Great Waterworks in Roman Greece

Aqueducts and Monumental Fountain Structures

Function in Context

Edited by

Georgia A. Aristodemou and
Theodosios P. Tassios

Archaeopress Roman Archaeology 35

ARCHAEOPRESS PUBLISHING LTD
Summertown Pavilion
18-24 Middle Way
Oxford OX2 7LG

www.archaeopress.com

ISBN 978 1 78491 764 7
ISBN 978 1 78491 765 4 (e-Pdf)

Cover: The monumental arcade bridge of Moria, Lesvos,
courtesy of Dr Yannis Kourtzellis
Creative idea of Tasos Lekkas (Graphics and Web Designer, International Hellenic University)

Contents

Preface

Georgia A. Aristodemou and Theodosios P. Tassios

The achievements and development of ancient hydraulic engineering that resulted to an efficient use of natural resources and provided ancient civilizations with the means to improve their living standards, attracted the interest of specialists of various disciplines quite early in scholarly research.

Series of international conferences of water management and hydraulic engineering (e.g. the Cura Aquarum International Conferences, Symposia held under the auspices of the Frontinus Gesselschaft, International Water Association Symposia [IWA], and many more) being held the last four decades, have indeed promoted the exchange of scientific knowledge about the latest findings and theoretical approaches and stimulated an interdisciplinary discussion about water management in antiquity, involving Archaeologists, Historians, Hydraulic Engineers, Civil Engineers, Geologists and many other scientists.

Scholarly discussions and relevant publications include (among others) large-scale water transfer structures, such as aqueducts and their complex systems, or monumental fountains, such as nymphaea – and geographically expand throughout the Mediterranean region.

Regarding the Eastern part of the Roman Empire, however, modern research seems to focus more on the Asia Minor, the Levantine and the North African Provinces rather than the Greek provinces.

During the last few years a newly launched discussion on water archaeology in Greece seems to emerge from exactly this need to promote research on this particular field and on this geographical region. However, there is still a definite and considerable gap in modern archaeological research and bibliography, regarding water management and monumental water structures in Roman Greece. The few and scarce publications on aqueducts from the Greek territories usually refer to the technological aspect of these imposing monuments. Though they have increased in number there is still a lack of a concise publication that would provide the archaeological community with gathered knowledge and data upon these significant monuments – on grounds of archaeological information, engineering progress and technological development. Similarly, studies related to monumental fountains and nymphaea in the Greek provinces were until now confined to their architectural and typological study, without expanding to other issues, such as decoration, interaction with their surrounding environment, and their impact upon the local communities.

Based on the above, the collective volume that we here present consists the very first presentation of great waterworks in Roman Greece and aims not only to fill an essential scientific gap but also aims to bring together a wide body of experts along with their knowledge from the newly emerged and expanding field of water technology and water archaeology in Roman Greece. The main goal that this multi-author volume attempts to succeed is to show that large scale waterworks (aqueducts and nymphaea)

- were a striking novelty in the Greek Provinces, both in form and function
- drastically changed the architectural landscape of their surrounding environment
- improved the living standards
- increased the water supply, in quantity and in quality
- introduced the concept of luxury in the urban landscape, and
- formed the link between utility and design, political ambitions and regional development

The invited papers deploy along a wide geographical area, covering the Roman provinces of Macedonia and Thrace, Epirus, Achaia, the Aegean islands and Crete. The chronological frame of the presented subjects extends throughout the Imperial period (1st century BC – 4th century AD).

This multi-author volume is structured in two parts (Part I: Aqueducts; Part II: Monumental Fountains) and evolves around four major axes: Architecture – Engineering – Historical Sources – Decoration.

Within this framework the invited authors addressed issues such as the following:

- Archaeological remains of aqueducts and monumental fountains in Roman Greece
- Technological development of urban water supplies (i.e. castella, aqueducts, fountains, nymphaea, etc) in Roman Greece.
- Function in Context: The interaction and spatial affiliation of aqueducts and nymphaea within the urban environment of Greek provincial cities.
- Sponsors and Benefactors
- Written sources and epigraphical data
- Aesthetics and Sculptural Decoration of Fountains and Nymphaea
- Religious aspects of water use
- Social, Cultural and Political issues of urban water supplies and water consuming structures in the Greek provincial cities.

As editors of this demanding volume, we were glad to invite and honored to welcome a wide range of authors, all expert scientists in the field of Water Archaeology in Roman Greece. Thirteen (13) studies were gathered in the present volume, all presenting us with new and valuable knowledge that addresses the needs of modern scholarship in regards of Roman hydraulic archaeology in Greece.

We are confident that it will soon be a valuable asset for senior and young scholars, as well as for students of a large variety of disciplines, such as Archaeology, History, Architecture, Engineering, Urban planning, etc.

With the increasing worldwide awareness of the importance of water resources management in the ancient civilizations, collective studies such as this, not only will enlighten and project the multifaceted significance of the archaeological remains regarding water management technology of the Roman period in the Greek regions, but they will also reveal the enormous impact of Roman technological heritage in the Greek territories.

As editors of this volume we would like to cordially thank each and every one of our invited authors, for providing us not only with their studies and their expertise, but for contributing to this volume with supreme academic professionalism, consistency and patience.

Furthermore, we extend our appreciation towards Archaeopress and, especially David Davison and Rajka Makjanic, who accepted and undertook the publication of the volume with a genuine academic interest and enthusiasm.

Lastly, we convey our warm thanks to all those people (family, friends and colleagues), who have supported us during the long preparation period of this volume and contributed in many way to its completion.

Dr Georgia A. Aristodemou
Researcher of Roman Archaeology
School of Humanities
International Hellenic University, Thessaloniki, Greece
E-mail: g.aristodemou@ihu.edu.gr;
garistodemou@gmail.com
www.hum.ihu.edu.gr

Dr Theodosios P. Tassios
Civil Engineer
Professor Emeritus, National Technical University,
Athens, Greece
President of the Association of Ancient Greek
Technology Studies
Member of the Academy of Sciences of Turin
E-mail: tassios@ntua.ua

Introduction I.
Roman Aqueducts in Greece

Theodosios P. Tassios

'Aquae [...] est enim maxime necessaria et ad vitam et ad delectiones, et ad usum cotidianum ...'
Vitruvius, *De Architectura*, 8.1.1

'Water is an absolute necessity for life, for pleasure and for daily use...'
Vitruvius, *Of Architecture*, 8.1.1

Preface

Vitruvius, the first Roman technical writer (1st century BC) in his book, *De Architectura* (Chapter VIII) offers several recommendations for the construction of water-supply works. To start with and regarding the methods of catching of waters, he mentions:

(i.) The tapping of aquiferous underground layers by means of several wells, interconnected through subterranean conduits gathering all [waters] into one place (8.1.6).[1]
(ii.) The waters originated from the rain (rivers, mainly, 8.2.6), without any additional technical description.
(iii.) The springs (8.1.6), again without any additional technical description.

Thus, only the underground catching by means of 'wells and fan-of-galleries' is technically described.

As Vitruvius continues (8.6.1-3): 'Water can be conducted in three ways: by flow in masonry channels, lead pipes and terracotta pipes. [...]. If in **channels**, the construction must be as solid as possible, and the stream bed must have a uniform slope of no less than ~ 5 ‰. The channel is to be vaulted over, so that the sun does not touch the water at all. [...]. If there are hills along the course between the city and the water source [...], an **underground** channel is to be dug, with the uniform slope described above. If the bed-rock is tuff or hard stone, the channel is to be cut directly in it; but if it is earth or sand, a vaulted channel with floor and walls is to be built in the tunnel. [...]. Vertical shafts are to be cut from the surface every one *actus* (~ 35 m).'[2]

Subsequently, Vitruvius describes in detail aqueducts made of lead pipes, before giving some specifications regarding terracotta pipes. He omits only stone and timber pipes.

Generally speaking, the earlier Roman aqueducts do follow the descriptions of Vitruvius – who, surprisingly enough, does not clearly mention water-supply-bridges in his text.

An exception might be a short and indirect comment of his (8,6.5): 'If [between the source and the city] exist valleys ('intervalla') then it is necessary to elevate the conduit by means of a substructure so that a smooth flow is achieved'.[3] This comment consists a rather inadequate reference to a possible bridge that is. This is perhaps due to the fact that during the 1st century BC, Pax Romana was not yet well established, and the risk of inimical destruction (or water contamination) of surface aqueducts was perhaps still probable, as it was during the Greek era. Another possible influence of previous Greek technical rules

[1] This is the typical Greek tapping, as for example behind the earlier krēnē, known as the 'krēnē of Theagenēs', in Megara, dated in 6th century BC, cf. Augerinou 2017: 44-48.
[2] Translation copied from Humphrey *et al.*, 1998: 295.
[3] Translation based on the views of P. Lefas 1998: 187.

on Vitruvius' opinions, may be his subsequent recommendation regarding the use of 'inverted siphons' when a ravine ('vallum') is to be crossed (8.6.5): 'But when long ravines exist, then we direct the conduct to the slopes [and] when it reaches the depth [of the ravine], we construct a not high substructure, so that a smooth flow is achieved in the greater possible length; [...] this is what the Greeks call κοιλίαν'.[4] In this passage of Vitruvius, a tendency to avoid the construction of a high and long bridge could be observed, preferring the adoption of the Greek solution of an inverted siphon of pipes under pressure, instead of a very high bridge.

Nevertheless, the aqueducts built in the mainland and island Greece during the Roman period cover all types of construction (including water conduit bridges) – with the exception of lead pipes, which seem not to have been used so frequently. The chronology of construction of these aqueducts (after the 1st century AD) explains the important developments made in Roman technology after the time of Vitruvius.

1. Roman Aqueducts in Greece

Aqueducts built during the Roman times in the mainland of Greece, the Islands and in the Greek Cities in Asia Minor, were (partly at least) a development of Greek (Archaic, Classical and, especially, of Hellenistic) Engineering. This simple fact may be helpful in comprehending some of the Roman technologies initially used, as well as some of the characteristics of local technicians in Roman dominions. This may be for instance the case with the use of strong stone-pipes in inverted siphons (as in the Hellenistic aqueducts indicated by Hodge)[5] or the use of lead pipes (as in the Hellenistic aqueduct under pressure in Pergamos).[6]

It is worth noticing, however, that these remarkable examples of Greek hydraulic engineering are taken from Greek cities outside the mainland. Having this in mind, it can be noted that, since mainland Greece had not much benefitted of the significant Hellenistic developments in Alexandria and Asia Minor (both in economic and technological terms), the influence of the Roman hydraulic technology was much more clearly observed in Greece than in the Greek regions abroad.

Scope of this first Part of the present volume is the description of significant aqueducts built in Greece during the Roman period. There are of course several equally important Roman aqueducts too, as shown in a broader catalog here below (cf. Table 1). It is however thought that the examples selected in this book are in fact the most representative of the entire period. We welcome the contributions of **Asimina Kaiafa-Saropoulou** on the ***Vaulted-Roof Aqueduct Channels in Roman Macedonia***, of **Constantinos Zachos and Leonidas Leontaris** on the long awaited ***Aqueduct of the Actian Nikopolis***, of **Manolis Manoledakis** regarding the ***Water Supply of Roman Thessaloniki***, giving emphasis on the Hortiatis' Aqueduct, of **Eustathios Chiotis** and his treatise on the renowned ***Hadrianic Aqueduct of Athens and the Underlying Tradition of Hydraulic Engineering***, of **Yannis Lolos** on the ***Hadrianic Aqueduct in Corinth***, of **Yannis Kourtzellis, Maria Pappa and George Kakes** and their study on the unique ***Roman Aqueduct of Mytilene***, of T**elauges** N. **Dimitriou** and his presentation on the ***Roman Aqueduct of Samos***, and lastly, of **Amanda Kelly** and her enlightening study about an ***aqueduct through the Highlands***, presenting us ***the Water Supply for elevated Lyttos*** **(Crete)**.

The above mentioned authors of the respective chapters are to be complimented for the detailed presentations they offered and the scientific methodology they followed in their research. I am sure that the scientific value of these chapters will be greatly appreciated by their readers, as it was by me.

In what follows, an attempt is made to summarise the main characteristics of the aqueducts presented in Part I, in an effort to move from the specific to the more general.

Table 1: An abbreviated catalog of Roman aqueducts in Greece (based on Lolos 1997: 302, modified)

[4] 'κοιλία' in Greek stands for the Latin 'venter' (=belly, in English).
[5] Hodge 1992: 33, l. 36.
[6] Garbrecht 1987: 26.

No	Location	Period
1.	Dion (Macedonia)	2nd century AD
2.	Philippi (Madeconia)	2nd century AD
3.	Hortiatis (Thessalonikē)	(1st century AD)
4.	Samothrace (Thrace)	(Roman)
5.	Demetrias (Thessaly)	(Roman)
6.	Glyky (Epiros)	(?)
7.	Goritsa (Thessaly)	(?)
8.	Nikopolis (Epiros)	1st and 2nd century AD
9.	Athens	2nd century AD
10.	Delion (Boeotia)	(?)
11.	Eleusis	(Roman)
12.	Lamia (Central Greece)	(Roman)
13.	Marathon (Attica)	2nd century AD
14.	Megara	(?)
15.	Piraeus	(Roman)
16.	Corinth	2nd century AD
17.	Argos (Peloponnese)	2nd century AD
18.	Gytheion (Peloponnese)	(Roman)
19.	Hermionē (Peloponnese)	(Roman)
20.	Kalabryta (Peloponnese)	(?)
21.	Loukou (Peloponnese)	(Roman?)
22.	Mantineia (Peloponnese)	(?)
23.	Megalopolis (Peloponnese)	(Roman)
24.	Olympia (Peloponnese)	2nd century AD
25.	Patras (Peloponnese)	(Roman)
26.	Sikyon (Peloponnese)	(Greek + Roman)
27.	Sparta (Peloponnese)	2nd century AD
28.	Kos (Aegean)	(Roman)
29.	Leros (Aegean)	(Roman)
30.	Mytilēnē (Aegean)	2nd century AD
31.	Samos (Aegean)	(4th century AD?)
32.	Chersonēsos (Crete)	1st / 2nd century AD
33.	Eleutherna (Crete)	(Roman)
34.	Elyros (Crete)	(Roman)
35.	Gortys (Crete)	(Roman)
36.	Inia (Crete)	(Roman)
37.	Knossos (Crete)	1st to 2nd century AD
38.	Koufonisi (Crete)	(Roman)
39.	Lyttos (Crete)	2nd century AD
40.	Basilikē (Crete)	(Roman)

2. Baths, Fountains or Individual consumers

The scope of these relatively large aqueducts cannot always be identified. It all depends on Archaeological evidence based on the remains of public baths, public fountains and large water reservoirs, serving an internal network or directly the city's Agora.

In the cases described in this volume, most of these scopes are served, but in some cases information on these matters is rather missing. On the other hand, many of the authors in this Part (I) of the volume are indeed providing us with valuable historical details regarding the activities of the inhabitants of the cities concerned, as a way to estimate (though indirectly) their consumption needs in water, as well as to propose possible chronologies for the construction of the discussed aqueducts.

3. Water catching

In most of the Roman aqueducts presented in this book, natural springs are yielding their waters to the conduit, directly or after a small reservoir. In Macedonia, however, catching directly from rivers was also effectuated. Finally, tapping of underground water by means of interconnected wells (or also along a permeable tunnel-work) is described in some important cases. The authors participating in Part I of this volume are making a commendable effort to identify the specific areas where water catchments were located each time. And this is a particularly difficult task because, among other reasons, the level of the underground water horizon and the capacities of the aquiferous zones have drastically changed through time. Furthermore, it is worth noticing that, in some cases, the initially drawn water quantities were further increased by means of smaller transversal or parallel secondary aqueducts, constructed along the main aqueduct.

4. The Length of the Aqueducts

Several historical developments resulted in a considerable increase of the distances where water sources were sought during the Roman period – when compared to Classical Greek or even Hellenistic times: the needs in water supply were increased, both because of the city-population growth and because of the changes in bathing-culture. Moreover, better hydraulic technology and enhanced security measures taken regarding neighbor cities' hostilities of the past, have also contributed to the lengthening of aqueducts. The 1km of Peirene (Corinth, 7th century BC), 1,5km of Megara (6th century BC), the 2,5km of Samos (6th century BC), the 7,5km of Peisistration in Athens (6th century BC), the 11km of Naxos (6th century BC), were already surpassed by the Hellenistic period – even without the extreme case of the Pergamos aqueduct system (with one of its aqueducts being 25km long). In the present volume the records are different: Dion 10km, Samos 15m, Hortiatis 20km, Hadrianic aqueduct of Athens 20km, Lyttos 22km, Mytilene 33km, Nicopolis 70km, Corinth 85km. One can clearly detect a typically Roman 'magnification' in these cases.

5. Construction technology

As is described within this volume, traditional and modern construction methods were employed in the construction of aqueducts both in mainland and island Greece during the Roman period:

If the route of the aqueduct follows the topographic contours of the **ground**, it is possible to construct the conduit (almost on the surface of the ground) either as:

a) a simple masonry-revetted ditch on a strong soil, or as
b) a trapezoidal (or rectangular) conduit carved on a rocky slope.

More frequent however is the construction of low-depth **buried** conduits:

c) In a 'cut and cover' excavation, a rectangular masonry conduit is constructed, high and large enough to be visited and maintained; its cover may be a half-circular vault or a stone-slab. Soil of variable thickness covers the conduit.

If however, hydraulic or geomorphological conditions require an **underground** course of the aqueduct,

d) the typical 'shaft-and-gallery' system is followed (a shaft is excavated, and from its bottom the worker digs the almost horizontal gallery up to the bottom of the preceding shaft),[7] or
e) a more deep tunnel is dug (with revetted or naked walls).
f) Nevertheless, when an intermediate valley is offered as a possible shortening of the entire aqueduct, then
g) the solution of an elevated channel is adopted, on top of appropriate walls or, mainly, **bridges** with arcades.

Finally, if a relatively narrow ravine is to be crossed,

h) an '**inverted siphon**' (pipes under pressure) is constructed close to the bottom of the ravine, in order to avoid the construction of a very high bridge.

Due to the meticulous work of all authors of Part I, the readers of this volume may find collected in the following Table (2) all the construction technologies followed by the Roman aqueducts presented in this book.

Table 2: Construction Technologies

		On the Ground	**Underground**				
No	**Aqueduct**	**In ditch or carved on the rock**	**Cut and cover**	**Shaft and gallery**	**Tunnel**	**On walls or on bridges**	**With Inverted Siphons**
		a or **b**	c	**d**	e	**f**	g
1	Hortiatis			+		+	
2	Dion		+			+	
3	Nikopolis	+	(+)	+	+	+	
4	Athens			+			
5	Mytilēnē	+		+	+	+	
6	Corinth			(+)	+	+	
7	Philippoi Edessa		+				
8	Samos	+				+	
9	Lyttos	+		+		+	+

Some other technical details described in the corresponding chapters of this volume are as follows:

- manholes for cleaning (located approx. every 1 'actus' = 120 feet ~ 35m, or sometimes every multiples or submultiples of actus)
- deep wells for the aeration of deep tunnels,
- water towers installed at the juncture of abrupt directional changes (horizontally or vertically)
- impurities' settlement tanks (*piscinae liminariae*) are sometimes mentioned
- water reservoirs to feed an internal city-distribution network (*castella divisoria*) – although few of them have survived.
- secondary water reservoirs.

[7] For a discussion regarding the tendency for a broader use of the term 'qanat', see Chiotis 2017: 1-16 and Defteraios *et al.* 2017: 50-54, as well as the papers of M. Manoledakis and E. Chiotis in present volume.

6. Cross sections of conduits

It needs to be reminded that in the case of free water flow in open channels, the geometrically available cross sections do not necessarily coincide with the 'hydraulic' cross section; the geometry of these conduits is imposed by practical rules regarding:

- minimal dimensions for digging or carving the channel,
- space needed to apply the masonry-covers of the walls of the channel, and
- space needed for a person to walk in underground conduits for cleaning and possible repairs.

In many cases, however, terracotta or lead pipes are accommodated within such larger conduits, operating under pressure or even under free flow.

Based on the detailed information offered by our Authors, the following summarized Table 3 may be prepared, regarding the cross sections of conduits of Roman aqueducts in Greece.

Table 3: Typical cross-sections of conduits (indicative dimensions)

Open channels[(a)] (carved on rocky slopes or built on top of retaining walls)	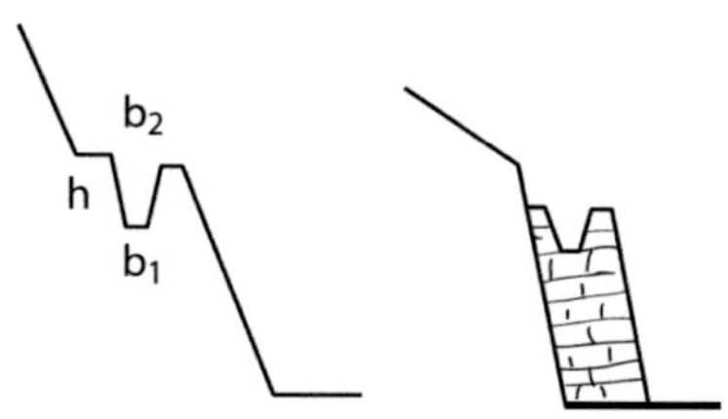
Masonry revetted channels[(b)]	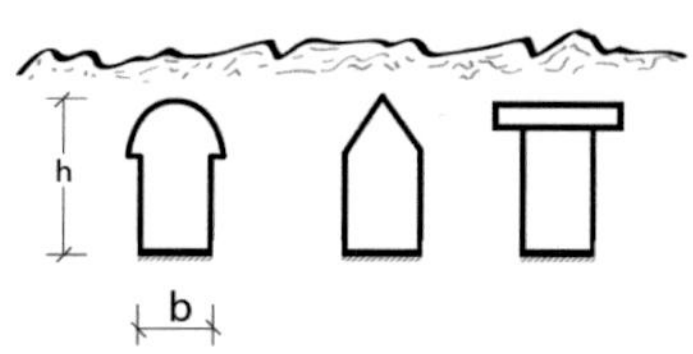
Channels on walls or bridges[(c)]	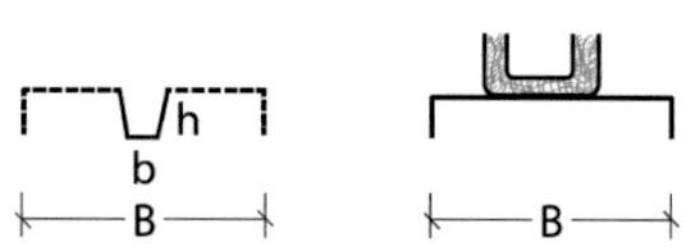
Conduits including pipes[(d)]	e.g. 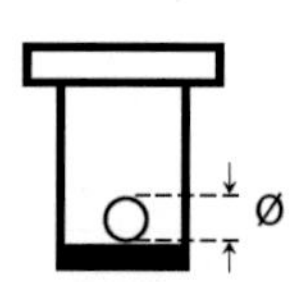
(a) b_1 ~ 0,20 to 0,30 m , b_2 = 0,30 to 0,40 m , h ~ b (In the aqueduct of Corinth b_1 > 1,00 m)	
(b) b ~ 0,40 to 0,80 (1,30) m , h ~ 1,00 to 1,80 m (In Corinth, Eleutherna and Elyros, the width was larger than 1 m.	
(c) b ~ 0,35 to 0,60 m , B ~ 2,00 to 5,00 m , h »	
(d) Ø ~ 0,20 m	

7. Some characteristics of bridge-aqueducts

It seems that the geometrical data of the structures were decided as a function of the height of the bridge, the span of the arcades, the existence of buttresses and (mainly) the employed masonry-building technique (ashlar or rubble masonry). That is why a considerable variability is observed in the wall-thickness of the bridge-aqueducts described in this volume:

	max. wall thickness	max. height
Lyttos	2,4 – 4,5 m	8,10 m
Hortiatis	5,00	20,10
Corinth	2,90	11,20
Mytilene	2,50	25,0
Samos*	7,00	21,7
Nikopolis	(3,00)	7,80

Dimitriou 2003: 206, drawing 19.

The case of the Mytilene aqueduct is however particular; it seems that this aqueduct is the most 'monumental' of the preserved Roman bridges in mainland and island Greece and one of the two with three 'floors' (zones) of arcades.[8] Despite its rather insufficient inspection level, it was the only one with such a high slenderness, since the 'height-to-thickness' ratio equals to 5, whereas in all other aqueducts the applicable ratio is lower than 4.

Details regarding the foundations of the piers of these bridges are not easily available, at least on the basis of Archaeological excavations, which as understandable, are very difficult to conduct.

This short paragraph cannot be closed without a brief reference to the '**substitute**' of a water-bridge, that (in the case of a deep and long ravine) can be an '**inverted siphon**', as it is described by Vitruvius (8.6.5). Among the aqueducts presented in this Volume, it seems that only in the case of Lyttos an inverted siphon was identified near Terazi, confirmed also by one stone-pipe (perforated diameter 0.22m, minimum wall thickness 0.15m) found *in situ* by Oikonomakis (1984).[9] Assuming a tensile strength of that soft rock equal to $f \sim 2\ N/mm^2$ and following simple calculations[10], it may be estimated that this pipe was able to withstand a hydrostatic pressure of 70m of water, with a safety factor equal to 2. It is obvious that the actual hydrostatic pressure in this case was considerably lower.

For the sake of completeness on this subject, it is very interesting to note that in the island of Lesbos (cf. the aqueduct of Methymna, mod. Molyvos), a series of such perforated stone cubes were found (internal diameter 8cm), with recessed male-female joints.[11]

8. Building Materials

At this point it is considered appropriate to present a brief summary of building materials employed in the Roman aqueducts described in this Volume.

- Blocks:
    - Stones (ashlar or rubble)
    - Bricks (thick and large)
- Mortars:
    - Binders + Sands
- Binders:
    - Lime
    - Lime and crushed ceramics
    - (Lime and puzzolan?)
    - Clay
- Plastering and pointing material:

[8] The other is the aqueduct of Samos, at the 'Yianniou-bridge' (Dimitriou 2003: 206).

[9] Oikonomakis 1984: 66-69.

[10] Tassios 2006: 21.

[11] I am indebted to Dr Yannis Kurtzellis for the bibliographical information, and to Prof. Yannis Lolos who mentioned this archaeological find to me.

- Fine grain hydraulic mortar. Special care was taken against possible cracking at the corners of vertical walls and the bottoms of conduits.

Arcades and vaults were made on timber-formwork, radially arranging large thick bricks or flat stones with thick intermediate mortar layers.

Remnants of local quarries and lime-kilns are occasionally identified along the course of the aqueducts.

9. Water capacity of the aqueducts

As it is well known, the water capacity of a conduit depends

(i.) on its hydraulic cross section
(ii.) on the roughness of the walls of the channel
(iii.) on the 'hydraulic gradient' of the aqueduct, and
(iv.) on the various water losses, such as evaporation, leakage through joints; intermediate illegal water-connections may sometimes be also important.

Regarding the hydraulic cross section, in ancient open channels we generally ignore the average flow **depth**; and this may result in very large errors in calculating both the velocity and the capacity of the channel.

On the contrary, regarding the roughness of the internal walls of the channel, we may make better estimates – assuming however a certain level of cleaning and maintenance care (including also measures against possible sprouting of plants in the conduit).

Finally, geomorphological and hydraulic conditions may shape the value of the hydraulic gradient 'J' (i.e. the ratio between the altimetric difference between the 'starting' and the terminal point of the aqueduct, over the length of the aqueduct between these two points). The selection of these two reference points should however be carefully made, so that no turbulent stretches are included between them. Evidently, this is not an easy task for us regarding a dry ancient channel.

Due to all these difficulties and uncertainties, our attempts for an *a posteriori* calculation of the possible water-capacities of ancient conduits are rather alleatoric.

Nevertheless, our interest to acquire a broader understanding of the function of ancient aqueducts necessitates the knowledge of an even roughly estimated margin of their water-capacity. Such knowledge can also be used to check – or complete – our information on the number of inhabitants of the ancient city, as well as their 'water culture'.

From this point of view, it is evident that the valuable data presented by all authors contributing in this Volume, constitute a sound basis and provide the 'impetus' for further research on the field of water archaeology, emphasizing on aqueducts in the Greek provinces of the Roman Empire.

Theodosios P. Tassios
Civil Engineer
Professor Emeritus
National Technical University, Athens, Greece
President of the Association of Ancient Greek Technology Studies
Member of the Academy of Sciences of Turin (Italy)
tassiost@central.ntua.gr

Bibliography

Augerinou, P. 2017. The Archaic underground aqueduct in Megara, in Chapter 4: An Updated Appraisal of Ancient Underground Aqueducts in Greece. In: A. N. Angelakis, E. D. Chiotis, S. Eslamian and H. Weingartner (eds) *Underground Aqueducts Handbook*: 44-48. Boca Raton, FL: CRC Press.

Chiotis, E. 2017. Underground Aqueducts: Types, Definitions and General Conclusions. In: A. N. Angelakis, E. D. Chiotis, S. Eslamian and H. Weingartner (eds) *Underground Aqueducts Handbook*: 1-16. Boca Raton, FL: CRC Press.

Defteraios, P., Chiotis, E. and Mamassis, N. 2017. Underground evidence on the tunneling technique of the Hadrianic aqueduct of Athens, in Chapter 4: An Updated Appraisal of Ancient Underground Aqueducts in Greece. In: A. N. Angelakis, E. D. Chiotis, S. Eslamian and H. Weingartner (eds) *Underground Aqueducts Handbook*: 50-54. Boca Raton, FL: CRC Press.

Dimitriou, T. 2003. *Το Ρωμαϊκό υδραγωγείο της Σάμου*. Athens: 'N. Dimitriou' Samos Cultural Institute.

Hodge, T. 1992. *Roman aqueducts and water supply*. London: Duckworth.

Humphrey, J. W., Oleson, J. P., Sherwood A. N. 1998. *Greek and Roman Technology: A Sourcebook. Annotated Translations of Greek and Latin Texts and Documents.* London: Routledge.

Garbrecht, G. 1987. Die Wasserversorgung antiker Pergamon. In: Frontinus Gesellschaft e. V. (eds) *Die Wasserversorgung antiker Städte: Pergamon. Recht / Verwaltung. Brunnen / Nymphäen. Bauelemente:* 11-48. Mainz-am-Rhein: von Zabern Verlag.

Lefas, P. 1998. *Μετάφραση και σχολιασμός του Βιτρουβίου, Περί Αρχιτεκτονικής*, τ. I (Βιβλία 1-5), τ. II (Βιβλία 6-10). Athens: Plethron Editions. (2001, 2nd Edition).

Lolos, Y. 1997. The Hadrianic aqueduct of Corinth. *Hesperia* 66.2: 271-314.

Oikonomakis, N. E. 1984 .Το Υδραγωγείο της Λύκτου. *Lyktos* 1: 66-99.

Tassios, T.P. 2006. Selected topics of water technology in Ancient Greece. In: A. N. Angelakis and D. Koutsogiannis (eds) *1st IWA International Symposium Water and Wastewater Technologies in Ancient Civilizations. 28-30 October 2006*: 3-26. Herakleion.

Introduction II.
Roman Monumental Fountains (Nymphaea) in Greece

Georgia A. Aristodemou

Monumental fountain structures, built by Emperors and wealthy patrons in cities throughout the Roman Empire, have always been striking features of Roman architecture. These imposing - often multi storied - fountain structures, not only became a trend in the Roman East, but they contributed greatly to the formation of a luxurious urban environment. During the last four decades scholars studying nymphaea focused mainly on producing a relevant terminology and establishing their architectonic typology, along with their evolution in form and design. Only the past few years, research extended to issues such as their sculptural decoration, their aesthetics, their interaction with their surrounding environment and the impact nymphaea had upon the local communities.

Research has been greatly developed, especially regarding the eastern part of the Roman Empire. However, modern scholarship has more to present about the Asia Minor, the Levantine and the North African provinces rather than the Greek provinces.[1]

There is a definite and considerable gap in modern archaeological research, discussion and bibliography, regarding monumental fountain structures in Roman Greece.[2]

This volume attempts to fill this gap, by gathering studies from scientists, all dedicated to the study and development of fountain archaeology in Greece during the Roman period. It is self-evident that not all monumental fountains were included in this collective work. One may observe, for example, the absence of the great Nymphaeum of Herodes Atticus located in the sanctuary of Zeus in Olympia. The main reason is that this monument was fully and in details published and presented as early as 1984 by Renate Bol.[3] Any addition to the information would not be significant. Or, the absence of the renowned Corinthian fountains, which were all thoroughly presented a few years ago by Betsey Robinson.[4] On other reason is that even though numerous fountains dated in the Roman period are excavated nowadays during rescue or systematic excavations in mainland and island Greece, unfortunately - and perhaps due to the lack of knowledge, or scientific resources - they either remain unpublished or they are poorly understood and scatterly presented. As a result the archaeological community is deprived from the opportunity of a better comprehension of these important monuments and their evolution in the Greek territories.

Under the above mentioned limitations, we decided to present here three kinds of studies; theoretical approaches and overviews (on sculpture, aesthetics, contextualization), discussions of specific monuments, under a new consideration (Messene, Gortyn), and entirely new presentations of so far unpublished monuments (Nikopolis).

Six papers were gathered in this Part (II) of the volume to discuss various aspects of the monumental nymphaea in Roman Greece. All six papers consider monumental nymphaea and luxurious fountains as

[1] For example, Dorl-Klingeschmid, Cl. 2001. *Prunkbrunnen in kleinasiatischen Städten. Funktion im Kontext.* Studien zur antiken Stadt 7, Bayerische Akademie der Wissenschaften, München; Uğurlu, Nur Banu, 2009. *The Roman Nymphaea in the cities of Asia Minor: Function in Context.* Saarbrücken; Richard, J. 2012. *Water for the City, Fountains for the People. Monumental Fountains in the Roman East: an Archaeological Study of Water Management.* Studies in Eastern Mediterranean Archaeology (SEMA), 9. Brepols.

[2] Longfellow, B. 2011. *Roman Imperialism and Civic Patronage Form, Meaning and Ideology in Monumental Fountain Complexes.* Cambridge University Press; Robinson, B. A. 2011. *Histories of Peirene: A Corinthian Fountain in Three Millennia.* Princeton: ASCSA; Aristodemou, G. 2012. *Ο Γλυπτός Διάκοσμος Νυμφαίων και Κρηνών στο Ανατολικό Τμήμα της Ρωμαϊκής Αυτοκρατορίας [Sculptural Decoration of Monumental Nymphaea in the Eastern Provinces of the Roman Empire].* Thessaloniki : Kornelia Sfakianakis Publications. Rogers, D. K. 2015. *Water-Display and Meaning in the High Roman Empire.* Unpublished PhD dissertation, University of Virginia.

[3] Bol, R. 1984. *Das Statuenprogramm des Herodes-Atticus Nymphaeums.* Olympische Forschungen, Band XV. Berlin: Walter de Gruyter.

[4] Robinson, B. A. 2001. *Fountains and the Culture of Water at Roman Corinth.* Unpublished PhD dissertation, University of Pennsylvania; Robinson, B. A. 2011. *Histories of Peirene: A Corinthian Fountain in Three Millennia*, ASCSA, Princeton.

reflections not only of the architectural aesthetics developed in the Greek provincial cities of the Roman Empire, but also of the social, political and religious life of their citizens. They are the following:

Dylan Kelby Rogers (*Shifting Tides: Approaches to the Public Water-Displays of Roman Greece*), is the new entry in scholarly discussion regarding the role of nymphaea into their urban environment. Since the study of Roman water-displays is constantly evolving (with research being conducted on the terminology of the structures, their typologies, their sculptural programs, and their impact on the social and physical landscapes surrounding them), he is approaching the subject under a new aspect, focusing on 'aesthetics'. His present paper focuses on a survey of the terminology used and the historiography therein, so that water-displays that until now have not received a great deal of attention (in religious, civic, and entertainment-related contexts) should also be considered. With his work, he poses several and significant new ways to advance the study of public Roman water-displays in Greece.

Georgia Aristodemou (*Fountain Figures from the Greek provinces: monumentality in fountain structures of Roman Greece as revealed through their Sculptural Display Programs and their patrons*), focuses on the definition of the term 'fountain figures', namely sculptures with a visible water outlet ascertaining a spouting function or sculptures that regardless their function derive from an elaborated fountain. Next, she discusses the presence and function of sculptures within the fountain structures based on their direct (organic) or indirect (thematic, regional) relation with water and nature. Imperial and private portraiture are also discussed since they underline the Imperial power and the prestige of local Elite. Issues about sculptural assemblages and the way these were carefully arranged to highlight the prestige of the local communities and sponsors are also presented. The significance of choices on a local, regional, imperial level allow us to study sculptural display programs of monumental nymphaea as a reflection of the patrons and the city's status and power.

Constantinos Zachos and Leonidas Leontaris (*The Aqueduct of Actian Nikopolis*), are participating with the most recent and complete presentation of the aqueduct of Nikopolis (the largest aqueduct of Roman Greece). Even more important is the fact that they supplement their study with a preliminary but extremely significant description of the twin monumental nymphaea of Nikopolis – two fountain structures that were directly supplied by the aqueduct and for many years have engaged scholarly discussions regarding their initial architectural form and their chronology (1st or 2nd century AD). We are privilege to have the excavators of these nymphaea reaching to concluding remarks regarding important questions about the nymphaea's building phases, their connection to the aqueduct, their dating and their association to the Hadrianic building program. It is needless to comment their valuable contribution to the archaeological discussion about monumental nymphaea in Roman Greece.

Susan Leigh (*The Monumental Fountain in the Athenian Agora: Reconstruction and Interpretation*), presents us with an up to date reconstruction and interpretation of the Monumental Fountain in the Athenian Agora, the so-call Southeastern Nymphaeum. In the last five years the architecture of and various issues regarding the monumental fountain in the Athenian agora, a building not preserved above foundation level, leaving all possible reconstructions largely hypothetical, have undergone renewed scrutiny. S. Leigh provides us with a fresh look at this extraordinary edifice after presenting earlier discussions and recent reconstructions (especially regarding the existence – or not – of a second floor). More specifically, she reconsiders the evidence and possibilities for the architecture and decorative program of the Athenian fountain, and the degree to which its reconstruction based on the Olympia structure is likely. Additionally, she postulates the possible water technology utilized in the structure, a topic largely ignored in previous studies. She also discusses the meanings behind the siting of the 'nymphaeum', its imperial connections, and how the monument and its supply aqueduct visually changed the southeastern Agora space and the important ceremonial approach to the Acropolis, the Panathenaic Way. This focused restudy allows the building to be better understood within its context in the Imperial Greece as well as within Hadrian's program of Athenian Euergetism.

Mario Trabucco della Torretta (*New water from old spouts: the case of the Arsinöe Fountain of Messene*), reconsiders an old monument, the Arsinöe fountain in Messene – an urban monumental water facility originally built at the end of the 4th century BC and transformed several times afterwards, both using the same architectonic members and adding new ones. In his paper, he proposes a new analysis of the two Roman phases of the specific fountain. The first Roman transformation of the building during the second half of the 2nd century BC is examined regarding its architectural characteristics and set in the broader context of the so called 'Romanisation' of the province of Achaia. Trabucco della Torretta is comparing the evidence from the Arsinöe fountain with similar cases in Greece and the other provinces (Asia Minor, Sicily), whereas he is posing questions on how the Romanisation process – a process accomplished throughout the Roman world also through the establishment of Roman architectural models in the newly annexed provinces – can be detected at the provincial landscapes. In accordance to this, he examines the second large-scale refurbishment of the fountain (during the second half of the 1st century BC), underlining the transformation of urban monumental fountains from plain and utilitarian buildings to places of display and propaganda, and their connection with their urban environment.

Brenda Longfellow (*Reflecting the past: the nymphaeum near the so-called Praetorium at Gortyn*), discusses the single-storey Pi-shaped nymphaeum (F25) near the so-called *Praetorium* in Gortyn, which was built by an unknown patron in the second half of the 2nd century AD. At this time, it was adorned with an extensive sculptural group that included seated muses, fountain sculptures of gods, and portrait statues that presumably portray the patron of the monument and members of his family. Standing in the civic center and facing the so-called *Praetorium* across a large street that traverses the city, the fountain was highly visible in the cityscape, where it provided a convenient source of potable water as well as an eye-catching artistic water display. Sometime between the 5th and 8th century AD, the open-air settling basin was enclosed and the edifice transformed into a barrel-vaulted reservoir, where, however, the 2nd-century sculptural group was re-erected along the colonnaded façade wall of the transformed monument.

Longfellow's paper examines the ancient and late-antique phases of the monument in order to answer the question of why a late-antique patron would re-erect the sculptural group at all. According to her, the sculptural group's resurrection should be attributed to the connections of the late-antique patron to the specific community – as well as to the late-antique imperial legislation that repeatedly restricted the removal of statues and other architectural elements from public buildings and urban centers across the Empire.

....

It is now comprehended that monumental nymphaea are centrally located in major cities of the Roman East and are usually financed by the Emperor, the city, or the local Aristocracy. Financing the construction of a nymphaeum and (or) its sculptured decoration is a politically motivated action, whether the benefactor is the Emperor himself, the city, or a private individual. Evidently, monumental nymphaea are actually products of specific historical and social conditions. In this Part (II) of the collective volume an attempt was made, to reconstruct the social, political as well as cultural role that these edifices held in the Greek provinces. The main goal was not only to fill an essential scientific gap but also to launch the discussion among experts regarding fountain archaeology in Roman Greece.

Dr Georgia A. Aristodemou
Researcher of Roman Archaeology
School of Humanities
International Hellenic University, Thessaloniki, Greece
E-mail: g.aristodemou@ihu.edu.gr; garistodemou@gmail.com
www.hum.ihu.edu.gr

PART I: AQUEDUCTS

Vaulted-roof aqueduct channels in Roman Macedonia

Asimina Kaiafa-Saropoulou

Abstract

During the Roman period, water management systems, which were based on the previous Greek hydraulic technology, were characterized by a great technical progress. One of the innovations applied in aqueducts were associated with the widespread use of vaulted roof channels which conveyed water by gravity, whether above or below ground. In Macedonia, following the new technological achievements, the sufficiency of water in cities depended also on open-channels, either with flat or vaulted covers. Vaulted roof aqueduct channels are detected in Dion, Amphipolis, Edessa, Philippi and the eastern suburbs of Thessaloniki (Thermi). Slight variations in materials and construction methods can be identified, as well as minor differentiations in their dimensions. They all present built -masonry with vertical sides and paved bottoms, with slight inclination to regulate the water flow and avoid erosion. The inner side walls and the floors were waterproofed so as to prevent leakages. Square cross-section manholes and shafts for maintenance and cleaning purposes supplemented the water projects. All of the above features undoubtedly confirm the deep hydraulic knowledge of the Macedonian engineers and their ability to supply Macedonian cities with both impressive and effective aqueducts, based on the same engineering principles applied throughout the Roman Empire.

Key Words: Macedonia, aqueducts, channels, vaulted covers, masonry.

Introduction

Since early in antiquity, in Macedonia, water management has been strongly related to the environmental and geological conditions, hydrographic status and urban profile of its settlements. Based on the geological features, the main aquifers for groundwater exploitation in Macedonia are developed in alluvial deposits (porous aquifers) and in terrestrial facies of Neogene deposits, as well as in carbonate rocks (Karstic aquifers). Karstic aquifers in carbonate rocks discharge through many springs.[1]

Apart from the numerous natural springs and streams all over the Macedonian territory, major rivers constituted permanent boundaries and provided sources of fresh water for a great number of settlements that were sited on their banks. Moreover, the abundance of water in the entire Macedonian region was a privilege recognized already in antiquity, and this is the reason why Macedonians accorded their rivers with great respect. Both rivers and numerous permanent streams were included in the Macedonian myths or were worshiped as gods, while they were also depicted on the reverse of the Macedonian coins having a human form. For example, the Strymon River, who was animated and personified, was evolved through time into a god and was given temples and sacrifices. It was of no surprise that the river was also depicted on the Amphipolis coinage, as a young man either standing crowned with reeds or reclining.[2]

In the ancient Greek literature, Euripides (*Bacchae*, 571-575) mentions the 'rushing waters' of the Axios River along with the father of all rivers, the river Lydias, that 'with his sparkling waters, offers the joy of comfort in living to all the mortals' (*...τ'ὠκυρόαν Ἀξιὸν....Λυδίαν πατέρα τε τὸν τᾶς εὐδαιμονίας βροτοῖς ὀλβοδότατον ...ὕδασιν καλλίστοισι λιπαίνειν...*). On the other hand, Herodotus (VII, 127, 1), within the topographic analysis of the region, reports that 'Lydias and Haliacmon are mingling their waters together in one and the same stream', ending on the west coast of the Thermaikos Gulf (*...Λυδίεώ τε ποταμοῦ καὶ Ἁλιάκμονος, ... ἐς τώυτὸ ῥέεθρον τὸ ὕδωρ συμμίσγοντες.)*. The crucial role of the Axios River to the hydrography of the area is verified very early in antiquity as deducted by the Homeric text reference (B, 849-850) that its '...water whereof floweth the fairest over the face of the earth' (....*ἀπ' Ἀξιοῦ εὐρὺ ῥέοντος, Ἀξιοῦ οὗ κάλλιστον ὕδωρ ἐπικίδναται αἶαν...*).

1 Kaiafa *et al.* 2014: 177.
2 Kaiafa 2008, II: 117; Samsaris 1976: 16-21.

The existence of numerous springs, torrents and rivers in Macedonia is reflected in many of the Macedonian toponymics. In several cases names of small settlements, cities or even entire regions, etymologically indicate the existence of water.[3] For example, the name of the city called Veroia (Βέροια) apart from the Phrygian term BE (i.e. bring) and the word ΡΟΥΣ (i.e. water flow). This analysis undoubtedly reflects the main characteristic of the area where the city was located, that is the presence of abundant flowing water.[4] The name of the nearby located Edessa derives from the Phrygian word VEDU, which stands for water – in other words, a city founded on water sources.[5] Furthermore, the names of the cities Leibithra (Λείβηθρα), meaning channel and conduit, and Pimbleia (Πίμπλεια) meaning spring and/or rainwater, both indicate the water availability of the region where they are located.[6] In addition, the name Mygdonia (Μυγδονία), the region extending from the east valley of the Axios River until Mount Vertiscos and the Strymonic Gulf, is a compound word, consisting of the letter -MY- which in the Phoenician alphabet stands for water and -ΧΘΩΝ, that means land, referring to the hydrogeological picture of the entire area.[7]

Bearing these in mind, it is worth mentioning that during the prehistoric period in Macedonia water was the factor that affected the most the population's movements and settlements in a place. That is the reason why sites of both the Neolithic and the Bronze Age were located next to springs, permanent streams or along rivers and lakes. Organised hydraulic networks did not exist in prehistoric Macedonia, as was the case in the Minoan-Mycenaean world, due to the fact that Macedonian settlements, unlike those in southern Greece, were not classified as urban since then.[8] Well structured settlements with a central authority rule appeared in Macedonia from the 7th century BC and developed during the 6th century BC. The development of water supply and waste water systems in Macedonia during the Classical period were contemporary with the emergence of urban centers, the implementation of the Hippodamian system and the stabilization of architectural types; not only the Macedonian water management systems were in accordance with the technological achievements of that period, but they reached a comparable level with the contemporary hydraulic infrastructure of southern Greece.[9] The aqueducts in Pydna, Stageira and Amphipolis were based on the use of clay pipes buried under the soil or protected in underground tunnels carved into the rock, not to be exposed to enemies. They were always following the topographic contour lines.[10] In Olynthus, water was transported through an underground aqueduct via closed clay pipes, with cleaning and venting holes on their walls. In its last part, pipes were laid in a vaulted tunnel with cleaning shafts. They were supplying a fountain at the edge of the town, with a cistern with permanent taps on the front side.[11]

During the Roman period water management systems were characterized by great technological progress. Innovations on aqueducts were associated with the widespread use of vaulted roof channels which were conveying water through gravity, either above or below ground. The water efficiency of Macedonian cities depended also on open-channels, according to new achievements and applications, either with flat or most impressive vaulted covers (Figure 1). Vaulted roof aqueduct channels have been found in Dion, Amphipolis, Edessa, Philippi and the Kalamaria valley, east of the city of Thessaloniki (Figure 2).

Roman vaulted-roof aqueducts in Macedonia

Romans, based on the Greek knowledge, improved hydraulic technology, developed high engineering skills and were able to expand these technologies on large scale projects so as to improve urban supply and drainage systems. This is definitely related to the fact that in the Roman period baths and latrines played a major role in the lifestyle and demanded continuous water flow. As far as water transfer systems,

[3] Kaiafa 2008, II: 121-122.
[4] Girtzy 2001: 69; Chionides 1960: 69.
[5] Girtzy 2001: 62.
[6] Kaiafa 2008, II: 121-122; Kotzias 1948-49: 25-26; Girtzy 2001: 100.
[7] Kaiafa-Saropoulou 2014a: 557 and 2014b: 277.
[8] Kaiafa 2008, II: 123-127; Kaiafa and Karadedos 2011: 480; 2013: 29.
[9] Kaiafa 2008, II: 127; Kaiafa and Karadedos 2011: 480-486; 2013: 29-31.
[10] Besios and Pappa 1998: 140; Pikoulas 1997: 610-611; 2001: 107-109, 210-211; Sismanides 1995: 453-455; 2003: 45-46.
[11] Crouch 1993: 171-176; Robinson 1946: 96, 103-114.

Macedonia: Roman vaulted roof aqueduct channels					
Aqueducts	***Amphipolis***	***Dion***	***Edessa***	***Kalamaria***	***Philippi***
Date	Roman	2nd c. AD	Late 3rd c. AD	3rd c. AD	2nd c. AD
Side walls	Part-worked stone masonry with bricks and many marble members	Waterproofed lime masonry	Porous lime masonry	Lime part-worked rubble masonry	Waterproofed lime masonry
Floor	Unknown configuration	Bricks in a single row	Bricks in a double row	Bricks lined	Lined with clay slabs
Vault roof	Brick made vaulted roof	Stone built arch	Porous stone built arch	Brick and stone made vaulted roof	Stone made arch
Manholes	Not found in the uncovered section	Not found in the uncovered section	Not mentioned	Inspection shafts prominently projecting some cm up above the rounded top of the channel	Quadrangular manholes, built of stones and mortar
Dimensions	Width 55cm High 1,00m	Width 55cm High 1,00m	Width 70cm High 1,00m	Width 43cm- 55cm High 1,10m- 1,70m	Width 80cm High 1,20m
Comments	Scanty remains, incomplete research	Last part above ground		Many parts interspersed in Kalamaria valley (eastern suburb of Thessaloniki), from Toumba until Vasilika	

Figure 1. Roman vaulted roof aqueduct channels from Macedonia. General construction features and differentiations.

Figure 2. Region of Macedonia. Known aqueducts of the Roman period.

Roman innovations were associated with the widespread use of open channels, and vaulted roof in many cases, which were conveying water by gravity, either above or below ground.

The Macedonians, following the achievements of Roman hydraulic engineering, exhibit an equally high level of technological expertise regarding water management – and, especially, water supply systems. Regarding typology, Macedonian aqueducts seem consistent with the evolution of Roman hydraulic technology – as viewed in the rest of the Graeco-Roman world. There is no doubt that the water efficiency of Macedonian cities depended not on closed clay pipes, but on open channels; there, the water was flowing by gravity, from the source to the locations where it was needed. According to new achievements and applications, these aqueduct conduits had either flat or quite impressive vaulted covers. Free flow flat slab-covered aqueduct channels were preferred in Veroia; they were made of porous limestone from the Mount Vermion and dated at the end of the 1st century AD.[12] Similarly, a trapezoid shaped cross section is observed in the open channel of the Roman aqueduct of Pydna; it is constructed of bricks placed vertically at the sides, and also has a brick-constructed floor covering. In Kesareia (Kozani region), the waterproofed channel of the Roman aqueduct is also flat covered, with side walls of rubble and pebbles embedded in mortar for the bottom. Finally, sidewalls stone constructed with plaster and with flat slab coverings formed the aqueduct channel discovered in the Polichni suburb. It was carrying water from the Lebet springs to the Roman city of Thessaloniki.[13]

Despite the fact that the flat slab coverings of aqueduct channels were considered more adequate for their maintenance and cleaning, the use of the vaulted roof is also represented by several examples in Macedonia. Nevertheless, the high level of knowledge of the Macedonian engineers during the Roman period, and their implementation capacity is reflected more in vaulted culvert pipes. Vaulted roof aqueduct channels were already discovered in Dion, Amphipolis, Edessa, Philippi and the Kalamaria valley, in the southeast of Thessaloniki. They are all impressive curved covered conduits, rectangular in cross-section, usually subterranean – the visible terminal point of the aqueduct of Dion is an exception.[14]

In Dion, water was collected from sources in the foothills of Mount Olympus, about 6 miles away from the city and was flowing by gravity through an underground vaulted stone built conduit (Figure 3 a-b). In its terminal part it was built on an arcade and carried the water to the main distribution tank (*castellum*) located in the west side of the city wall. This compact aboveground water construction contained five arched openings, three of which correspond to two roads – one of them being bidirectional. The other two openings are above fortification ditches (Figure 3 c). From the main distribution tank, water was supplying the city via three secondary *castella* and pipes of clay or lead.[15] The aqueduct of Edessa has been identified in different plots within the modern city. Its arched stone canal, well waterproofed, was constructed at the end of the 3rd century AD (Figure 4).[16]

On the other hand, Philippi, due to the poor water quality of the area, received water mainly from karstic springs, about 8km northwest. A barrel vaulted aqueduct channel with cleaning manholes was carrying water in a central tank located at the city's southwestern slope. It was a quite expensive construction of the 2nd century AD (Figures 5, 6).[17] Similar structural features have been recognized in the remains of the Roman aqueduct of Amphipolis, with the vaulted covering of the channel detected both in situ, in a fragmentary preserved section of the pipeline (Figure 6).[18]

A series of vaulted water conduits have been revealed in the Kalamaria valley, southeast of the city of Thessaloniki, indicating the existence and operation of an important and extensive aqueduct (Figures 7, 8). Its location and function were benefited by the soft soil slopes of the region. The aqueduct was located during salvage excavations at three locations (Toumba, Pylaia, the American Farm School premises), as well as at various places in Thermi village. Finally, it was uncovered during a road construction, southwards of the public (country) road of Galatista – Vasilika. The aqueduct was constructed with rough stones and with manholes on its arch. It was probably built by emperor Galerius at the end of the 3rd century AD, and it was initially placed either in the Anthemountas valley or near the Thermi dam; its endpoint is located in Thessaloniki, about 150m north of Kalamaria's Gate in the eastern city walls, in the area of the Evangelistria cemeteries.[19]

[12] Kaiafa 2008, I: 184-193; Petkos 1989: 205-206.
[13] Valavanidou 2000: 138; Gala-Georgila 2015: 62-69; Kaiafa 2008, I: 391-393, Kaiafa 2010: 29-32.
[14] Kaiafa 2008, II: 178-182; 2013: 32; Kaiafa-Saropoulou *et al.* 2015: forthcoming.
[15] Kaiafa 2008, I: 238-247; 2014b: 182; Kaiafa and Karadedos 2011: 489-491; Karadedos 1985: 33-41; 1990α: 217-230; 1990β: 144-165; 1994: 155-166; 2000: 51-68)
[16] Kaiafa 2008, I: 52-53; Kaiafa-Saropoulou *et al.* 2015: in press.
[17] Kaiafa 2008, I: 516-519; Kranioti 1989: 475-482.
[18] Kaiafa 2008, I: 426-430.
[19] Doulgeri 1998: 219-220; Kaiafa 2008, I: 307-309, 394-396; Tsimbidou-Avloniti 2010: 375-384; Kaiafa-Saropoulou 2014b: 277-282; Gala-Georgila 2015: 69-76; Yannopoulos *et al.* 2017: 468..

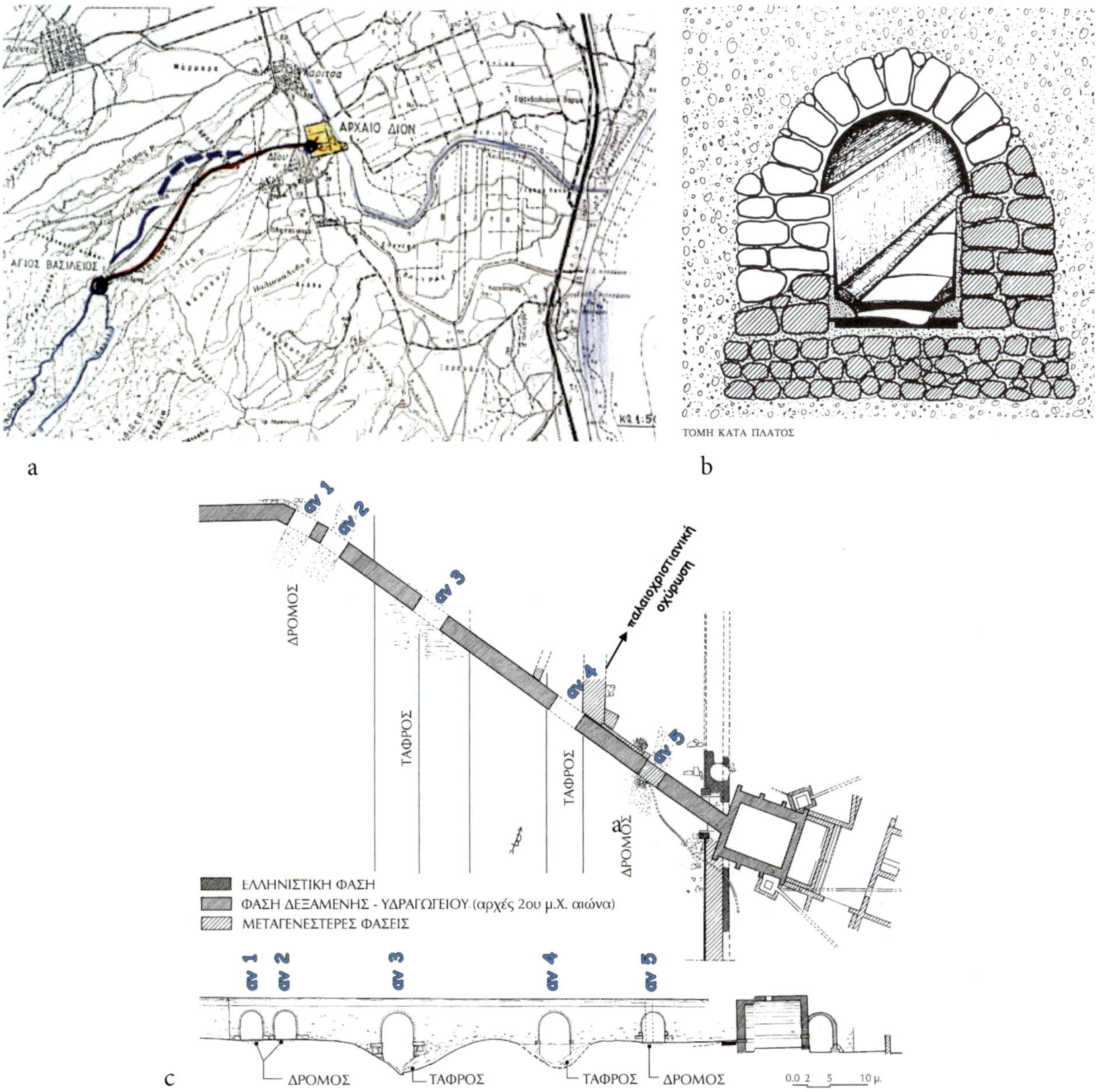

Figure 3. Dion. Aqueduct channel. a) the route form the sources to the distribution tank (after Karadedos 1990b: 149, fig. 6); b) cross section; (after Karadedos 2000: 58 plan 7); c) terminal part of an arcade, ground plan and section (after Karadedos 2000: 59 plan 8)

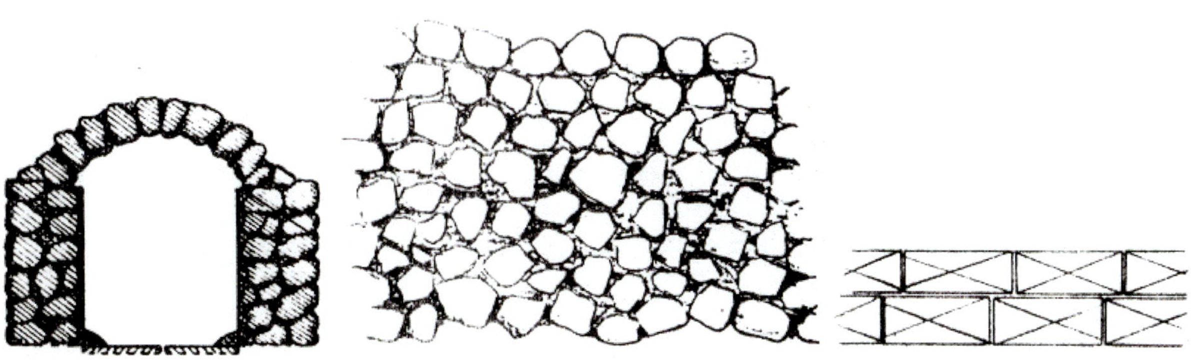

Figure 4. Edessa. Aqueduct, cross section, bed configuration and masonry (after Petsas 1975: 309 plan 15)

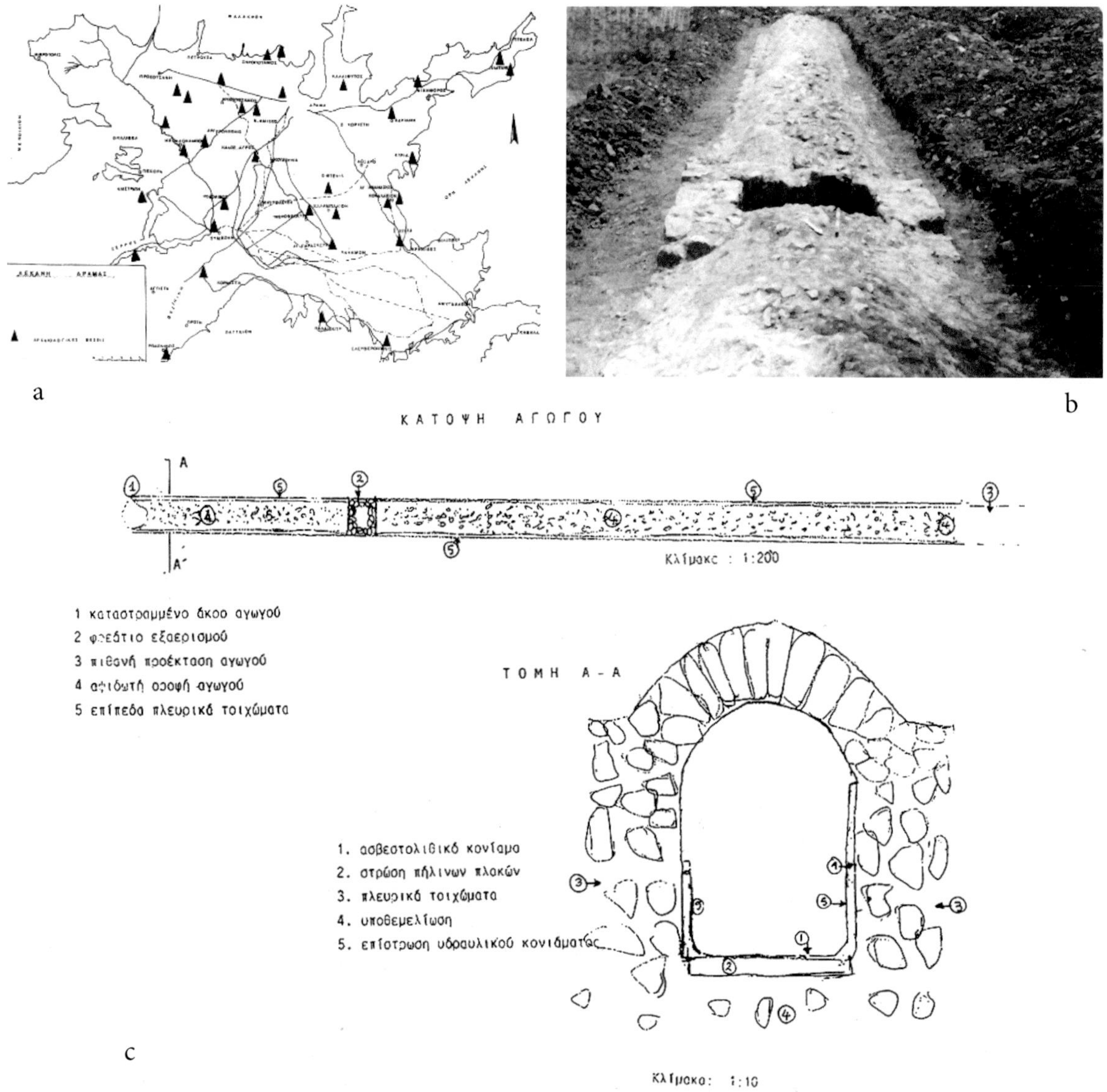

Figure 5. Philippi. Aqueduct channel. a) the sources in Kefalari; b) manhole (after Kranioti 1989: 481, fig. 3); c) ground plan and cross section (after Kranioti1989: 476, plan 2)

General construction features and differentiations

Regarding their construction, slight variations in materials and construction methods of the walls and arch forms can be identified, as well as small differentiations in their dimensions. These differentiations are not related to the exact time of their construction. In cross section all conduits were rectangular. Their sizes naturally varied, but the measurements from Philippi (80cm wide X 1,20m high), Dion (55cm wide X 1,00m high) and Kalamaria aqueducts (43cm-55cm wide X 1,10m-1,70m high) provide us with the general proportions (Figures 3 b; 5 b; 8 a-c). They were all masonry built with vertical sides and paved bottoms and with a slight gradient in order to regulate the water flow and avoid erosions. The waterproofed lime rubble masonry is a common feature in all examples. The Amphipolis aqueduct channel differentiates since a large amount of both bricks and marble members were used for the vaulted cover, together with stones and lime mortar (Figure 6). Likewise, bricks were also used for the covered construction of the Kalamaria aqueduct. The brick arch has an external shell made of small stones and mortar.

An important feature of all known aqueduct channels was their waterproofing by a mixture of quicklime and crushed brick, which made them impermeable to leaks, or seepage (Figures 3 b; 5 c). The hydraulic material was used to cover the inner side walls and the floor, but not the surface of the semi-cylindrical cover. Several layers were laid each time, while larger cement quantities were applied on the junction

Figure 6. Amphipolis. Remains of the aqueduct channel (after Malama 2001: fig. 2)

of the vertical walls with the conduit's bottom, this way forming a thick layer of convex section. These practices, applied so as to prevent leakages, were also particularly common in the similar structures throughout the Empire.[20] In the case of Philippi, underneath the hydraulic (cement) coating of the vaulted channel, a thick layer of plaster -which smoothed the surfaces of the side walls- was spread, so as the surface to be continuous and uniformed.

The configuration of the bed with bricks is applied to the reservoirs of Philippi and Edessa. Conduit briquettes at the Dion aqueduct exhibit a slight curvature on their surface. Bricks are usually placed on a substrate of a hydraulic material, and depending on the floor width, were laid in a single or a double row. This way, the narrow bottom of the Dion duct (55cm width) allowed placement of the blocks in only one single row (Figure 3 b and c), while two rows of elongated bricks were necessary to cover the wider surface of water flow in the case of the Edessa aqueduct, which reaches 70cm (Figure 4). At Philippi, the width of the aqueduct pipe was measured at 80cm. On the other hand unfortunately, the configuration of the channel bottom of the Amphipolis aqueduct remains unknown, since its research is still pending. Finally, the vaulted ducts of the Kalamaria aqueduct were constructed of clay lined with hydraulic mortar on the bearing surface, the width of which varied from 0.43m to 0.55m.

Two variations are identified regarding the way the arched cover crowns the sidewalls: the first one is represented by the channels of the Kalamaria aqueduct; its walls and arch form a single inseparable unit of the same density. On the other side, the semi-cylindrical arch of the Dion duct is less thick than its vertical sides (Figure 3 b and c). The vault recessed back from the tops of the vertical side walls, thus forming a ledge. It was probably constructed of wooden formwork and it seems it was deposited upon the solid rectangular structure, this way ensuring an internal height of 1.00m. Similarly, the arch of the Edessa aqueduct channel was also constructed by means of formwork, and is visibly lighter than the double thick vertical sides, upon which it rests (Figure 4). Finally, the arched construction of the Philippi aqueduct, which is 0.20 m. higher than the previous two, was raised from the side walls in a way to form a uniform flat ridge on both long sides (Figure 5 b and c).

Furthermore, some of the vaulted aqueducts were equipped with manholes or inspection shafts, thus providing practical confirmation of the high level of hydraulic technology applied at the Macedonian region. These features are detected only on top of vaulted channels and served for maintenance and cleaning purposes. When the conduit was roofed by flat cover slabs, these could simply be removed to provide access for cleaning. In every case the openings were square, unlike the circular cross-section openings which were common to the rest of the Empire.[21] Regarding the aqueduct channel of Philippi, its course is interrupted by square shaped manholes, built of stones and mortar, and having the same width as the conduit on its outer dimensions (Figure 5 b and c). Inspection shafts perforated the vaulted cover

[20] Hodge 1992: 98.
[21] Hodge 1992: 102.

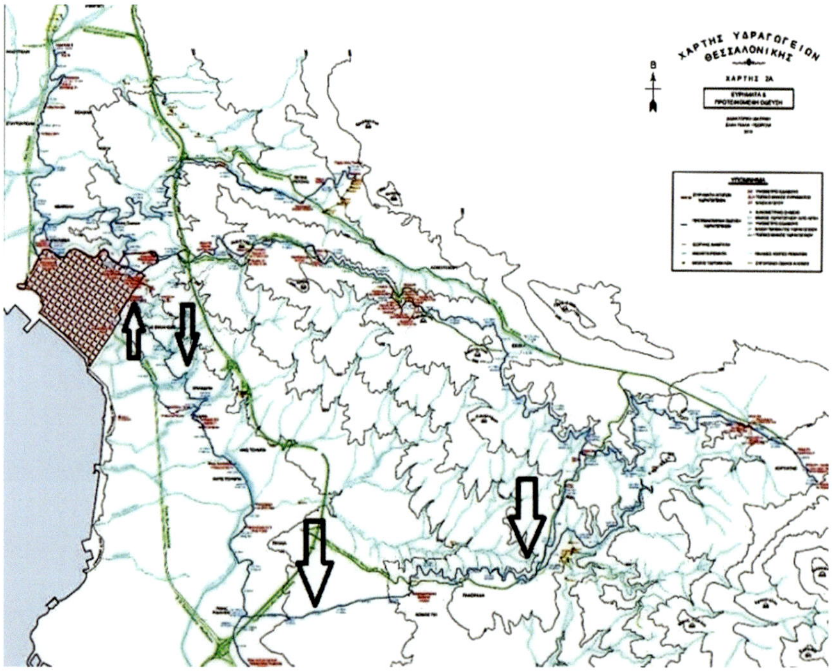

Figure 7. Kalamaria valley. Known aqueducts at Toumba, Thermi, Pylaia. The route from the sources to the city (after Gala – Georgila 2015: map 2)

a

b

c

Figure 8. Kalamaria valley. a) Aqueduct channel with an inspection shaft; b) channel (after Doulgeri 1998: 222, fig. 1.3); c) cross section (courtesy of the former 16th Ephorate of Antiquities of Thessaloniki)

of the Kalamaria aqueduct – at least in the parts running through Toumba, Pylaia and Thermi (Figure 7). They were prominently projecting some centimeters above the rounded top of the channel, resembling the conning tower of a submarine (Figure 8 a, b and c). They were constructed of few scattered brick fragments from the masonry and were covered by monolithic slabs. This type of structures, which were slightly protruding from the arched top, were also found in the channels of Gier aqueduct in Lyon, and also in the aqueduct conduits of Cologne and Trier, but also in those in the eastern provinces of the Roman Empire, such as the Madradag aqueduct in Pergamon.[22] Regarding the aqueducts in Macedonia region, no constant distance between the manholes is measured, since the intervals between them vary from 11.30m, 22.00m or even 33.00m. This disorderly repetition is observed not only on the Kalamaria aqueduct, but also on other main water channels all over the Roman Empire, though Vitruvius (*De Architectura*, VIII, 6, 3) specifies the ideal interval to be one actus, which is 35m. On the other hand, according to Pliny (*Historia Naturalis*, XXXI, 3.1), the most appropriate distance is 2 actus, i.e. 75m.[23]

Within the same framework, in Macedonia region, care for maintaining the proper water gradient in the free flow ducts was with certainty reflected at the Dion aqueduct channel.[24] Its final section was made of a continuous stone wall with a pivoting path to halt the momentum of the water flow, which permeated the jambs of the gate of the Hellenistic fortification ending in the central tank of the inner city. Slight gradient in order to regulate the water flow and avoid erosions is also observed in the aqueduct of Kalamaria, where the slope is estimated at 2.2 ‰ -1.2 ‰.[25]

Conclusions

Based on the above discussion, it is evidenciated that Macedonian engineers acquired a deep knowledge on hydraulic technology. Thus, they were capable to supply their cities with both impressive and effective aqueducts, using the same engineering principles which were diffused throughout the Roman Empire. The sufficiency of Macedonian cities in water was also depended on open-channels, constructed in accordance with the new technological achievements and applications – either with flat stone slab roofs or with the more impressive vaulted covers. Vaulted roof aqueduct channels are already found in Dion, Amphipolis, Edessa, Philippi, as well as in the wider area east of Thessaloniki (Kalamaria Valley) (Figure 2).

They are all rectangular in cross-section, following a subterranean course. The terminal part of the Dion aqueduct constitutes an exception, since its water channel was carried upon an arcade, so as to maintain the water level high enough to provide efficient service within the city.

Slight variations in materials and construction methods of the walls and arch forms can be detected, as well as small differentiations in their dimensions. These differentiations are not always related to the initial construction period. They were all masonry built with vertical sides and paved bottoms, with slight gradient in order to regulate the water flow and avoid erosions. The vaulted ceilings were usually formed of stones or bricks and sometimes were less thick than the vertical walls on which they were standing. Water proofed lime mortar generally covered the inner side walls and the floors, but not the surfaces of the semi cylindrical covers, so as to prevent leakages. In addition, square cross-section manholes and inspection shafts supplemented the water projects, for maintenance and cleaning purposes.

Bibliography

Besios, M. and Pappa, M. 1998. Νεολιθικός οικισμός Μακρυγιάλου. *To Archaeologiko Ergo stē Macedonia kai stē Thrakē (=AEMTh)* 8: 137-146.

22 Burdy 1991: 36-38; Fahlbusch 1982: 170-171; Hodge 1992: 94, 101-102.
23 Kaiafa 2008, II: 68-69, note 54.
24 Kaiafa 2008, II: 68, 181.
25 Gala-Georgila 2015: 71.

Burdy, J. 1991. Some directions of future research for the aqueducts of Lugdunum (Lyon). In: Tr. Hodge (ed.) *Future currents in Aqueduct Studies*: 29-44. Francis Cairns Publications.

Chionides, G. 1960. *Ιστορία της Βέροιας*, vol. I. Veroia.

Crouch, D. 1993. *Water Management in ancient Greek cities*. Oxford: Oxford University Press.

Doulgeri, E. 1998. Σωστικές ανασκαφές υδραγωγών στη Θέρμη και Τούμπα Θεσσαλονίκης. *To Archaeologiko Ergo stē Macedonia kai stē Thrakē (=AEMTh)* 12: 217-222.

Fahlbusch, H. 1982. *Vergleich antiker griechischer und römischer Wasserversorgungsanlagen*. Mitteilungen des Leichweiss Institut für Wasserbau Heft 73.

Gala-Georgila, E. 2015. Δρόμοι του νερού και οργάνωση του χώρου στη Θεσσαλονίκη κατά τη Μέση και Ύστερη Βυζαντινή περίοδο. Unpublished PhD dissertation, Aristotle University of Thessaloniki.

Girtzy, M. 2001. *Historical Topography of Ancient Macedonia, Cities and other settlement - sites in the late Classical and Hellenistic period*. Thessaloniki: University Studio Press.

Hodge, Tr. 1992. *Roman aqueducts and water supply*. London: Duckworth Archaeology.

Kaiafa, M. 2008. Συστήματα ύδρευσης και αποχέτευσης κατά την ελληνιστική και ρωμαϊκή περίοδο στη Μακεδονία. Unpublished PhD dissertation, vols. I-II, Aristotle University of Thessaloniki.

Kaiafa, M. 2010. Ελληνιστική και ρωμαϊκή Θεσσαλονίκη: ζητήματα υδραυλικών υποδομών. *Thessalonikeon Polis* 8: 23-36.

Kaiafa-Saropoulou, A. 2014a. Supplying water in Hellenistic and Roman Macedonia: the art of the wells. In: I. K. Kalavrouziotis and A. N. Angelakis (eds) *Wastewater and Environment: Traditions and Culture, 1st IWA Regional Symposium on Water 22-24 March 2014*: 557-568. Patras: Hellenic Open University, Patras, Greece.

Kaiafa-Saropoulou, G. A. 2014b. Water management and sanitation technologies in Hellenistic and Roman Thessaloniki. In: *10th International Hydrogeological congress, with emphasis on Groundwater depended ecosystems* Vol. 1: 277-286. Thessaloniki.

Kaiafa, M. and Karadedos, G. 2011. Η διαχείριση του νερού στην αρχαία Μακεδονία. In: Δ. Β. Γραμμένος (επιμ.) *Μακεδονία, από τον 7ο αι. ως την ύστερη αρχαιότητα. Μελέτες και λήμματα για την τρίτη εκθεσιακή ενότητα της μόνιμης έκθεσης του Αρχαιολογικού Μουσείου Θεσσαλονίκης*: 477-500. Thessaloniki: Zitros Publications.

Kaiafa, M. and Karadedos, G. 2013. Έργα ύδρευσης και αποχέτευσης στο Μακεδονικό χώρο, In: K. Βουδούρης (*επιμ.*) *Υδροτεχνολογίες στην Αρχαία Ελλάδα/* Hydrotechnologies in Ancient Greece, organized by the Aristotle University of Thessaloniki's Integrated Water Management Centre (ΚΕΟΔΥ), one-day conference, 23 March 2013: 29-38. Thessaloniki: AUTh.

Kaiafa, A., Papanikolaou, E., Melfos, V., Papacharalabou, Chr. and Voudouris, K. 2014. Sanitation and wastewater and stormwater management in ancient Kingdom of Macedonia. In: A. N. Angelakis and J.B. Rose (eds) *Evolution of Sanitation and wastewater Technologies through the centuries, Chapter 8*, IWA: 177-192. London: IWA Publishing.

Karadedos, G. 1986. Ύδρευση και αποχέτευση στο αρχαίο Δίον. In: *Οι αρχαιολόγοι μιλούν για την Πιερία*, Καλοκαίρι 1985: 33-41. Pieria: Prefecture Committee for Popular Education (Ν.Ε.Λ.Ε.).

Karadedos, G. 1990a. Το υδραγωγείο και η κεντρική δεξαμενή του αρχαίου Δίου. *To Archaeologiko Ergo stē Macedonia kai stē Thrakē (=AEMTh)* 4: 217-230.

Karadedos, G. 1990b. Υδραυλικές γνώσεις και τεχνολογία των αρχαίων. *Επιστημονική Επετηρίδα της Πολυτεχνικής Σχολής, Τμήμα Αρχιτεκτόνων*, vol. ΙΒ': 143-172. Thessaloniki.

Karadedos, G. 1994. Η Ύδρευση του αρχαίου Δίου από τα νερά του Ολύμπου. In: *Πρακτικά Ζ΄ Πανελληνίου Συνεδρίου 'Ο Όλυμπος στους αιώνες'*: 155-166. Elassona: Cultural Organization of the Municipality of Elassona.

Karadedos, G. 2000. Υδραυλικές γνώσεις και τεχνολογία – Η περίπτωση του Δίου. In: Κ. Σουέρεφ (επιμ.) *Υδάτινες σχέσεις, Το νερό ως πηγή ζωής κατά την αρχαιότητα*: 51-68. Thessaloniki: University Studio Press.

Kotzias, Ch. N. 1948-1949. Λείβηθρα, Πίμπλεια, Πιερίς, Η πατρίς του Ορφέως. *Archaeologike Ephemeris*: 25-40.

Kranioti, L. 1989. Αρχαίος αγωγός στο Κεφαλάρι Δράμας. *To Archaeologiko Ergo stē Macedonia kai stē Thrakē (=AEMTh)* 3: 475-482.

Malama, P. 2001. Νεότερα στοιχεία από το ανατολικό νεκροταφείο της Αμφίπολης. *To Archaeologiko Ergo stē Macedonia kai stē Thrakē (=AEMTh)* 15: 111-126.

Petkos, A. 1989. Το δίκτυο υδρεύσεως της Βέροιας. *Archaeologika Analekta Athenon* XXII: 197-210.

Petsas, F. 1975. Χρονικά Ἀρχαιολογικά. *Makedonika* 15: 171-355.

Pikoulas, G. 1997. Το υδραγωγείον της Αμφιπόλεως. In: Μορφωτικός Όμιλος Κομοτηνής (επιμ.) *Πρακτικά 2ου Διεθνούς Συμποσίου Θρακικών Σπουδών, Αρχαία Θράκη*, vol. 2: 605-611. Komotini: Centre of Thracian Studies.

Pikoulas, G. 2001. *Η χώρα των Πιέρων, συμβολή στην τοπογραφία της*. Athens, Municipality of Pieria, Kavala, and The National Hellenic Research Foundation/ Research Centre for Greek and Roman Antiquity (Ethnikon Hidryma Ereunōn/ Kentron Hellēnikēs kai Rōmaïkēs Archaiotētos [EIE/KEPA]).

Robinson, M.D. 1946. *Excavations at Olynthus. Part XII, Domestic and Public architecture* (The Johns Hopkins University Studies in Archaeology 36). Baltimore: The Johns Hopkins Press.

Samsaris, D. 1976. *Ιστορική Γεωγραφία της Ανατολικής Μακεδονίας κατά την αρχαιότητα* (Society of Macedonian studies Nr. 49). Thessaloniki.

Sismanides, K. 1995. Ανασκαφή αρχαίων Σταγείρων 1992. *To Archaeologiko Ergo stē Macedonia kai stē Thrakē (=AEMTh)* 6: 451-465

Sismanides, K. 2003. *Αρχαία Στάγειρα, η πατρίδα του Αριστοτέλη*. Athens: Ministry of Culture. Archaeological Receipts Fund.

Tsimbidou-Avloniti, M. 2010. Πυλαία 2001-2010, Ανασκαφικές ψηφίδες από μια άγνωστη περιοχή. *To Archaeologiko Ergo stē Macedonia kai stē Thrakē (=AEMTh)* 24: 375-384.

Valavanidou, A. 2000. Η υδροδότηση των πόλεων: η περίπτωση της Θεσσαλονίκης. *Thessalonikeon Polis* 1: 133-142.

Yannopoulos, S., Kaiafa-Saropoulou, A., Gala-Georgila, E., Eleftheriadou, E. 2017, Evolutionn of water supply infrastructures of the Thessaloniki ity, Hellas through centuries. *European Water* 58: 465-472.

Dr Asimina Kaiafa-Saropoulou
Archaeologist
Independent Researcher of Water Archaeology
Thessaloniki, Greece
minakasar@gmail.com

The aqueduct of Actian Nicopolis[1]

Konstantinos L. Zachos and Leonidas Leontaris

ἤθελόν πώς ποτε πάντα μαθὼν ἐπανελθεῖν,
ἀλλὰ πολλοῦ πόνου χρεία καὶ οὐδεὶς οὐδὲν
πέμπει καὶ ἐν Νικοπόλει σαπρῶς λούει τὰ βαλανεῖα
καὶ ἐν οἴκῳ κακῶς καὶ ὧδε κακῶς.
Επίκτητος, Διατριβαί, Περὶ ἀνομολογίας, 2.21.14.

I wish I could learn everything before my return;
but this requires much labor,
and nobody sends me anything.
The baths are very bad at Nicopolis;
and things go very ill both at home and here

Epictetus, *Discourses, Of inconsistency*, 2.21.14.

Abstract

From the abundant sources of the Louros River to the modern village of Haghios Georgios, to the walls of Nicopolis, the Aqueduct travelled a distance of about 50 km. Its construction was a significant achievement in hydraulic engineering of the Imperial period in mainland Greece, comparable to the great aqueducts of Corinth and Athens. This paper is presenting the most recent research conducted at the aqueduct remains which managed to provide answers to important questions regarding the aqueduct's chronological and building phases, its course and function as the city's water supplier.

The paper also provides a preliminary but significant description on the twin monumental Nymphaea of the city, that were directly supplied by the aqueduct.

Keywords: Nicopolis, Louros, Aqueduct, Waterbridges, Nymphaea.

Introduction

In memory of his victory at the naval battle of Actium (2 September 31 BC), Augustus, following the tradition inaugurated by Alexander the Great which was continued by Hellenistic rulers and adopted by Roman generals like Pompey, founded Nicopolis, the 'city of victory'.[2] The location selected for the city's foundation was not near the site of the battle. Rather, it was about 8.5km away, in a straight line to the north, at the isthmus of the peninsula of Preveza. The geopolitical advantages on grounds of maritime communications with the Italian peninsula and the coasts around the fish-farming area of the Ambracian Gulf, whence roads led to the inland regions of Epirus, the fertile expanses of the peninsula of Preveza, and water sources in the plain of Nicopolis were the primary factors determining the selection of this location.[3] Augustus had camped in this area during the months intervening between his arrival at the coast of Epirus and his departure a few days after the battle. The plain of Nicopolis is bounded on the north by the Michalitsi hills. Augustus had installed his *praetorium* on one of these hills, overlooking the Ambracian Gulf and Ionian Sea. Infantry and cavalry forces had camped further down as far as the bay of

[1] This contribution to research on Nicopolis is dedicated to the memory of Alexandros Mantis, who excavated the region of the Nymphaea of Nicopolis early in his career. ΑΛΕΞΑΝΔΡΕ ΑΓΑΘΕ ΧΑΙΡΕ! Our warm thanks to Anthi Angeli, Head of the Ephorate of Antiquities of Preveza for facilitating us during preparation of this study, to Nikos Choinas for his suggestions, and to Angelika Douzougli and Ioulia Stamou for their support. Our warm thanks also go to Professor Emeritus Theodosios Tassios and Dr Georgia Aristodemou for their kind invitation to contribute to the present volume and to Deborah Brown-Kazazis who undertook the English translation of our text.

[2] Jones 1987.

[3] In the plain of Nicopolis, exploitation of the aquifer found at a shallow depth (phreatic zone) was until recently accomplished with wells, many of which can be found even today scattered across the archaeological site, built from ancient construction materials. For natural sources of water in the region see Zachos 2007a: 274-275.

Monolithi, where the fleet had anchored. The encampment was fortified by long walls extending from the hill on which the command post was established down to the coast, thus ensuring direct communication with his ships.[4] The location of Augustus's camp and the city itself acquired a sacred character in harmony with the religious symbolism which the new world order embedded in the narrative of events at Actium. The episode of the 'good omen', the ruler's encounter with the muleteer Eutychos at dawn on the day of the battle occurred near Augustus's headquarters. The meeting foretold the foundation of Nicopolis,[5] which was created through a synoecism of Greek residents from the surrounding areas and Roman veterans.[6] Excavation to date has encompassed only a small part of the extensive ruins of Nicopolis. However, the resulting evidence amply supports its characterization as an *urbs Romana.*[7] Nicopolis's city plan and the construction of its buildings were executed on the basis of a programmatic planning after Roman models, which premeditated the apportionment of space in rectangular building blocks on either side of the two main streets, the *decumanus maximus* which intersects the perpendicular *cardo maximus,* and the erection of public buildings at specific locations.[8] Streets, at least the main ones, were paved and flanked by sidewalks. Beneath the streets, a system of drains was dug out, to collect wastewater and channel it towards the sea (on the Ambracian Gulf coast at the site of today's Mazoma lagoon). The *Forum*, the center of political and administrative life, was located westwards of the *cardo maximus,* near the Odeum, though archaeological evidence on this remains insufficient. Roman practices were also applied as regards construction materials and building methods. The city's programmatic foundation plan also included a surrounding brick wall. This wall, with a length of about 5 km, enclosed an area of 150 hectares. The Augustan wall, an impressive infrastructure work due to its size (which entailed a tremendous outlay), reflected the generosity and power of Nicopolis's imperial founder, suggesting the Greek archaic and classical tradition of the *hero-founders* (*ktistēs*).[9] As in most urban centers of the Roman Empire, the infrastructure of Nicopolis included also its preserved aqueduct – although according to the evidence presented below this was not erected in the Augustan period.[10] The aqueduct of Nicopolis met not only the needs of the city itself, but also those of the city's north *suburbium*, called by Strabo the *Proasteion*, where the infrastructure for the quadrennial Actian Games was built. Athletes from the wider Empire, but mainly from the eastern provinces, participated in these Games.[11] Nicopolis had three harbors: one on the Ionian coast, named Komaros;[12] a second one at the coast of the Ambracian Gulf in the bay of what is today the Mazoma Lagoon, and a third at Vathy Bay, 4 km directly to the south of the city.[13] Nicopolis's aqueduct did not supply this third harbor, the city's main commercial port, nor the settlement which developed along the west coast of the bay.[14] Rather, water was provided by other source of water near the Vathy area.

The water supply at the springs of Haghios Georgios and the bridges on the Louros river

Water supplied to the Nicopolis aqueduct came from the springs of Haghios Georgios, located near the homonymous village in the prefecture of Preveza (Figure 1: point 1), which today gush forth over the slopes of the limestone hill of Isioma at an absolute altitude of 111.82m (Figure 2: point 1). These same springs, whose average water flow today is around 2.8 m^3/sec,[15] currently supply the cities of Arta, Preveza, and Lefkada in addition to a number of villages.[16] The remaining water coming from modern

[4] Dio Cassius, 50.12.4

[5] Plutarch, *Antonius,* 65.3; Suetonius, *Augustus,* 96.2.

[6] Purcell 1987; Ruscu 2006; Isager 2009.

[7] See in brief Zachos 2015: 46-49.

[8] Andreou 2007; Zachos 2007a: 288-295.

[9] Zachos 2007a: 273-288.

[10] See Zachos 2015, passim.

[11] On the Actian games (*Aktia*), see Tidman 1950; Sarikakis 1967; Lämmer 1986; Zachos 2008; Pavlogiannis *et al.* 2009.

[12] Strabo, *Geography*, 7.7.5; Dio Cassius 50.12.4.

[13] On the coastal evolution of Mazoma Lagoon and Ormos Vathy, see Jing and Rapp 2003: 164-177.

[14] Wiseman 2001: 49-50; Stein 2001: 66-70.

[15] The hill of Isioma belongs to the limestone formations of Pantocrator (Upper Triassic-Lower Jurassic). The limestone, which was off-white to ash-colored, appears throughout the length of the Ionian Zone. The stone is solid, no-layered, and its thickness reaches a depth of 1000m. The gushing spring at Haghios Georgios is attributed to a difference in permeability of the limestone at Pantocrator, see Katsanou 2012: 29, 170.

[16] The total population supplied with water from these springs today is around 70,000.

Figure 1. The course of Nicopolis's aqueduct.

water tapping flows into the Louros River, the ancient Oropos,[17] which flows beneath the springs and into the Ambracian Gulf. The initial source of the Louros lies several kilometers to the north on the eastern slopes of Mt Tomaros. Along its course it is fed by other springs like those of Haghios Georgios as well as by tributaries.

[17] Based on a – now lost – Hellenistic inscription incorporated into the masonry of the east row of the aqueduct piers at the springs of Haghios Georgios, which was examined and published by Leake (1847), various scholars have suggested that the ancient name of the river was Oropos [ρωπωποταμω/καθιερ(ω)σανευχα]. On the above discussion, Karatzeni 1997: 239-246. Del Barrio Vega (1998) proposed another reading of the inscription, dating it to the time of Julian in accordance with the benefactions of the Emperor to the city mentioned in the panegyric of Claudius Mamertinus (see below, note 64), though there is no evidence for this.

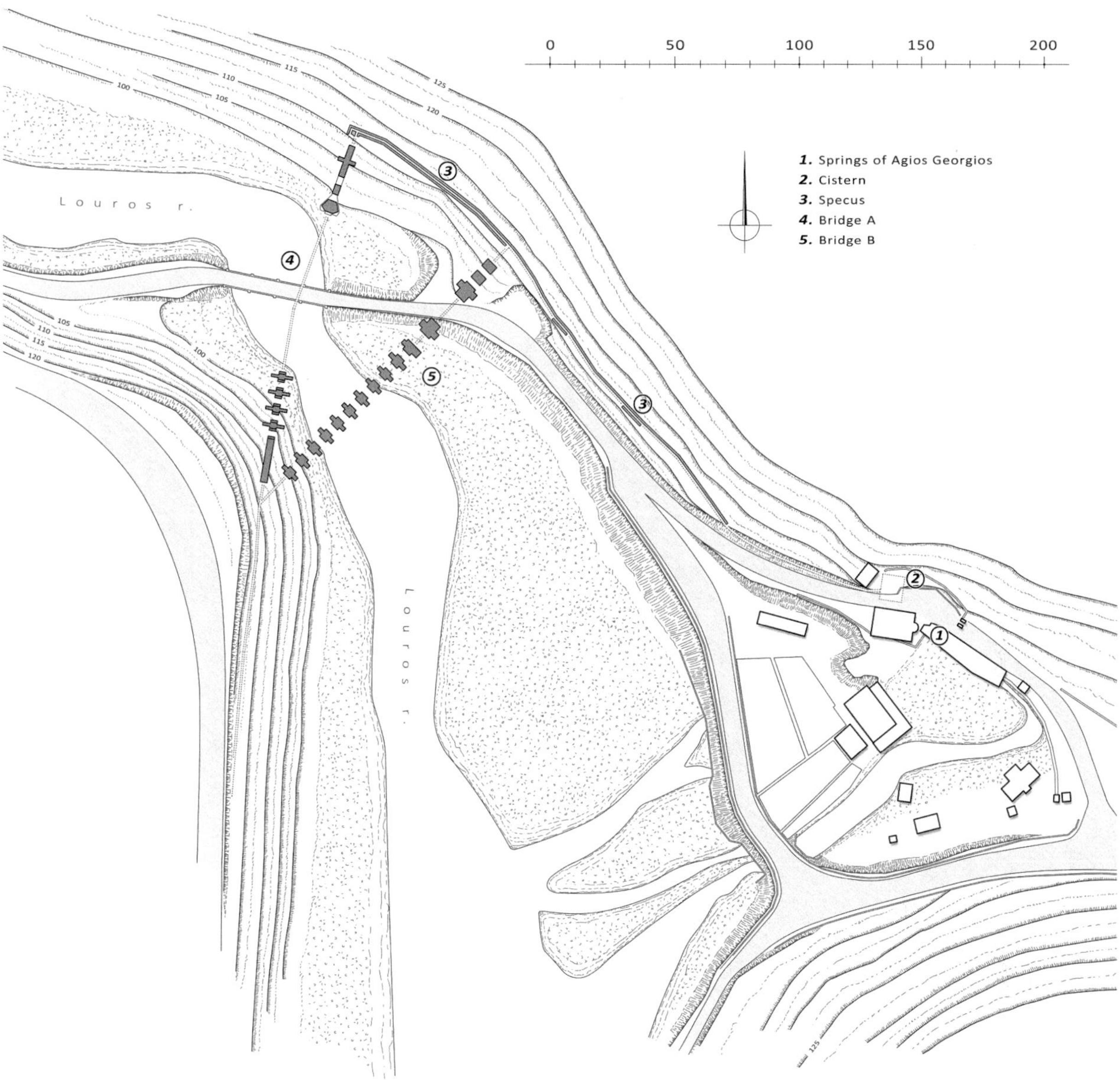

Figure 2. Topographical plan of the area of the source of the aqueduct.
(drawing: L. Leontaris).

At Haghios Georgios the water was accumulated in a rectangular collecting basin situated at the place where the church of Haghios Georgios stands today.[18] The dimensions of the basin, traces of which were identified in 1978 below the paving of the church courtyard in the course of work being carried out there,[19] are estimated at 9 x 10m, although these remain undocumented by archaeological evidence.[20] (Figure 2: point 2) From the basin, water was channeled into the conduit (*specus*) along the length of the slope of the hill, remains of which are preserved to a length of about 230m (Figure 2: point 3). This was a conduit built into the rock in the *opus incertum* system, which followed the contour lines of the ground after appropriate hewing of the rock to achieve the desired incline of the channel floor. Where appropriate, and depending on the natural state of the rock, the carved walls also functioned as the walls

[18] Pouqueville (1826: 238), who visited the site in the first decades of the 19th century, saw the ruins of the basin, which he described as follows: '*On remarque à quelques pieds au-dessous de leur chute un bassin ou château-d'eau dans lequel elles étaient reçues pour les élever à la hauteur d'un canal creusé dans le rocher, qui les versait à quarante toises le là sur un double aqueduct construit en brique*'.

[19] Aggeli *et al.* 2015: 46.

[20] Riginos 2013: 99, note 29.

Figure 3. Bird's-eye view from the East of the area of the source.

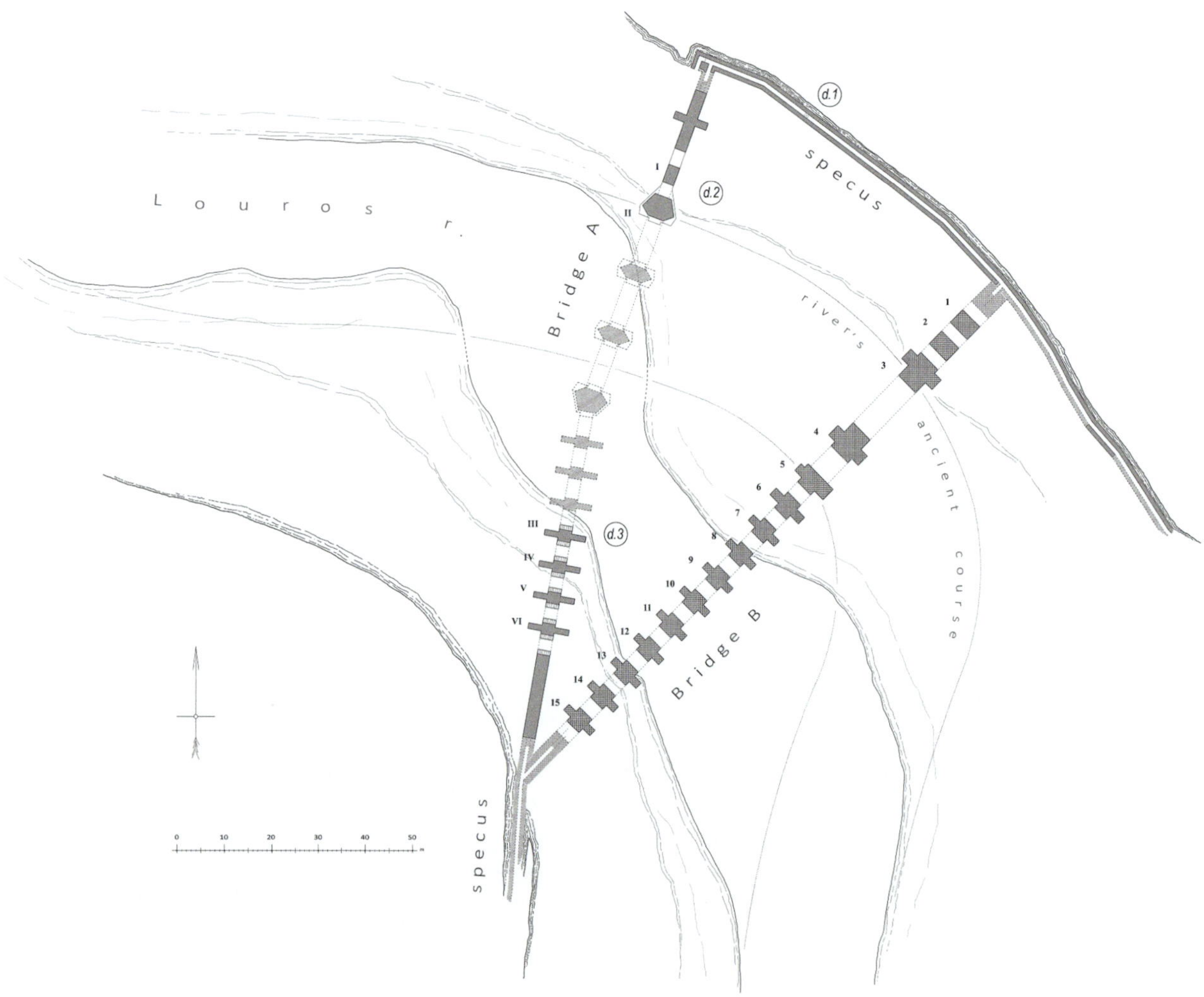

Figure 4. Ground plan of the aqueduct's bridges at Haghios Georgios (drawing: L. Leontaris).

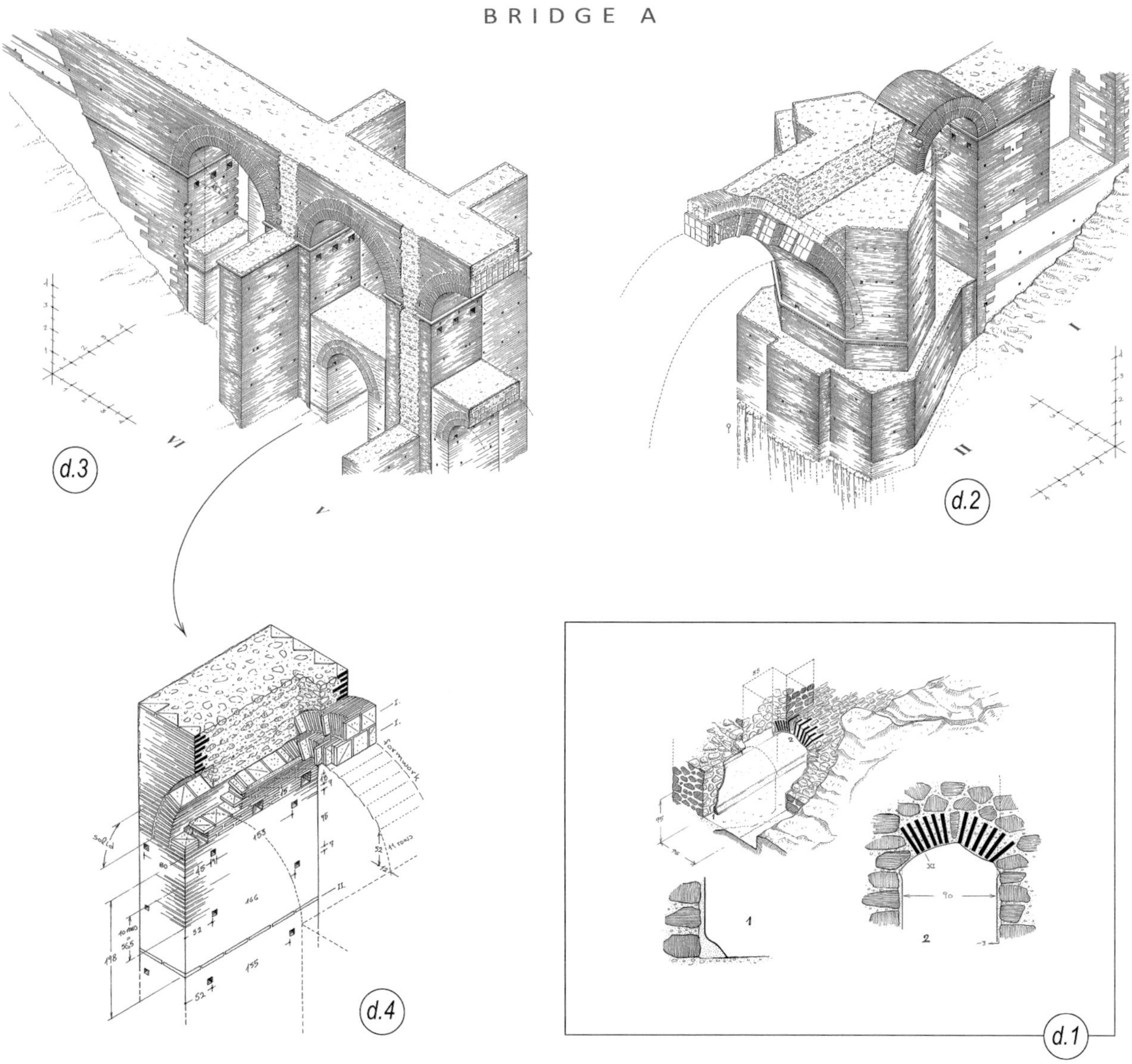

Figure 5. Drawing details describing the construction of the specus (d.1), north part (d.2), south part (d.3) and supporting vaulted constructions (d.4) of bridge A (drawing: L. Leontaris).

of the conduit, which had a width of 0.90m and maximum height of 1.20m, beginning its course from an absolute altitude of about 112m[21] to terminate at its westernmost end 0.50m below its original altitude, showing a gradient of 0.2%.[22] The built section of the conduit was covered by a pointed barrel vault of radially-positioned stones and bricks (Figure 5: d.1). Inspection shafts of roughly square plan (0.90 x 0.85m) were opened at intervals; there are traces of one of these 25m from the western end of the conduit (Figure 4: d.1).[23] The conduit was coated in solid, fine-grained hydraulic mortar which is preserved at many points. At the two bottom corners, the mortar was reinforced forming a continuous quarter round strip protecting the construction against cracks and water penetrating to the mass of the cement (Figure 5: d.1.1).[24] From the hill slope, the conduit made a right-angle turn and continued an aboveground course

[21] According to the preserved evidence, ancient tapping/catchment was carried out above where it is today. This indicates the depletion of the aquifer over the course of time.

[22] The conduit terminated at an altitude of 33.40m in the north nymphaeum of Nicopolis's West Gate (see relevant chapter below).

[23] According to Vitruvius (*De Architectura*, VIII, 6, 3: *puteique ita sint facti, uti inter duos sit actus*), the ideal distance between two shafts is one *actus*, i.e. approximately 35m (*c.* 120ft). According to Pliny (*Historia Naturalis*, XXXI, 3.1), the most appropriate distance is 2 *actus*, i.e. approximately 75m. Concerning the conduit at the hill slope, also Merkouri 2017: 538-542.

[24] A similar configuration is presented by the conduits of Roman aqueducts throughout the Empire, see Hodge 1995: 95-97. In the Greek region, this configuration is observed in the aqueducts of Corinth (Lolos 2010: 17-18), and Dion (Karadedos 1990: 220).

Figure 6. View of bridge A from the northeast.

along two rows of piers that bridged the Louros gorge. Beside the points where the conduit turned, sluices for relieving and regulating water pressure were created, as is indicated by the grooves in the side walls of the conduit to receive vertical sliding sluice gates. The two bridges (*henceforth, the north bridge will be referred to as bridge A, and the south one as bridge B)* which ran from NE to SW (Figure 2: points 4 and 5; Figures 3; 4), which were about 70m apart at the points where they were connected to the conduit on the hillside, converge at an acute angle of about 35° and terminate in a joint conduit on the opposite slopes of the gorge.[25] The conduit's route from this point to a distance of 1.5km in a straight line up to the location of Kokkinopilos has not been located. It appears to have run along the western slopes of the Louros gorge, generally coinciding with the route of the Ioannina-Arta national road, whose construction destroyed (or covered) any traces of the conduit.[26] Bridge A was configured as follows (Figures 4; 5; 6): at the beginning, the conduit was carried on a solid wall-*substructio* (thickness 2.40m, length 13m) reinforced by buttresses projecting 2.50m and resembling a cross. It is founded on steeply-sloping rock, and built in the *opus mixtum* construction system. Apart from the corners which connect the *substructio* to the buttresses, all the other edges of the construction are reinforced by brick masonry,[27] while horizontal zones of five brick courses run along the face of the wall at intervals.[28] The wall continues as far as the base of a pier which is rectangular in plan, and whose superstructure is reinforced with brick in the *opus testaceum* system. The wall and pier are joined at the top by a single arch built with a double row of brick voussoirs (0.29 x 0.42/0.43 x 0.05/0.06 decreasing to 0.04m) set radially and in correspondence to each other. On the faces of the pier, metopes are configured in masonry (*opus incertum*), which for better cohesion with the brickwork are in an ⊐⊏-shape.[29] At the arch springer, an impost moulding is formed with three brick courses running around the pier. At the same level, another identical in section moulding decorates

[25] The English traveler W. Leake, who visited the Preveza region in the early 19th century, identified and roughly described the ruins of the aqueduct of Nicopolis from the outskirts of the city to the Michalitsi hills, into a gorge behind the hills, to the Lamari plain in the region of the modern-day village of Archangelos, to Kanza on the slopes of a hill, and to the springs of Haghios Georgios, where he sketched the ground plan of the two rows of piers. See Leake 1835, I: 244, 256, 258-261; IV: 45, 47.

[26] Doukellis *et al.* 1995: 215.

[27] The bricks (0.27/0.28 x 0.27/0.28 x 0.035/0.040m) installed were more often cut in two half pieces and having their one side visible, with a horizontal joint measured about 0.025-0.37m between courses. More rarely they were employed cut into triangles – mainly in the brick courses of the piers. On the techniques used in cutting bricks, see Lugli 1957: 542-547.

[28] These courses did not pass through the thickness of the constructions; thus, they were not bonding courses. Their most likely role was related to artisans' constant efforts to create levels in the stonework of the *opus incertum*, see Lugli 1957: 514. Furthermore (and this is not a coincidence) most of the rows of putlog holes were placed above these horizontal brick courses. This view has also been sustained for other Roman constructions, see Vitti 2013: 102-103.

[29] Zachos and Pavlidis 2010: 147.

the solid wall at the starting point of the bridge.[30] On the inner face of the pier below the brick moulding, three rectangular putlog holes are preserved. These are at the same height as three similar ones on the inner face of the opposite solid wall, and were intended for securing the scaffolding for the vault. The next pier (henceforth, *pier II*, see Figures 2; 3; 4 and 5: d.2), which is hexagonal in plan, is built of bricks in the *opus testaceum* system. The same construction system is used for the high built podium, which is polygonal in plan, upon which this pier is founded. The polygonal podium abuts on the base of the previous pier without being organically bonded to it, forming a construction joint.[31] At their top, however, the two piers are bonded by a single arch formed by two successive courses of radially-arranged bricks. No visible traces of the following piers, which would have been in today's riverbed, are preserved. The first of these would have been strong, probably supported on a high base, and connected to the previous pier by a large arch with a diameter of around 11m, as may be inferred from the surviving elements of pier II.

On the opposite bank, four cross-shaped piers have been identified (overall dimensions 7.80 x 3.00m) along with a long wall-*substructio* (Figures 3 and 5: d.3). The preserved piers are built almost entirely of *opus testaceum*[32] and connected at their top by a single arch consisting of a double series of bricks. At the height of the springline, a moulding consisting of three brick courses runs around the construction. The long solid wall which ends below piles of stones on the slopes of the gorge is built in the *opus mixtum* system like the corresponding wall on the opposite bank. In the intervals (4.10m) between the cross-shaped piers in this part of the bridge, arches and reinforcement walls built in the *opus testaceum* system have been identified; their sides abut the piers, without being organically bonded to them (Figures 5: d.4, and 6). These were support structures for the bridge, apparently postdating the appearance of failures caused by some natural phenomenon.

The two surviving parts of bridge A are not on the same axis, but they converge at about the center of the distance they bridged in the modern-day riverbed, forming an obtuse angle of *c.* 170°. This finding is rather unlikely to have been owed to an error in alignment of the bridge, and therefore merits further investigation. In what follows we will endeavor to interpret this choice, using a number of assumptions in the absence of other obvious remains.

As noted above, in contrast to the other surviving elements, pier II (Figure 5: d.2) modifies the shape of its ground plan by being, at the same time, the first structural element of the construction marginally placed within the river bed. Its shape recalls the pier of a river bridge, and the large arch it supported strengthens this observation.[33] Given the opening of the first arch and the distance to the point of union of the two sides of the bridge, it is estimated that there were three arches (each with an opening of around 11m), two intervening central piers (with a width ranging from 2.80-3.70m),[34] and one more pier similar to pier II, which would have been sufficient to neutralize the thrust of the large end arch and at the same time allow the conduit (*specus*) to change course. With this alignment, the bridge would have crossed the river at right angles to the flow of water, minimizing the width of the construction that

[30] The bricks belonging to the middle course of the moulding were beveled on their visible side, specially-configured in the workshop where they were produced. At Nicopolis, corresponding examples of bricks made of a special mould are also recognized at the baseline of the city walls, see Zachos 2007a: 277-287 and Vitti 2013: 222-223; they were also used on Mausoleum 4 in the north necropolis, see Georgiou 2007: 317-318.

[31] The presence of this construction element should in all probability be attributed to the different requirements for the piers' foundations: at this point, at least macroscopically, it appears that the solid rocky ground of the hill was replaced by the marshy river bed. Thus, the rock was carved out to receive the overlying construction, while the massive pier was probably founded on piles. The break in structural continuity at foundation level must have aimed at the different response of the two sections in case of geological activities or thrusts from the course of the river, so that the possible collapse or other damage to one section would not carry away the other part with it. Its unusual shape, designed to sustain the hydrostatic pressure, is also pointed out by Malacrino (2004: 115).

[32] Metopes of *opus incertum* are preserved on the narrow sides of some of the piers, as they are on the constructions on the opposite bank.

[33] The presence of an elaborate dentil cornice consisting of four courses of bricks on the south side of pier II above the podium might be attributable to aesthetic reasons due to the passing of a road below the large arch and along the riverbank. However, the deep sinking of the ground at this point would have made the crossing of the road safe only during the summer months. In any case, the placement of pier II with the acute angles of its ground plan set against the river current aimed at successfully confronting the hydrostatic pressure from the overflowing water. Its non-canonical shape was most likely owed to the fact that in some sense it formed an abutment.

[34] The Roman engineers determined the width of the central piers on the basis of an empirical rule, according to which, the ratio of the pier width to the arch opening had to be between 1/3 and 1/4, cf. Korres 2014: 168.

Figure 7. Drawing of the south elevation of Bridge B (drawing: S. Pantazis) and view from the northeast.

was exposed to hydrostatic pressure (Figure 4).[35] It should also be noted that the zig-zag course of the conduit contributed to halting the momentum of the water being transported, a common practice in the construction of large aqueducts as a measure for maintaining a steady flow.[36]

Bridge B (Figures 2; 3; 4 and 7), with a length of 120m, was supported by fifteen piers along the width of the gorge basin. Thirteen of these have a cross-form plan, since buttresses projecting 2m were attached to their main body (3.80 x 3.50m). The brickwork piers were installed at intervals of 3.00-3.10m, and their height reaches 14m. They were built in the *opus testaceum* system (square bricks 0.28/0.29 x 0.35/0.40m) and linked by double arches above and below. The higher arches were built of two successive courses consisting of brick voussoirs (0.38/0.39 x 0.38/0.39m) arranged radially, in contrast to the lower arches, which were formed of a single row (0.58/0.59 x 0.58/0.59 x 0.05 decreasing to 0.035m). Simple bricks were employed for the arches' cores, similar to those used for the construction of the piers' brick masonry. At the springline, brick mouldings which surrounded each pier were formed. Two large arches (openings of 13.50 and 6.15m, respectively) at a distance of about 27m from the northern end of the bridge disrupt

[35] When one observes the gorge's geomorphology, despite various anthropogenic interventions which have obviously altered the river's course (dam 1.2 km south of the point under study; modern carriageway bridge and road towards and from the neighboring village of Haghios Georgios), it would appear that its original course laid perpendicular to the axis of the north section of the northwest bridge, and that it turned at a location about at the point of the southeast bridge. The ancient bed of the Louros river has been a subject of study by other scholars as well, who have expressed similar proposals for the river's course, cf. Doukellis *et al.* 1995: 215.

[36] Cf. the zig-zag course of the aqueduct of Dion, Karadedos 1990: 221.

the harmonious repetition of equal intervals of the remaining arches. They were most probably associated with the bridging of the ancient river bed.[37] In the masonry of the piers that follow southwards of the above-mentioned large arches, one can detect successive bonding courses of *bipedales* (0.59/0.60 x 0.59/0.60 x 0.057/0.06m), which normally enter the wall thickness to a depth equal to one brick length.

Figure 8. Interior view of the tunnel at Kokkinopilos.

At the site of Kokkinopilos (Figure 1: point 2), the conduit has been identified on the right side of the Ioannina-Preveza highway in the form of a tunnel carved into limestone (Figure 8), above which there are thick deposits of *terra rossa* – after which the site was named; 'Kokkinopilos' (=κόκκινο+πηλός) stands for red coloured clay.[38] Only the entrance of the tunnel, which is on the road side, has been identified; its exit, which is estimated to be about 400-500m away from the entrance, remains unidentified. Above the tunnel and at regular intervals of *c.* 40-50m, the ruins of six (out of a total of eight) rectangular shafts (*putei*) opened in the layers of the *terra rossa* and the underlying rock, have been found; these were used for inspection, maintenance and ventilation.[39]

The course of the aqueduct

According to investigations occasionally carried out from the 1960s to the present by numerous researchers, the course of the conduit from its exit at the tunnel of Kokkinopilos to the city has largely been determined.[40] The three ways of construction of the aqueduct already described from the point of water catchment to Kokkinopilos (built on the hill slopes; carried over rows of piers-bridges; opened as a tunnel in limestone) were each time repeated according to (and depending on) the geomorphology of the terrain throughout its course. In addition, especially on loose soils, the conduit continued underground, built of *opus incertum,* within a trench which was afterwards covered.[41]

[37] See above, notes 33-35.

[38] Harris and Vita-Finzi 1968: 539-541; on the date and interpretation of the deposits of *terra rossa*, see Runnels and van Andel 1993.

[39] The shafts were built in the *opus mixtum* system, i.e. rubble with layers of bricks at intervals (shaft´s outer side 1.85m. in length; inner conduit 0.90-0.95m.), see Dakaris 1964; Doukellis *et al.* 1995: 213-218, fig. 3 ; Merkouri 2017: 546-548.

[40] For descriptions of the course of the conduit, see Dakaris 1961-1962: 194, drawing 1; 1964; Andreou, E., 1978; 1979; 1980; Andreou, I. 1987a; 1987b; 2007, Andreou, I. 2007: 235-237; Doukellis *et al.* 1995; Wiseman 1992: 4; 1993: 13; Malacrino 2004: 111-114; Zachos 2007a: 284-285; Riginos 2013. Within the framework of the National Strategic Reference Framework (NSRF) 2007-2013, the Ephorate of Antiquities of Preveza carried out extensive excavation and enhancement work at five points along the course of the aqueduct, see Riginos 2013; Aggeli *et al.* 2015; Merkouri 2017. Excavation and other evidence will be published by the excavators.

[41] For the 'cut and cover' principle, Hodge 1995: 93-94.

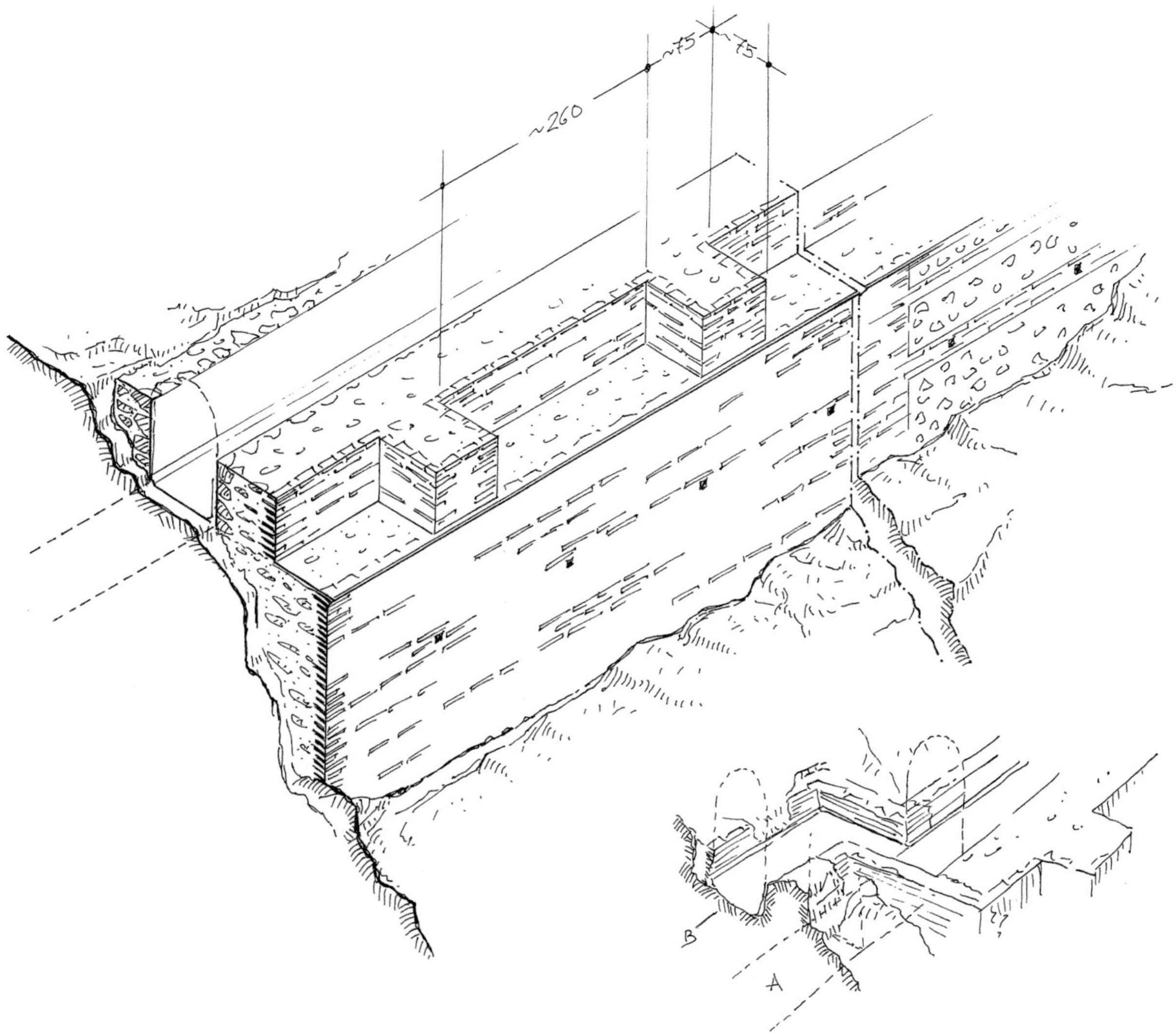

Figure 9. Axonometric drawing of the wall at Mt Valaoritis near the village of Stefani (right side). Bottom right corner: the restored specus bypassing (B) the earlier course (A) (drawing: L. Leontaris).

After Kokkinopilos, the conduit crossed a rivulet (the Dichalo Rema), the width of which did not exceed 2-3 m., upon an arched bridge – of which only few ruins are preserved with deposits of calcium carbonate (sinter) caused by water leakage. Traces of the conduit are found next north of the acropolis of Kastro Rizovouni,[42] and on the east side of the valley of Xiropotamos, from the village of Rizovouni to the village of Galatas southwards. The conduit is detected again, this time in an underground course through loose soil, along the west side of the Xiropotamos valley near the southern edge of the Thesprotiko village. The point at which it bridged the Xiropotamos remains unidentified (Figure 1). Doukellis (*et al.*) suggested that the conduit followed a peripheral course around the valley, as far as Thesprotiko village, positioning the bridge (at its narrowest point), at the height level of the Papadates village.[43] Five hundred meters northwest of the Stefani village (Figure 1: point 3), the channel followed an underground course into a tunnel about 20m long through a rocky elevation before continuing above ground on the slopes of Mt Valaoritis. A section of its above-ground course was reinforced on its outer face (to a length of *c.* 60m) by a buttress wall (*substructio*) as a result of the terrain's geomorphology. This was an *opus mixtum* structure which displays the same structural features (dimensions of bricks; thickness of horizontal joints; mortar) as the bridge A at Haghios Georgios. Some notable features include the reinforcement of the outer face of the wall with bricks in the *opus testaceum* system for about 30m, and the appearance of buttresses at the

[42] This acropolis has been identified as ancient Batiae, a colony of the Eleans in Kassopaia. Cf. Strabo, *Geography*, 7.7.5; Dakaris 1971:183-191.
[43] Doukellis *et al.* 1995: 218.

conduit's level (Figure 9). After the retaining wall, the original conduit appears to have gone out of use, probably due to the collapse of the supporting wall, and a new parallel conduit was constructed north of the former after further carving of the rock, as attested firstly by the traces of carving on the rock, and secondly by the surviving built constructions (Figure 9, bottom right corner). Further to the west, the channel continued its course upon the slopes of Mt Valaoritis. At the site of Skala Lourou, near the church of Haghia Paraskevi the conduit is preserved intact up to its roof over a notable length. Further to the south, it has been identified on a retaining wall (*substructio)* built of *opus incertum*. It then follows a northwest course. Near the village of Sfinoto, it travelled along a viaduct, as shown by the remains of three piers. It then followed an underground course in a southerly direction along the slopes of Mt Zalongo, crossing an ancient walled settlement identified as the Elatreia, one of the Elean colonies in Kassopaia.[44] It then crossed over a stream called Antorema upon an arched aqueduct bridge, before continuing along an underground course to the eastern slopes of Mt Zalongo, running parallel to the provincial Oropos-Kamarina road. It is then again identified running an underground course on the slopes of a sandy hill at the site of Paliomana. Northeast of the village of Archangelos and about 600m from the hill of Haghios Athanasios at the site of Kamares (Figure 1: point 4), the conduit follows an above-ground course upon a series of piers which crossed a small stream; however, the length of this series appears to have been greater, extending across the width of the entire valley. After the hill of Haghios Athanasios, it followed an underground course, crossing the village from northeast to southwest. At the southern entrance of the village, where a WWII memorial monument is built (Figure 1: point 5),[45] it surfaces and continues upon a series of piers of unequal height that follow the terrain.

Nineteen piers have been found at the site of Kamares (Figure 10).[46] Today, due to the wild vegetation, only eleven are visible, arranged in a Π – formation, along with the destroyed body of another pier. Of the eleven visible piers, two (1.90 x 1.80m) deploy in a parallel line and at a distance of 3.50m from the line of the remaining nine piers; the twelfth pier was located further to the west, off the route of the long side of the Π-formation. Nine of the piers are arranged in a row (two outermost piers: 2.90 x 2.65m;[47] remaining piers: 2.35 x 1.90m). Some of the piers reach a height of 6-7m, at intervals of 3.50m, and were joined, both to the two parallel piers and to each other, by a single arch consisting of brick voussoirs (0.29 x 0.53 x 0.045 decreasing to 0.035m). Their exterior faces were built of brick (0.28/0.29 x 0.28/0.29 x 0.035/0.04m) in the *opus testaceum* system, reinforced at the corners by larger bricks (0.29 x 0.42 x 0.035/0.04m). The south side of the two parallel piers was not visible, as ascertained by the imprint of an earlier structure to which the eleven piers mentioned above were attached. It appears that at some point and at this location the initial aqueduct construction faced intractable technical problems which necessitated its bypass. These technical problems probably became even more extensive later, leading to the continuation of the bypass westwards with the erection of a new series of piers. Certainly, one cannot exclude the possibility that the bypass at this location was also connected with the site of intermediate water distribution in this area. Identification of additional evidence through a future excavation could clarify the role it played.

At the southern entrance of the village of Archangelos (Figure 10), one can detect a notable number of piers of unequal height (again due to the terrain's geomorphology). Four of these are preserved to a notable height, up to the arch's springer. The piers are built of *opus testaceum* and present a similarity in construction to the vaulted reinforcement constructions of bridge A at Haghios Georgios. The conduit (built in the *opus mixtum* system) that crosses the settlement of Archangelos, has brick courses with similar bricks (0.29 x 0.29 x 0.037/0.04m), which confirms that they were built at the same time.

From the village of Archangelos the conduit followed a southwestern course, either underground and built in loose soil, or, above ground and carried upon rows of piers. Eastwards of the village of Kanali, the conduit crossed a small stream atop a bridge. On the slopes of the Michalitsi hills, which, as noted above,

[44] Cf. Strabo, *Geography*, 7.7.5; Dakaris 1971: 172-173, fig. 43.

[45] A commemorative monument was erected to honor the memory of those who had fallen at the battle at Archangelos (July 6th, 1944).

[46] Riginos 2013: 101.

[47] The two outermost piers had incrementally decreasing length along their height to serve as sufficient buttresses for the change in course of the row of piers.

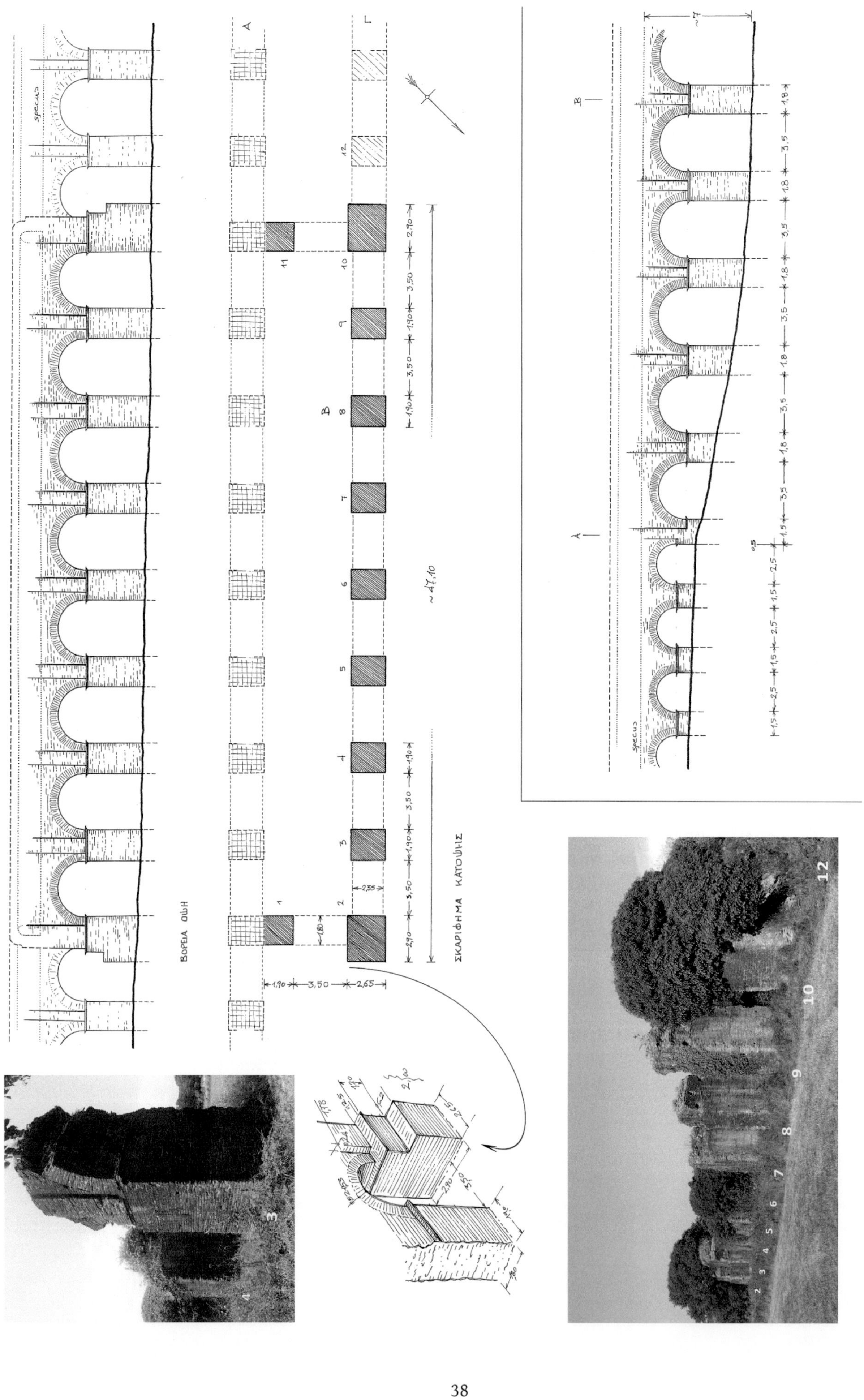

Figure 10. The Aqueduct at Kamares near Archaggelos village: restoration drawing (ground plan and north elevation), sketchy detail of northwest corner and photographs. Bottom right: part of the aqueduct at the Heroon of Archangelos (restoration plan of its west elevation) (drawing: L. Leontaris).

bounded the plain of Nicopolis on the north, it surfaces and continues upon a row of piers. About 1,4km from the city walls, the row of piers diverges into two branches (Figure 1: point 6). At this point, four piers are preserved – they apparently formed an intervention to replace already-existing ones, since they were constructed with bricks of various types, and at points their masonry also incorporates large broken-off pieces of stone blocks, as *spolia.*[48] One branch of the aqueduct continues southwards, while the other makes its way to the east, towards the *Proasteion* (the suburb) to supply water to the Actian Festival's infrastructures. The *Proasteion's* infrastructure included two bath complexes: one is associated with the Gymnasium westwards of the Stadium; the other, known as the North Thermae, is located further south, on the outskirts of the North Necropolis. Salvage excavation conducted at a plot in the modern city of Nicopolis northwest of the Stadium and at a higher altitude brought to light part of a strong structure built of concrete with a brick revetment in the *opus testaceum* system. This is probably the remains of a distribution tank (*castellum divisorium*) which collected and distributed water to the installations of the *Proasteion.* South of the Stadium, into the fields, the remains of a row of piers (1.20 x 1.20m), that carried the conduit to the North Thermae, are preserved. In Donaldson's early 19th century ground plan, which was published by Leake, the row of piers is noted as terminating at a rectangular tank which apparently served the baths exclusively.[49] According to the plan, the basin consisted of seven parallel barrel-vaulted chambers with a transverse chamber set crosswise. Donaldson's plan also notes the continuation of the row of piers to the southwest of the Baths; it probably carried the conduit *extra muros,* to another important building complex, as its terminal point.

Excavations conducted at the Victory Monument of Actium which Augustus erected on the sacred hill of Apollo, have uncovered sections of lead pipes in the west part of the north portico on the upper terrace, as well as lower, at the eastern terminus of the first retaining wall. These belonged to the monument's water supply network. Water was required for the sacred garden on the upper terrace, as well as for the rituals which took place at the outdoor sanctuary.[50] Due to the altitude difference between the site of the monument and the conduit of the aqueduct on the western slopes of Michalitsi, the possibility that the monument was supplied with water from the aqueduct must be excluded; in all likelihood, water came from some local spring located in the hills.

From the hills of Michalitsi, the main branch of the conduit followed a southerly course parallel to the imaginary line of the unfinished northwest wall, making its way toward the west Gate (Figure 1: point 7). Just a few meters from the Gate, the conduit's course turned slightly eastwards and entered the city, where it split into two branches (Figures 11-12). One branch supplied water to the first (north) of two facing nymphaea on either side of the *decumanus maximus*, nearly touching the Gate, while the second continued its course and after passing above the Gate was once against divided into two branches. Of these, one fed the second (south) nymphaeum, while the other continued its above-ground course on a series of piers not far from, and parallel to, the wall. At the height of the southwest Gate, the conduit passed over the *cardo maximus* and continued to make its way eastwards, terminating beside the wall, *c.* 500m from the southeast Gate.

The nymphaea at the West Gate

The two nearly-opposite nymphaea which framed the *decumanus maximus* at the western entrance to the city, (Figures 11-12) consist *terminal display fountains.*[51] They belong to the category of fountain buildings with a Π-shaped aedicular façade, being defined on three sides by walls (here, 2.30m thick).[52] The back wall and the lateral wings framed a rectangular basin (dimensions 12.50 x 8.80m) fronted with a parapet

[48] At some points, brick voussoirs from other previous constructions were incorporated into the brickwork rather haphazardly and clumsily, as simple building materials.

[49] Leake 1835, end of Vol. I. According to Wilson's (2001: 85, fig. 7.2) classification of the tanks in North Africa, the tank at the North Thermae of Nicopolis belongs to the second type.

[50] For the sacred garden at the Victory Monument of Augustus, see Zachos 2001a: 57-59, fig. 33; 2001b: 38; 2003: 81, fig. 24; 2007b: 413.

[51] For terminal fountains, Aicher 1993.

[52] On aedicular façade with lateral wings, Letzner 1990: 150-155, 201-203 (Typus IX B: Fassaden), 155-158, 204-206 (Typus X: Fassadenbauten mit Eckrisaliten); Dorl-Klingenschmidt 2001: 48; Aristodemou 2012: 37-38.

Figure 11. The Nymphaea at the West Gate: bird's-eye view of the complex of Nymphaea, Aqueduct, and Walls of Nicopolis.

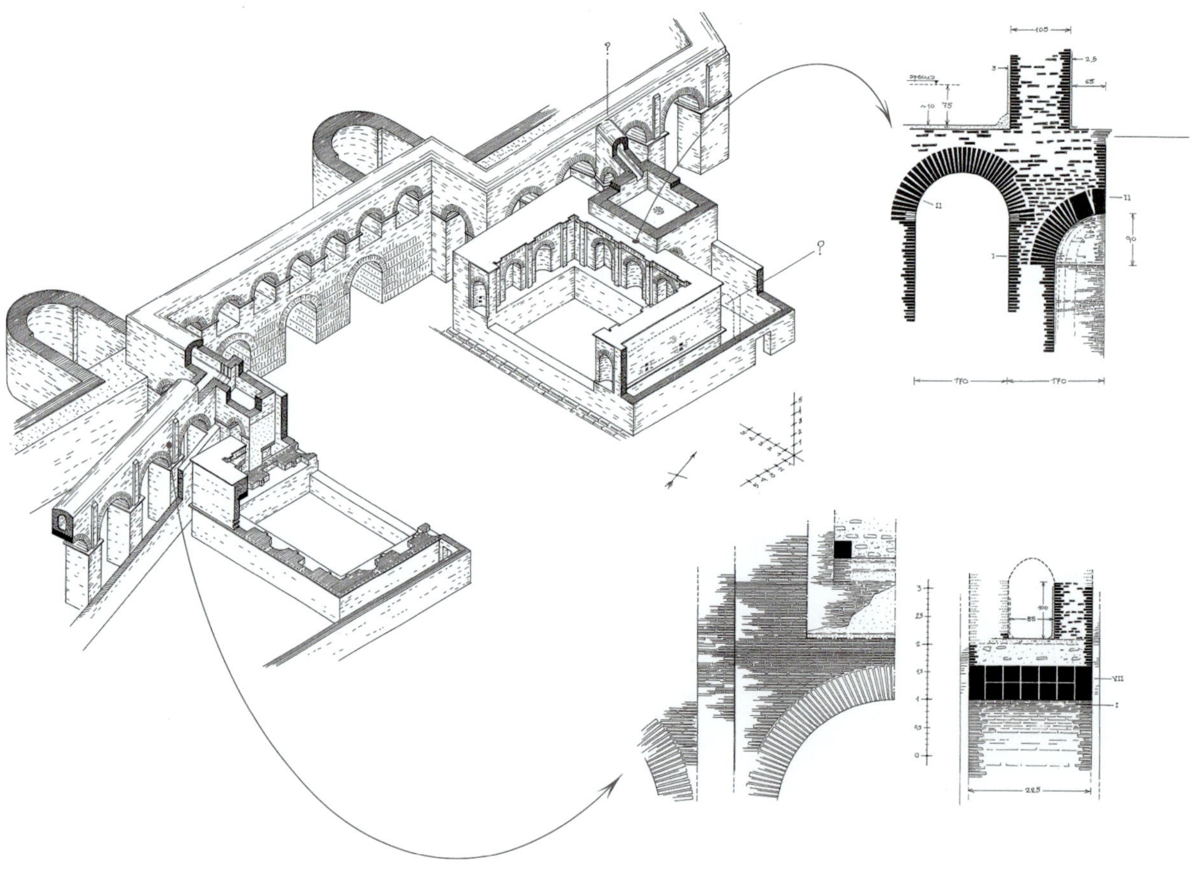

Figure 12. Axonometric plan of the complex of Nymphaea, Aqueduct, and Walls of Nicopolis (drawing: L. Leontaris).

wall 1.60m in height – of which only ruins are preserved at the side walls of the Π. The floors of the basins of the two nymphaea were covered by rectangular slabs. Seven rectangular and semi-circular niches were opened alternaly on the façade and lateral walls of the Π (as well as on the front sides of its ends), to a height of 1.85m above the basin's floor. Within each niche, a pair of successive rectangular shaped holes corresponds to the monument´s water outlets. These outlets either carried faucets, of a plain or more decorated form, or, possibly, indicate the spots where once fountain statues stood.[53] Nymphaea of a similar type (cf. the Nymphaeum of Herodes Atticus in Olympia) normally carried a second storey which also had niches decorated with statues of the Imperial family, of the benefactors who had probably sponsored the construction and/or the decoration of the monument, as well as of the divinities belonging to the local pantheon.[54] However, the preserved traces of their superstructure indicate that the nymphaea at the West Gate had only one storey.[55] More specifically, the east side of the aqueduct pier, which tangents to the south nymphaeum was configured with bricks from the level of the first storey of the nymphaeum and above. On the level below, the structure simply rests against the nymphaeum, creating a construction joint and employing the west wall of the earlier built nymphaeum, as a formwork (mould). The appearance of this structural differentiation along the vertical axis (height) – which, it should be noted, is not related to a chronological difference between the two constructions (cf. same bricks and construction method, joining of the two by a shared, thinner wall) – testifies that, from a certain height upwards, the ancient builder could work without obstacles, unhindered. Moreover, the water tank of the north nymphaeum rests partly on the masonry of the monument, leaving a minimum of space (0.65m) available to the south face of the north wall to continue the masonry upwards. The wall framing the water basin was plastered on both its interior and exterior sides and was, like the south nymphaeum, built in the *opus testaceum* system. The total height of the two nymphaea is estimated to have been about 7m. The inner walls and facades carried polychrome marble revetments, as shown by the holes of the supporting nails preserved on the walls. The niches of the nymphaea also carried elaborate revetments, in which the imprints of the hexagonal marble tiles are preserved on the remains of the mortar.

The north nymphaeum was surrounded by a wall on its north side and arched passageways on its east and west sides. Behind it rose the water tower (*castellum aquae)* linked by an arch to one of the aqueduct piers, thus creating a vertical construction joint. The remains of the conduit that supplied the rectangular tank of the water tower are preserved upon the arch. The conduit, with an absolute floor altitude at 33.40m, preserves remains of its coating of compact hydraulic mortar. The tank's floor is inclined and paved in hydraulic mortar and has a large funnel-shaped opening at its center. Through gravity, water was channeled through this opening and circulated through a network of lead pipes that terminated at the water outlets of the nymphaeum's niches. No other drainage is detected around the perimeter of the water tower's tank. The three sides of the south nymphaeum were enclosed by a wall for aesthetic reasons, as well as to protect its pipes, which were supported on the exterior surfaces of the walls. The secondary conduit which was diverged from the primary one (and supported on an arch), terminated in a small water tank at the south nymphaeum. The two arch pillars rested on piers which rested against the Augustan wall and the water tower, respectively. The water supply of the two nymphaea was controlled by vertical movable sluice gates -as shown by the three vertical channels carved in the hydraulic mortar of the south nymphaeum's conduit branch. The central sluice gate would probably carry a metal screen to filter the water.[56] The tanks of the water towers located above the two nymphaea were exclusively intended for supplying the nymphaea and were not distribution tanks (*castella divisorium*) as was previously suggested. However, it is not impossible that section of the pipes network under the funnel-shaped opening of the north tank, might run underground far from the nymphaeum and supply other buildings with water. This hypothesis, however, cannot be verified without excavation. Further investigation is also required

[53] Longfellow 2011: 131-134; Aristodemou 2012: 196, mainly 256. See also G. Aristodemou's paper in present volume.

[54] On the sculptural decoration of monumental nymphaea (subjects and contextualization) see Aristodemou 2011 and Aristodemou 2012: mainly 101, 199. Also G. Aristodemou's paper in present volume.

[55] On the previous discussion regarding the two nymphaea, see Vokotopoulou 1975; Walker (1979: 138-148, mainly 140) was the first to suggest that there was a second, smaller story which was probably recessed from the line of the ground floor, while Zachos and Georgiou (1997: 588-592) too, considered the possibility of a second floor; Also, Chrysostomou and Kephallonitou 2001: 29-33, figs. 22-25; Zachos 2007a: 283-285; Longfellow 2011: 131-134; Aristodemou 2012: 56, 71, 182, 256; Zachos 2015: 107-110.

[56] On the way sluice gates operated in Roman aqueducts, Schram 2014: 238-242.

to decipher the network of lead pipes that carried water to the pairs of holes opened in the niches of the Π-shaped walls of the nymphaea, within which faucets (or sculpture) were adjusted. In all probability, these pipes were supported externally, resting upon the thick brick cornice that surrounded the exterior walls of both nymphaea at a slightly lower level of that of the holes.

Intra-urban water distribution

The actual *castellum divisorium* that distributed water to the city's various sectors has not been identified at Nicopolis.[57] Nevertheless, its location can be almost accurately defined not far from the wall and westwards of the southeast Gate – in the area where the aqueduct's bridge becomes lost and its altitude is about 32m.[58] Further to the east, the ground rose smoothly, terminating at the city's highest area, known as Asyrmatos (the area is located outside the Early Christian walls, and it is where the Basilica D was built). Due to the great altitude difference, this area could not be supplied with water by the aqueduct.

A third nymphaeum, the so called, the 'small nymphaeum', has been discovered on the west side of the paved *cardo* that passes westwards of the *domus* of the *ekdikos* Georgios. It was rectangular in plan, built out of bricks and divided in four vaulted chambers. The first and northern chamber was accessible and might have been used for rituals (or sacral purposes in general); the remaining three and smaller chambers had their lower section formed as a small basin supplied by water through lead pipes set in the wall mortar; these pipes are now removed, but their imprints are visible. Water was channeled into the lead pipes from the roof of the building, where thick layers of hydraulic mortar are preserved in various levels.[59]

The general rule was that water was transported from the *castella divisoria* upon rows of piers, or through underground pipes, to *secondary castella* in the form of brick-built water towers carrying a lead storage tank at their top. Rome, for example, had 247 secondary *castella*. From these, a network of lead and clay pipes distributed water to its final recipients, including houses, public baths, etc.[60] The role of these secondary *castella,* to our understanding of Roman urban water supply networks, represents a significant gap in scholarly research.[61] Most of the known secondary *castella* have been found in Pompeii, where they are situated at strategic points, normally at the corners of intersecting roads.[62]

At Nicopolis, a *secondary castellum* and fountain complex has been found strategically situated at the northwest corner of the intersection of the *cardo maximus* and *decumanus maximus,* about 370m eastwards of the West Gate.[63] It was approximately square (dimensions: 3.90 x 3.75m; preserved height *c.* 4.40m) and rested on a stone base. On the west side of the tower there is an inherent vaulted chamber with a trapezoidal ground plan, and its south side open. On the chamber's north wall, a rectangular niche is formed having two rectangular openings in its walls into which faucets would have been fitted. On the east wall of the water tower thick layers of precipitates of calcium carbonate (sinter) (result of the overflow of the basin at its top) are preserved. The building is constructed of bricks in the *opus testaceum*

[57] In Buthrotum (Epirus), at the back of the Nymphaeum located southwards of the Great Basilica, a single-chambered vaulted distribution tank was built to supply the lowest in altitude part of the city, Wilson 2013: 90-91. In the Greek territories, *castella divisoria* were identified in Attica (Leigh 1997), in Argos and in Dion (Karadedos 1986; 1990).

[58] At the site of Haghios Georgios considerable evidence survives regarding the conduit's inclination: at the first section and in a length of 230m the gradient is about 0.2% (see above, note 22), while, overall and up to the beginning of the carved tunnel at Kokkinopilos (estimated length of the conduit 2000m, distance between locations in a straight line 1.550m, altitude difference 1.90m) it is 0.1%. Within the fortified area of Nicopolis, it is estimated that with a similar inclination of 0.1%, the water could follow a zig-zag course of 1350m with natural flow until it encountered a higher obstacle.

[59] Zachos 2015: 111-112; Pavlidis *et al.* 2017.

[60] So far, in Nicopolis, two houses are known to have received piped water: the *domus* of Manius Antoninus and the *domus* of the *ekdikos* Georgios. At the first house the pipes supplied the kitchen installations, while at the second house piped water was used to supply the bath of the house and a flashing fountain within a triclinium, Kyrkou 2006: 49; Pavlidis 2015: 41, 45. Terracota pipelines were found below the pavement of the *cardo,* at the east of the domus of Manius Antoninus.

[61] Hodge 1995: 294.

[62] Olsson 2015: 15-17, 30-35. A water tower was discovered at the southern side of a street at Dion in Macedonia. Its lead pipe was running lengthwise and terminated at a lead, truncated cone-shaped siphon, Karadedos 1986: 37-39.

[63] Zachos 2015: 106. This monument has not been fully excavated. A small-scale excavation was carried out in 1952 by S. Dakaris, cf. Dakaris 1953.

system, and resembles the structure of the nymphaea at the west Gate and the neighboring aqueduct piers.

Remains of a second water tower with a rectangular ground plan (2.50 x 1.00m) built in the *opus testaceum* system are preserved at the west corner of the intersection of the *decumanus maximus* and the *cardo,* flanking to the west the *domus* of the *ekdikos* Georgios.

The location of these two water towers on the north side of the *decumanus maximus* and at a distance of 470m from each other suggests the implementation of a model for dispersing the locations of water towers – which needs to be further clarified with a targeted program of field surveys, near-surface geophysical prospection systems, and excavation test trenches, given that the basic sections of the city´s grid have – until today – been sufficiently defined through excavation.

The date of the aqueduct

Literary and epigraphic testimonies concerning Nicopolis, do not record the foundation of the aqueduct. Moreover, the lack of adequate excavation data makes the determination of its chronology even harder. The only literary reference to the aqueduct is that of the praetorian prefect *per Illyricum* Claudius Mamertinus, at his panegyric speech delivered before the Senate in Constantinople on January 1st, AD 362, upon assuming the office of *prefectus*; he then extolled the benefactions and interest of the Emperor Julian in Epirus. In referring to the wretched state into which Nicopolis had fallen, he notes that among other things its aqueduct had been destroyed.[64] The date of the aqueduct remains an open question in contemporary research, by which it is dated either to the reign of Augustus or to that of Hadrian or even later.[65] Here, we endeavor to decipher the historical evolution of the aqueduct using two approaches. The first approach is to document the differences and groupings of the various construction and repair phases which are visible today on the basis of their common structural features. The second approach is to attempt to associate the construction groupings with historical evidence and the iconography of Nicopolis's coinage.

Based on the above approaches, one group includes the ruins of the bridge A at Haghios Georgios and those of the conduit built (as circumstances dictated) upon a bridge, on a row of piers, or on hillsides, etc., as far as the site of Skala Lourou. These employed (as previously described) the same construction techniques and materials. We consider that this section of the aqueduct's course represents the first building phase of the monument (Phase A), on which work appears to have remained unfinished, since building remains with similar characteristics have not been identified or excavated to date beyond the area of the sandy hillocks at the site of Paliomana.[66]

One second group corresponds to a main building phase. It is detected in the vaulted reinforcements and support walls in the intervals between the piers in the south section of bridge A, the conduit and row of piers which crosses the village of Archangelos, the row of piers which approaches and traverses the *extra muros* area of Nicopolis, the branch of the aqueduct which leads to the *Proasteion*, the aqueduct

[64] Mamertinus (*Grat. Actio Iuliano,* 9.2-3) with the characteristic exaggeration of panegyrics, presented the aqueduct as having been completely destroyed, apparently to emphasize the magnitude of the beneficence of the Emperor who repaired it: *Urbs Nicopolis, quam divus Augustus in monumentum Actiacae victoriae trophaei instar exstruxerat, in ruinas lacrimabiles prope tota conciderat: lacerate nobilium domus, sine tectis fora, iamdudum aquarum ductibus pessumdatis plena cuncta squaloris et pulveris. Certamen ludicrum lustris omnibus solitum frequentari intermiserat temporis maesti deforme iustitium*. (transl.: The city of Nicopolis, which the divine Augustus had built like a trophy as a monument to victory of Actium, had fallen almost totally into lamentable ruins: the houses of the nobles were torn apart, marketplaces were without roofs, everything was full of dirt and dust since the aqueduct had long since been destroyed. The unseemly cessation of business during that sorrowful time had allowed the public games customarily observed at every *lustrum* to lapse, Nixon and Rodgers 1994: 408).

[65] Andreou, E. 1978: 184; Doukellis *et al.* 1995: 232-233; Lolos 1997: 304; Malacrino 2004: 114-121; Andreou, I. 2007: 235-237; Longfellow 2011: 49. In an earlier paper (Zachos 2013: 261-269), it was suggested that the bridge A should be dated to the reign of Nero and the bridge B to the reign of Hadrian.

[66] The underground conduit (*specus*) has been only fragmentarily identified in this region. The use of the *opus incertum* system for its construction does not in itself comprise secure evidence so as to date the construction of the aqueduct in its first building phases – since incorporated bricks have not been found in the construction and, in addition, the same construction system was also adopted in succeeding building phases of the monument (cf. below regarding the *specus* which traverses the village of Archangelos). Clarification of the relative dating of this section of the conduit will form the subject of future large-scale excavations in the area's dunes.

and nymphaea complex at the West Gate,[67] and the row of piers within the city south of this Gate. The aqueduct was completed during this construction phase (Phase B), and water reached the city from the springs of Haghios Georgios.

A third group with common features includes the bridge B at Haghios Georgios, a repair to the conduit at the village of Archangelos, and a pier preserved to a notable height 140m. south of the West Gate. They correspond to a third building phase (Phase C). The bypass at the site of Kamares Archangelou could be dated at about the same period, though not to the same building program – due to particular construction differences (Phase D: small scale local repairs). Reinforcement constructions to shore up part of the row of piers south of the West Gate,[68] were probably associated with the repair of the wall found nearby (Phase E: small-scale repairs confined to the area around the West Gate). During the same period, the intervals in the row of piers north of the West Gate were probably sealed.[69] Interventions to the piers in the Michalitsi hills, a vertical repair to a pier south of the West Gate, the continuation towards the west of the bypass at Kamares Archangelou and probably, the bypass of the conduit on the slopes of Mt Valaoritis after the collapse of its retaining wall-*substructio* belong to a sixth construction-repair phase (Phase F).

Hadrianic monetary issues by the Nicopolis mint are proving to be important for dating the aqueduct's building phases. The West Gate is represented on these coins and its depiction has been linked with the construction of the aqueduct as an Imperial benefaction of Hadrian.[70] The renovation/reconstruction of the West Gate itself during the reign of Hadrian is also connected with the aqueduct structures in the area of the West Gate. The Gate's reconstruction is confirmed by the ruins of the north tower, which survives to a considerable height. At the base of the tower, the masonry is the same as that of the Augustan wall, with thick bricks and limestone *caementa* in the *opus caementicium* of the core.[71] In the tower's upper part one observes an intervention with thinner bricks[72] and *opus caementicium* with brick pieces as *caementa*. The bricks used in the intervention on the tower's superstructure resemble those used on the two nymphaea and parts of the aqueduct near the Gate, which, in accordance with the hypothesis proposed by both Walker and Zachos, leads to a date at the early 2nd century AD for the second construction phase of the aqueduct. It was at that period when water from the springs at Haghios Georgios first reached the city. Moreover, the use of bricks as *caementa* in the *opus caementicium* or as reinforcement courses in the walls and vaults[73] is also a characteristic feature of building in the age of Trajan and Hadrian.[74]

Furthermore, the depiction for the first time on Hadrianic coins of Nicopolis, of a fountain structure in the form of the famous *Meta Sudans* fountain in Rome, would also argue in favor of the completion of the aqueduct during the reign of Hadrian.[75] This had a major impact at residents' public and private life due to the abundant water supplied to public buildings, mainly the baths, as well as to private residences. We may also draw indirect information about the completion of the aqueduct's construction during the reign of Hadrian from the passage in the *Discourses* of Epictetus (*Of inconsistency*, 2.21.14.) with which the present article begins. During the persecution of philosophers by Rome, under Domitian (AD 89), the philosopher Epictetus settled in Nicopolis, where he founded a school and remained until his

[67] In the past, the south nymphaeum was dated to the first half of the 2nd century AD, while the north nymphaeum was placed in the early 3rd century AD, cf. Zachos and Georgiou 1997: 588-592; Malacrino 2004: 109-110, 117-118. In addition to the similarity of the two nymphaea as regards their architectural form, which has already been noted by various researchers (cf. note 55), in the course of research for this paper, further similarities between these two nymphaea as regards their construction systems, the size of their bricks and the thickness of their joints lead us to conclude that they belonged to the same building program.

[68] Aggeli *et al.* 2015: 70-71.

[69] Zachos 2007a: 286-288.

[70] For monetary issues of Hadrian and Antinous, on which the western triple Gate is depicted, see Karamesini-Oikonomidou 1975: 86, no 44, pl.17; 87, no 57, pl.18; Kraay 1976: 239-241; Calomino 2011: 237-241. Regarding the association of the Gate's representation with the aqueduct and Hadrian, see Walker 1979: 145-148 and Zachos 2013: 261-269.

[71] Dimensions of bricks: 0.29 x 0.45 x 0.08/0.085m and 0.29 x 0.29 x 0.08/0.085m. For the north tower, cf. Zachos 2007a: 285-286.

[72] Dimensions of bricks: 0.295 x 0.295 x 0.037/0.04m.

[73] As observed in the mass of the wall structure in the renovated parts of the walls at the West Gate at the reinforcement constructions of the bridge A at Haghios Georgios – either in the stratification of the *opus caementicium*, or as through route horizontal courses having brick voussoirs (0.29 x 0.42/0.44 x 0.05 decreasing to 0.04m), or as thick intervals courses with pieces of bricks in the nymphaea and on the piers of the aqueduct from Archangelos to Nicopolis.

[74] Vitti, M. and Vitti, P., *in press*.

[75] See Longfellow 2011: 46-49; Zachos 2013: 269-270.

death around AD 130. Among Epictetus's students were the Emperor Hadrian and Flavius Arrianus, who preserved the teachings of his master in written form. The unsatisfactory state of the public baths in Nicopolis is mentioned by one of Epictetus's students – who probably came from a wealthy family and from a city with well-established and luxurious bath complexes.[76] It should be connected with the lack of water, reflecting the situation in the baths (*balneae*) before the arrival of water from the spring of Haghios Georgios – when public and private baths were supplied by water from local springs and wells.[77] The Emperor Hadrian must have also formed a personal opinion about the state of the baths at Nicopolis, since he visited the city more than once.[78]

As regards the first building phase of the aqueduct, the use in the cornices of bricks with a moulded made cross section leads to its dating in the 1st century AD – especially since in subsequent building phases this cross section was achieved by cutting the bricks mechanically on the worksite. Furthermore, the so called 'Mausoleum 4' (in the North Necropolis) which displays a structural resemblance (i.e. bricks of a similar thickness; thick horizontal mortar joints; mould-made beveled bricks) with the bridge A at Haghios Georgios, dates at the first half of the 1st century AD.[79] The aqueduct's initial construction phase cannot be dated to the reign of Augustus because no bricks were used in any of the (safely dated at Augustus' reign) buildings excavated at Nicopolis (Actium Victory Monument, Stadium, Theater) – with the exception of the walls, in whose construction particularly thick bricks were employed, which one would expect to be also used on the aqueduct, if it had been built in the same period.[80] The first building phase of the aqueduct, which remained unfinished, could be attributed to Nero, who sojourned in Nicopolis and was crowned a victor at the Actian Games, while the city was renamed ΝΕΡΩΝΙΚΟΠΟΛΙΣ in his honor so as to be presented as the second founder of the city after Augustus.[81]

It has reasonably been supported that the erection of bridge B at Haghios Georgios should be dated to the reign of Hadrian,[82] especially since through-and-through courses of *bipedales* have been found in its walls.[83] However, the use of through-and-through bonding courses of *bipedales* returned as a construction technique in the reign of Commodus and the Severan dynasty,[84] as is confirmed at Nicopolis itself, at a Severan repair phase of the Odeum.[85] It would therefore be justifiable to date the erection of bridge B at this period, like the interventions at Archangelos and Nicopolis which employed a similar technique.[86] The construction of vaulted reinforcements in the south section of bridge A at the starting-point of the aqueduct at Haghios Georgios, which presents unquestionable affinities with the nymphaea and the section of the aqueduct at the West Gate, is dated to the Hadrianic era.

The propping up of the row of piers south of the West Gate must have been carried out in the second half of the 3rd century AD, since structurally it resembles the generalized repairs of the Roman wall which have been associated with the Herulian invasion.[87]

---

[76] Dedousi 1982: 28.

[77] For the natural water sources in the area, see Zachos 2007a: 274-275.

[78] On Hadrian's visits to Nicopolis, Zachos 2013: 267-269.

[79] It may be noted here that a lamp from a child's cist grave in the chamber of the so called 'Mausoleum 4', dated to AD 50-57, is considered a *terminus ante quem* for the mausoleum's erection, see Georgiou 2007: 319. In addition, given that the surrounding wall of the funerary monument is organically connected with its burial chamber, below which an earlier (built in the *opus reticulatum* system) transverse wall was found which was discontinued by the construction of 'Mausoleum 4' (op. cit. 318), the dating of the monument to the period when the city was founded, i.e. at the age of Augustus and Agrippa, must be excluded.

[80] For the reasoning involved here, also Zachos 2007a.

[81] Zachos 2013: 263-266. Concerning the change in the city's name to ΝΕΡΩΝΙΚΟΠΟΛΙΣ and its coinage under Nero, also Calomino 2011: 230-233. Two inscriptions found at Nicopolis, one in the Theatre and the other at the Victory Monument of Actium are also attributed to Nero's presence in Nicopolis, Zachos 2015: 32, 80.

[82] Zachos 2013: 266-269.

[83] The appearance of *bipedales* bricks goes back to the age of the Emperor Domitian (Hesberg 2009: 36). Their generalized use as bonding courses in the region of Rome, however, has been linked to the Emperors Trajan and Hadrian (Vitti 2013: 8-9).

[84] Lugli 1957: 570-574.

[85] Zachos *et al*, 2015: 62-70; Leontaris and Stamou in press.

[86] Due to individual differences in their structures, the most likely thing is that they either occurred at different but closely-related dates or were executed by different crews.

[87] Zachos 2007a: 286-287, note 65; Zachos 2015: 34-35.

Finally, the sixth construction phase should be identified with the extensive repairs carried out by the Emperor Julian, the dispersion of whose preserved remains confirms the panegyric speech delivered by the prefect of the Illyricum Claudius Mamertinus.

The end of the aqueduct

The aqueduct of Nicopolis permanently ceased to function around the end of the 5th century, when the city shrank to one-fifth of its original area under the pressure of barbarian invasions.[88] New walls were built to protect Nicopolis, which was confined to the southeast sector of the Augustan city. Two of the walls' sides (western and southern) were erected *ab initio*, while the others were built atop the damaged walls of the old city. It would appear that the criterion for its connection to the aqueduct, the cost of whose maintenance would have been unsustainable for the city, did not prevail in the choice of location for the new city. Rather, in addition to other factors it seems that the existence of a spring in this area, under the domus of *ekdikos* Georgios, played an even more important role in selecting its location. The water of this specific spring, when colonel William Leake visited the ruins of Nicopolis in the early 19th century, created a small stream that flowed in the direction of the eastern side of the new walls.

Bibliography

Aggeli, A., Saltagianni, E., Merkouri, Chr., Giannakis, I. 2015. *The Roman aqueduct of Nicopolis*. Preveza.

Aicher, P. J. 1993. Terminal Display Fountains ('Mostre') and the Aqueducts of Ancient Rome. *Phoenix* 47(4): 339-352.

Andreou. E. 1978. Άγιος Γεώργιος Πρέβεζας, *Archeologikon Deltion* 33. Χρονικά, Β'1: 184.

Andreou, E. 1979. Ρωμαϊκό υδραγωγείο Νικοπόλεως, *Archeologikon Deltion* 34. Χρονικά Β'1: 246.

Andreou, E. 1980. Άγιος Γεώργιος Πρέβεζας, *Archeologikon Deltion* 35. Χρονικά Β'1: 323, plan 8.

Andreou, I. 1987a. Πρέβεζα, *Archeologikon Deltion* 42. Χρονικά Β'1: 318-320.

Andreou, I. 1987b. Το έργο της ΙΒ΄ Εφορείας Αρχαιοτήτων στη Νικόπολη. In: E. Chrysos (ed.) Νικόπολις Α': πρακτικά του πρώτου Διεθνούς Συμποσίου για τη Νικόπολη *(23-29 Σεπτεμβρίου 1984)*: 145-152. Preveza: Actia Nicopolis Foundation.

Andreou, I. 2007. Τοπογραφικά και πολεοδομικά Νικόπολης. In: K. L. Zachos (ed.) Νικόπολις Β΄. Πρακτικά του Δεύτερου Διεθνούς Συμποσίου για τη Νικόπολη *(11-15* Σεπτεμβρίου *2002), τ.I-II // Proceedings of the Second International Nicopolis Symposium (11-15 September 2002)*: I: 231-262, II:125-143. Preveza: Actia Nicopolis Foundation.

Aristodemou, G. 2011. Sculptured Decoration of Monumental Nymphaea at the Eastern Provinces of the Roman Empire. In: Trinidad Nogales Basarrate and Isabel Rodà de Llanza (eds) *XI Colloquium on Roman Provincial Art 'Rome and the Provinces: Models and Diffusion'. Mérida, Spain 18-21 May 2009* (Hispania Antigua, Serie Arqueológica 3): 149-160. Roma: L'Erma di Bretschneider.

Aristodemou, G. 2012. Ο Γλυπτός Διάκοσμος Νυμφαίων και Κρηνών στις Ανατολικές Επαρχίες της Ρωμαϊκής Αυτοκρατορίας // *Sculptured Decoration of Monumental Nymphaea and Fountains at the Eastern Provinces of the Roman Empire.* Thessaloniki: Kornelia Sfakianiakis Editions.

Chrysostomou, P. and Kefallonitou, F. 2001. *Νικόπολις*. Athens: Ministry of Culture. Archaeological Receipts Fund.

Dakaris, S. 1961-1962. Αρχαιότητες και Μνημεία Ηπείρου. Νομός Πρέβεζας. *Archeologikon Deltion* 17:194-195.

Dakaris, S. 1964. Αγ. Γεώργιος. *Archeologikon Deltion* 19. Χρονικά: 309.

Dakaris, S. 1971. *Cassopaia and the Elean Colonies* (Ancient Greek Cities 4). Athens.

Dedousi, Chr. 1982. Ο Επίκτητος στη Νικόπολη. *Ηπειρωτικό Ημερολόγιο* vol. Δ΄: 25-37.

Doukellis, P. N., Fouache, É., Dufaure, J., J. 1995. Le contexte géomorphologique et historique de l'aqueduc de Nicopolis. *Bulletin de correspondance hellénique* 119(1): 209-233.

[88] The abandonment of the aqueduct at Buthrotum in Epirus is dated to the late 4th – early 5th century, probably in the wake of a destructive earthquake, Wilson 2013: 95.

Dorl-Klingenschmidt, Cl. 2001. *Prunkbrunnen in kleinasiatischen Städten. Funktion im Kontext* (Bayerische Akademie der Wissenschaften. Studien zur antiken Stadt, Band 7). München: Dr. Friedrich Pfeil.

Georgiou, A. 2007. Βόρεια νεκρόπολη της Νικόπολης. Οργάνωση και ταφική αρχιτεκτονική. In: K. L. Zachos (ed.) Νικόπολις Β΄. Πρακτικά του Δεύτερου Διεθνούς Συμποσίου για τη Νικόπολη *(11-15 Σεπτεμβρίου 2002), τ.I-II // Proceedings of the Second International Nicopolis Symposium (11-15 September 2002)* I: 307-323. Preveza: Actia Nicopolis Foundation.

Harris, D. R. and Vita-Finzi, C. 1968, Kokkinopilos: a Greek badland. *The Geographical Journal* 134(4): 537-546.

Hodge, A. T. 1995. *Roman aqueducts & water supply.* Bristol: Classical Press.

Isager, J. 2009. Destruction or depopulation of cities in Pausanias: Nicopolis, Aetolia, and Epirus. *Proceedings of the Danish Institute at Athens* 6: 201-215.

Jing, Z. and Rapp, G. 2003. The Coastal Evolution of the Ambracian Embayment and its Relationship to Archaeological Settings. In: J. Wiseman and K. Zachos (eds) *Landscape Archaeology in Southern Epirus, Greece I* (Hesperia Supplement 32): 157-198.

Jones, J. E. 1987. Cities of Victory: Patterns and Parallels. In: E. Chrysos (ed.) *Νικόπολις Α΄: πρακτικά του πρώτου Διεθνούς Συμποσίου για τη Νικόπολη (23-29 Σεπτεμβρίου 1984)*: 99-108. Preveza: Actia Nicopolis Foundation.

Karadedos, G. 1986. ΄Υδρευση και αποχέτευση στο αρχαίο Δίον. In: *Οι αρχαιολόγοι μιλούν για την Πιερία*, Καλοκαίρι 1985: 33-41. Pieria: Prefecture Committee for Popular Education (Ν.Ε.Λ.Ε.).

Karadedos, G. 1990. Το υδραγωγείο και η κεντρική δεξαμενή του αρχαίου Δίου. *To Archaeologiko Ergo stē Macedonia kai stē Thrakē (=AEMTh)* 4: 217-230.

Karamesini-Oikonomidou, M.1975. *Η Νομισματοκοπία της Νικοπόλεως*. Athens: En Athēnais Archaeologikē Etaireia.

Karatzeni, P., Τοπογραφικά της χώρας των Αμβρακιωτών. In: Αφιέρωμα στον N.G. L. Hammond. *Μακεδονικά παραρτήματα 7*: 233 -249.

Katsanou, K. N. 2012. Περιβαλλοντική υδρογεωλογική μελέτη της λεκάνης του Λούρου ποταμού. Unpublished PhD dissertation, University of Patras, Department of Geology.

Korres, M. 2014. *Η στέγη του Ηρωδείου και άλλες γιγάντιες γεφυρώσεις*. Athens: Melissa Editions.

Kyrkou, Th.. 2006. *Η ΄Επαυλη του Μάνιου Αντωνίνου: Μια πολυτελής ιδιωτική οικία στη Ρωμαϊκή Νικόπολη* (Μνημεία Νικόπολης 2*)*. Athens: Ministry of Culture. Archaeological Receipts Fund.

Lämmer, M. 1986. Die Aktischen Spiele von Nicopolis. *Stadion* 12: 27-38.

Leake, W. M. 1835. *Travels in northern Greece* I-IV. London.

Leake, W. M. 1847. Greek inscriptions from Thessaly and Epirus. Letter from Mr. Lyons to colonel Leake, with some notes by the latter. *Transactions of the Royal Society of Literature of the United Kingdom* Second Series vol. II: 229-236.

Leigh, S. 1997. The 'reservoir' of Hadrian in Athens. *Journal of Roman Archaeology* 10: 279-290.

Leontaris, L. and Stamou, I., in press. Ωδείο Νικόπολης. Νέα δεδομένα για την αρχιτεκτονική μορφή και την κατασκευαστική τεχνολογία του. In: *To Archaeologiko Ergo stē Voreiodytikē Ellada kai sta nesia tou Ioniou // The Archaeological Research and Activity in NorthWest Greece and the Ionian Islands, 10-14 December 2014.* Ioannina.

Letzner, W. 1990. *Römische Brunnen und Nymphaea in der westlichen Reichshälfte.* München: Charybdis.

Lolos, Y. 1997. The Hadrianic Aqueduct of Corinth. *Hesperia* 66.2: 271-314.

Lolos, Y. 2010. *΄Υδωρ βασιλεύς Αδριανός εσήγαγε εκ Στυμφήλου. Το Αδριάνειο υδραγωγείο της Κορίνθου και η μεταφορά του νερού στα ρωμαϊκά χρόνια.* Athens: Piraeus Bank Group Cultural Foundation.

Longfellow, B. 2011. *Roman Imperialism and Civic Patronage: Form, Meaning, and Ideology in Monumental Fountain Complexes.* Cambridge: Cambridge University Press.

Lugli, G. 1957. *La tecnica edilizia Romana: con particolare riguardo a Roma e Lazio.* Rome: G. Bardi.

Malacrino, C. G. 2004. L'approvvigionamento idrico di Nicopoli e l'acquedotto presso Haghios Georgios. Una nuova attestazione di Opus Reticulatum in Grecia. *Rivista di archeologia* 28: 107-124.

Merkouri, Chr. 2017. Ρωμαϊκό υδραγωγείο Νικόπολης. Προκαταρκτική έκθεση από τις εργασίες ανάδειξης του μνημείου (2011-2014). In: *Σπείρα: Επιστημονική Συνάντηση προς τιμήν της Αγγέλικας Ντούζουγλη και του Κωνσταντίνου Ζάχου.* Ιωάννινα 1-3 Νοεμβρίου 2012: 535-556. Athens: Ministry of Culture. Archaeological receipts Fund.

Nixon, C. E. and Rodgers, B. S. 1994. *In Praise of Later Roman Emperors: The Panegyrici Latini: Introduction, Translation, and Historical Commentary, with the Latin Text of R. A. B. Mynors* (Vol. 21). Berkeley: University of California Press.

Olsson, R. 2015. The water-supply system in Roman Pompeii, Lund University Department of Archaeology and Ancient History, dissertation for a licentiate degree.

Pavlidis, E. A. 2015. Νικόπολις. Ο Οίκος του Εκδίκου Γεωργίου. Μνημεία της Νικόπολης, 5 // *Nicopolis. The domus of the Ekdikos Georgios* (Monuments of Nicopolis 5). Athens: Ministry of Culture. Archaeological Receipts Fund.

Pavlidis, E., Kleidas, E., Papavasileiou, K. 2017. Το μικρό Νυμφαίο της Νικόπολης. In: *Σπείρα: Επιστημονική Συνάντηση προς τιμήν της Αγγέλικας Ντούζουγλη και του Κωνσταντίνου Ζάχου.* Ιωάννινα 1-3 Νοεμβρίου 2012: 557-574. Athens: Ministry of Culture. Archaeological Receipts Fund.

Pavlogiannis, O. Albanidis, E. Dimitriou, M. 2009. The Aktia of Nicopolis: New Approaches. *Nikephoros: Zeitschrift für Sport und Kultur im Altertum* 22: 79-102.

Purcell, N. 1987. The Nicopolitan synoecism and Roman urban policy. In: E. Chrysos (ed.) *Νικόπολις Α: πρακτικά του πρώτου Διεθνούς Συμποσίου για τη Νικόπολη (23-29 Σεπτεμβρίου 1984)*: 71-90. Preveza: Actia Nicopolis Foundation.

Pouqueville, F. C. H. L. 1826. *Voyage de la Grèce.* Tome Second, Épire. Paris: Firmin Didot, père et fils.

Riginos, G. 2013. Το ρωμαϊκό υδραγωγείο της Νικόπολης στην περιοχή της Λάκκας Σουλίου. In: St. Mamaloukos, G. Riginos, M. Stork (eds) *Λάκκα Σουλίου ΙΙ. Πρακτικά του Επιστημονικού Συμποσίου «Λάκκα Σουλίου. Νέες ιστορικές και αρχαιολογικές καταγραφές», Θεσπρωτικό, 28 & 29 Ιουλίου 2012*: 93-120. Thesprotiko, Athens.

Runnels, C. and van Andel, T. 1993. A handaxe from Kokkinopilos, Epirus, and its implications for the Paleolithic of Greece. *Journal of Field Archaeology* 20: 191-203.

Ruscu, L. 2006. Actia Nicopolis. *Zeitschrift für Papyrologie und Epigraphik* 157: 247-255.

Sarikakis, Th. Chr. 1967. Άκτια τα εν Νικοπόλει. *Archaeologikē Ephemeris*: 145-162.

Schram, W. D. 2014. Dynamic Control Elements in Roman Aqueducts – A Reconnaissance Study. In: Ch. Ohlig and Ts. Tsuk (eds) *Cura Aquarum in Israel II: Water in Antiquity. Proceedings of the 15th International Conference on the History of Water Management and Hydraulic Engineering in the Mediterranean Region Israel 14-20 October 2012* (Schriften der Deutschen Wasserhistorischen Gesellschaft 21): 229-246. Siegburg: Papierflieger Verlag.

Stein, C. A. 2001. In the shadow of Nicopolis: Patters of settlement on the Ayios Thomas Peninsula. In: J. Isager (ed.), *Foundation and destruction of Nicopolis and Northwestern Greece: the archaeological evidence for the city destructions, the foundation of Nicopolis and the synoecism* (Monographs of the Danish Institute at Athens 3): 65-75. Athens/Aarhus.

Tidman, B. M. 1950. On the foundation of the Actian games. *The Classical Quarterly* 44 (3-4): 123-125.

Vega Del Barrio M. 1998. A propos d'une inscription perdue d'Haghios Georgios (Epire). *Bulletin de correspondance hellénique* 122: 501-509.

Vitti, M. and Vitti, P., in press. Experimental innovations in vaulting construction under Trajan and Hadrian. In: S. Keay and C. Triantafillou (eds) *Trajan and the City of Rome: Recent Work on the Trajanic Building programme.* Southampton.

Vitti, P., 2013. Roman vaulted construction in the imperial period. Organization of the construction and dissemination of the building techniques in the Peloponnese. Phd dissertation, Aristotle University of Thessaloniki.

Vokotopoulou, I. 1975. Νικόπολις-Νυμφαίον. *Archeologikon Deltion* 30. Χρονικά, Β' 2: 214-216.

Walker, S. E. 1979. The architectural development of Roman nymphaea in Greece, Unpublished PhD dissertation, University College London.

Wilson, A. 2001. *Urban water storage, distribution, and usage, Water use and hydraulics in the Roman City.* Dubuque: Kendall/Hunt Publishing Company.

Wiseman, J. 2001. Landscape archaeology in the territory of Nicopolis. In: J. Isager (ed.) *Foundation and destruction of Nicopolis and Northwestern Greece: the archaeological evidence for the city destructions, the foundation of Nicopolis and the synoecism* (Monographs of the Danish Institute at Athens 3): 43-64. Athens/Aarhus.

Wiseman, J. 1992. Archaeology and Remote Sensing in the Region of Nicopolis, Greece. *Context* 9 (3-4): 1-4.

Wiseman, J. 1993. Land, Sea and Aerial Surveys in northwestern Greece. *Context* 10 (3-4): 11-15.

Zachos, K. L., 2001a. *Το Μνημείο του Οκταβιανού Αυγούστου στη Νικόπολη. Το τρόπαιο της ναυμαχίας του Ακτίου.* Athens: Ministry of Culture. Archaeological Receipts Fund.

Zachos, K.L. 2001b. Excavations at the Actian tropaeum at Nicopolis: A preliminary report. In: J. Isager (ed.) *Foundation and destruction of Nicopolis and Northwestern Greece: the archaeological evidence for the city destructions, the foundation of Nicopolis and the synoecism* (Monographs of the Danish Institute at Athens 3): 29-43. Athens/Aarhus.

Zachos, K. L. 2003. The tropaeum of the sea-battle of Actium at Nicopolis: interim report. *Journal of Roman Studies* 16: 64-92.

Zachos, K. L. 2007a. Η οχύρωση και η πολεοδομική οργάνωση της ρωμαϊκής Νικόπολης: Νεότερα στοιχεία και παρατηρήσεις. In: K. L. Zachos (ed.) *Νικόπολις Β΄. Πρακτικά του Δεύτερου Διεθνούς Συμποσίου για τη Νικόπολη (11-15 Σεπτεμβρίου 2002)* τ.Ι // *Proceedings of the Second International Nicopolis Symposium (11-15 September 2002)* v.I: 273-298. Preveza: Actia Nicopolis Foundation.

Zachos, K. L. 2007b. Τα γλυπτά του βωμού στο Μνημείο του Οκταβιανού Αυγούστου στη Νικόπολη: Μία πρώτη προσέγγιση. In: K. L. Zachos (ed.) Νικόπολις Β΄. *Πρακτικά του Δεύτερου Διεθνούς Συμποσίου για τη Νικόπολη (11-15 Σεπτεμβρίου 2002)* τ.Ι // *Proceedings of the Second International Nicopolis Symposium (11-15 September 2002)*: I, 411-434. Preveza: Actia Nicopolis Foundation.

Zachos, K. L. 2008. *Άκτια: αθλητικοί αγώνες των αυτοκρατορικών χρόνων στη Νικόπολη της Ηπείρου.* Τ.Δ.Π.Ε.Α.Ε. Επιστημονική Επιτροπή Νικόπολης. Athens.

Zachos, K. L. 2013. Coins and the City: Architecture on the Coins of Nicopolis. In: K. Liampi, Kl. Papaevaggelou-Genakou, K. L. Zachos, A. Douzougli and A. Iakovidou (eds) *ΚΕΡΜΑ III, Νομισματική και Οικονομική Ιστορία στην Ήπειρο κατά την Αρχαιότητα. Πρακτικά του 1ου Διεθνούς Συνεδρίου, Πανεπιστήμιο Ιωαννίνων 3-7 Οκτώβρη 2007// Numismatic History and Economy in Epirus during Antiquity. Conference Proceedings, Ioannina 3-7 October 2007*: 257-281. Athens: 'Lydia Lithos' - Society for the Study of Numismatics and Economic History.

Zachos, K. 2015. *Αρχαιολογικός Οδηγός Νικόπολης. Περιδιάβαση στο ιστορικό, ιερό και αστικό τοπία. Μνημεία της Νικόπολης* 10 // *An archaeological guide to Nicopolis: Rambling through the historical, sacred, and civic landscape* Monuments of Nicopolis, 10. Athens: Ministry of Culture. Archaeological Receipts Fund.

Zachos, K. and Georgiou A. 1997. Περιοχή Νοτιοδυτικής Πύλης - Θέση Μπούφι (αγροτεμάχιο αρ.59, ιδιοκτησία ΥΠΠΟ). *Archeologikon Deltion* 52. Χρονικά, Β' 2: 588-592.

Zachos, K. L. and Pavlidis, E. A. 2010. Die frühen Bauten von Nicopolis. Bemerkungen zu den Bauphasen und techniken der Kaiserzeit. In: R. Asskamp and T. Esch (eds) *IMPERIUM - Varus und seine Zeit. Beiträge zum internationalen Kolloquium des LWL-Römermuseums am 28. und 29. April 2008 in Münster.* Münster: Aschendorff Verlag.

Zachos, K. L. Stamou, I. Leontaris, L. Soukantos, I. 2015. *Το Ωδείο της Νικόπολης*. Athens: Ministry of Culture. Archaeological Receipts Fund.

Dr Konstantinos Zachos
Emeritus Director of Antiquities
klzachos@gmail.com

Leonidas Leontaris
Architect Engineer
PhD Candidate, Aristotle University of Thessaloniki
leontaris4@hotmail.com

The water supply of Roman Thessaloniki

Manolis Manoledakis

Abstract

The city of Thessaloniki was so much increased in terms of both size and population in the Roman period that a new significant water supply construction was urgently needed. This need was covered by a huge ambitious aqueduct that carried water from Mt Kissos (modern Hortiatis) to the acropolis of Thessaloniki, and from there to the city quarters. It consisted of a series of underground tunnels driven deep inside a mountainside to tap its aquiferous stratum, and a monumental bridge, of a total length of *ca.* 20km. The aqueduct was constructed in the Roman period, but experienced several repairs and reconstructions in the Early Christian, the Byzantine and the Ottoman periods, as can be assumed by careful observations and research. In this paper we deal with the history of the Hortiatis aqueduct by examining the several construction phases of the monument as well as the references to it in the written sources.

Keywords: Thessaloniki, Hortiatis, aqueduct, qanat, springs

Introduction: the city's water needs

The location chosen by Cassander to build in 316/5 BC the new city, which he named after his wife, was not accidental. Besides the fact that it was at the cove of a bay within the Thermaic Gulf, and therefore offered a perfect and protected harbour, it also covered an area surrounded by an important mountain, Hortiatis, then called Kissos, and several streams that flew very close to its walls.[1]

During the first phase of the history of Thessaloniki, the needs for water of its inhabitants, who, as we learn from Strabo (7 frg. 21, 24) came from twenty six neighbouring small towns of which we know six (Apollonia, Chalastra, Thermi, Gariskos, Aineia, Kissos), were covered by the area's groundwater,[2] which was drawn from various wells scattered throughout the city. Evliya Çelebi mentions in the chapter of the eighth book of *Seyahatnâme* (1667) that Thessaloniki had 3060 such wells in his time. This figure is certainly exaggerated as far as the ancient times are concerned, but in any case, the Hellenistic city seems to have had water autonomy; little more can be said on this issue, as very few parts of this early city phase have been archaeologically researched.

In the Roman period, however, the situation changed. Already in the 1st century AD, Strabo (7 frg. 21) calls Thessaloniki 'metropolis of Macedonia', while a century later Lucian *(Lucius or the Ass* 46*)* would call it 'the largest city of Macedonia'. It seems that the city experienced a remarkable increase of population, as well as a consequent increase of its territory. It is no coincidence that it was chosen by Apostle Paul as one of the cities he visited in his second missionary journey. Besides, the city flourished significantly during the Imperial period, especially from the 1st century AD, when large public buildings appear, such as administrative buildings, baths, the theatre-stadium, probably also the first phase of the Agora.[3] It is obvious that the needs for water increased significantly and the city's groundwater was no longer sufficient to cover them.

As a result, the inhabitants of Thessaloniki turned to other solutions, and the one that was offered as particularly suitable was certainly the water supply of the city from the water sources of Mt Kissos

[1] For these streams see Dimitriadis 1983: 218-221.
[2] Cf. Lioutas 2000: 39, who estimates that the population of the Hellenistic city was no larger than 5000 people; Tamiolakis 2001: 13.
[3] Vitti 1996: 58; Allamani-Souri 2003: 81; Adam-Veleni 2003: 143, 147, 155-158; 2004: 226.

(modern Hortiatis). Raised 1201m to the east of Thessaloniki, this mountain has always been famous for its verdant environment and abundant waters. Many authors of all historical periods, from Xenophon (*Hunt.* 11.1) until John Kameniatis (*De exp. Thess.* 5), Evliya Çelebi (*Seyahatnâme*, book 8) and later travellers, described in more or less detail and with a glorifying mood the fertile plains of the mountain, its dense vegetation, its rivers and springs; During the winter large pieces of ice were formed on its numerous lakes, which 'resemble the eye of life', and were sold during the summer by the people of the mount in the city of Thessaloniki, even until the last century.[4]

All this indicates that Mt Hortiatis was an ideal solution for the city's water supply problem, a solution that was indeed soon used. The aqueduct of Hortiatis was built in the Roman period and for many centuries supplied the city of Thessaloniki as well as other neighbouring settlements with water. Both the route it followed and part of the construction itself are today known and visible near the entrance of the village Hortiatis. However, until recently, some questions remained unanswered – and some still need to be investigated – such as: the dating of its first construction phase, the dating of its repairs and subsequent construction phases, which sector of Thessaloniki it supplied with water, and whether it was earlier, contemporary or posterior in comparison with the other two aqueducts, those of Retziki and Lembet, that are also reported as having supplied Thessaloniki.[5] However, for the last two, much less is known. We will examine all these separately.

The fact is that the excavations in ancient Thessaloniki have revealed several pipelines of water supply, which are divided into three types: the built water pipes, the clay pipes, and the lead pipes.[6] Most of them were found in districts outside the ancient city, such as Toumba, Thermi, and Polichni.[7] These water pipes present two difficulties: the first relates to their routes, which, because of the small part that is in all cases preserved, cannot be safely restored and thus identified with the course of either one of those three aqueducts, or be regarded as a part a fourth one.[8] The second difficulty concerns the precise dating of the earliest of these pipes, which are generally dated to the Late Roman and/or the Early Christian period.[9] The presence of both water pipes and cisterns grows significantly, especially in the Upper Town of Thessaloniki, during the late Early Christian and the Byzantine periods.[10]

There is also an honorary inscription of the 2nd century AD from Thessaloniki,[11] which has been completed by P. Nigdelis in a manner suggesting that it was referring to the repair of a city water supply system from the area of Altia; the scholar connects Altia with the recorded in the sources -but still unidentified - location named Altos.[12] As Nigdelis points out, Altos cannot be connected to one of the three known aqueducts, but the inscription is an important testimony to the water supply of Thessaloniki from distant areas, already from the Imperial period.

Finally, it is suspected that the name 'Caracalla', which until the early 20th century was born by a region of Pylea, between Hortiatis and Thessaloniki, could refer to an aqueduct built in that region by Emperor Caracalla (AD 198-217),[13] but this cannot be proven, as there are doubts even for the very fact of Caracalla passing through the region of Thessaloniki.[14]

---

[4] For the environment of Mt Kissos and its dramatic change since Antiquity, see Manoledakis 2007: 85-91.

[5] Tamiolakis 2001; Lioutas 2000; Nigdelis 2006.

[6] Cf. Lioutas 2000: 40, with relevant images.

[7] Eleftheriadou 1992: 424; Makropoulou 1992: 424-425; *Bulletin de correspondence hellénique* (=BCH) 122, 1998: 871; Doulgeri 1998; Soueref 2000: 35; Lioutas 2000: 40-46; Nigdelis 2006: 68.

[8] Indicative is the case of the duct in Toumba, for which Eleftheriadou (1992: 24) says that it is Late Roman and irrelevant to the Hortiatis aqueduct, while Doulgeri (1998: 221) considers it Early Christian and having Mt Hortiatis as its starting point.

[9] See above, notes 7 and 8.

[10] Nalpantis 1991.

[11] *IG* X 2 1, 226; *Bulletin de correspondence hellénique* (=BCH) 46, 1922: 527.

[12] Nigdelis 2006: 65-70. For Altos, which is mentioned by Stephen of Byzantium (s.v. Altos) to be a village near Thessaloniki, according to Theagenes, see Cramer 1828: 238; Dimitsas 1874: 252; Papazoglou 1988: 203; Papakonstantinou-Diamantourou 1990: 104; 1997: 352. The fact that Altos is placed on some maps in the region of Philyron (e.g. on the map in Papakonstantinou-Diamantourou 1997: 354), is due to its arbitrary, as Papakonstantinou-Diamantourou (1997: 352), correctly points out, placement there by Dimitsas (1874: 252), although there is no evidence for this placement.

[13] Dimitriadis 1983: 423.

[14] Allamani-Souri 2003: 71, 86-87.

But let us see what we know about these three aqueducts of Thessaloniki.

The aqueducts of Retziki and Lembet

The aqueduct of Retziki starts near the homonymous village, so consequently it also carries water from Mt Hortiatis. The name of the region is a corruption of a Turkish noun meaning paradise, a name given precisely due to the abundant water and lush vegetation of the area. The water was transported initially through water abstraction galleries and afterwards in closed pipes, covered with stone plaques and coated internally with kourasani, and entered Thessaloniki passing through the Litaia Gate.[15] From there it ended up in a reservoir at the Monastery of the Twelve Apostles (today Church), which is dated from the Byzantine period.[16] G. Tamiolakis, the only one so far that has written in detail about this aqueduct, considers that it may be older than that of Hortiatis, which he dates to the early 4th century. However, not only is there no evidence to confirm such a view, but on the contrary the following data can contradict it: 1. All parts of the Retziki aqueduct that survive today, such as that in the area of Asvestochori, belong to the Byzantine period; in the best case, some may possibly date in the Early Christian period.[17] 2. The dating of the Hortiatis aqueduct in the 4th century AD has now been reversed, since new data, as we shall see in the following paragraphs, date the first construction phase of the monument in the 1st century AD. 3. The Retziki aqueduct was known as 'New Water' (Yeni Su)[18] and was mentioned as such in contradistinction to the Hortiatis aqueduct, suggesting that it was later than this.

The aqueduct of Lembet, like that of Retziki, supplied with water the western part of Thessaloniki. Starting from the homonymous area (modern Stavroupolis), it ended up in the reservoir of the Twelve Apostles church.[19] This aqueduct is considered a construction of the Byzantine period,[20] although parts of water pipes in Stavroupolis and Polichni dating to the Roman times have recently been discovered, which could be linked to this aqueduct.[21] In the Ottoman period it was restored by the Grand Vizier of Rumeli, İbrahim Makbul Paşa. It should be noted however that in the very few – and all of a later period – texts that mention this aqueduct, there is a confusion of information or even an identification with information referring to the aqueduct of Retziki, as sometimes it is the Lembet aqueduct that is called Yeni Su, and associated with the reservoir of the Twelve Apostles church.[22] This could mean either a connection of the two aqueducts, which were anyway situated very close to each other, or a confusion of the scholars probably because of this very neighbourhood.

The aqueduct of Hortiatis

It is certainly remarkable that none of these two aqueducts, of Retziki and Lembet, was mentioned by any medieval source or even by a modern traveller of Thessaloniki until the 19th century.[23] In stark contrast, as we shall see later on, the aqueduct of Hortiatis is mentioned in a variety of texts from as early as the 12th century. Moreover, the Hortiatis aqueduct provides several archaeological data that allow both a more credible than in the other two aqueducts dating of its first construction phase and the pin-pointing and dating of its subsequent repair phases. This is due to the two large parts of the aqueduct preserved today, namely the qanat of Haghia Paraskevi and the part of the aqueduct bridge at the entrance of the Hortiatis village, to both of which separate chapters will be dedicated here.

---

15 For the Retziki aqueduct, see Tamiolakis 2001: 29-35, who is cited by all the later authors writing about the aqueduct.

16 Hatzi Ioannou 1880: 66; Papageorgiou 1901: 38-39; Orlandos 1940: 377-383.

17 Tamiolakis 2001: 31-33, figs. 7-9.

18 Hatzi Ioannou 1880: 66.

19 Papageorgiou 1901: 38-39.

20 For the Lembet aqueduct, see Dimitriadis 1983: 426-432; Lazaridis 2001; Tamiolakis 2001: 36-39. According to data from the Ottoman period, the amount of water from Lembet that reached Thessaloniki in this era was greater than that of the Hortiatis waters; however the latter watered a bigger part of the city. See Dimitriadis 1983: 426, 433.

21 Lioutas 2000:45.

22 Tafrali 1913: 118-119; Letsas 1961:146; Spieser 1984: 13.

23 Cousinéry (1831: 106-107) refers to 'Urendgik', but only to connect it geographically with Mt Hortiatis (which he erroneously calls Disoron), to the mountain range of which it really belongs, and without mentioning any separate aqueduct of this mountain.

Figure 1. Panoramic view of the aqueduct bridge of Hortiatis.
In the upper right side there is the entrance of the homonymous village.

The route to Thessaloniki and the city's water supply

The part of the aqueduct that we can see today near the entrance of the Hortiatis village (Figures 1 and 6) is a very small piece of a huge construction of approximately 20km, which started from the homonymous mountain and ended up in the city's reservoirs. The area where the waters of the mountain were collected and started their route – initially underground – is the region of Haghia Paraskevi, near the homonymous chapel to the east of the village of Hortiatis, on the northwest slopes of the mountain at an altitude of 575-585m. A qanat was constructed there,[24] the largest among others in Hortiatis,[25] to collect the mountain water (Figures 2-5). It is a water abstraction construction with a history of many centuries, a few samples of which survive today in Greece. Some hundred meters further to the west of the qanat, water was exiting the soil surface and led towards the city of Thessaloniki on an aqueduct bridge,which is preserved today near the entrance of the Hortiatis village and is the monument that is today known as 'the aqueduct of Hortiatis' (Figure 1). This is a unique monument of this kind to be preserved in Central Macedonia. Afterwards, the water pipe goes underground again, with short over ground intervals, depending on the natural terrain of the area through which it passed, so that the desired inclination was maintained in any case.

[24] See in detail below.
[25] For the other qanats of Hortiatis, see Manoledakis 2007: 98, 116 and figs. 35 and 47.

The route of the Hortiatis aqueduct, as this is indicated by several, much smaller, parts of it that survive scattered between Hortiatis and Thessaloniki, shows that it entered the city from its so-called eastern part.[26] After the aqueduct bridge of the village, the water route continued towards northwest, to the junction of the roads to Asvestochori and Kavala, then towards Panorama; afterwards it turned northwest again, to the area of the modern 'Philippeion' Hotel, from where it sloped down to the Citadel of Thessaloniki, where a reservoir was situated.[27] From there, the Hortiatis water entered the city, where it was distributed to fountains, baths and reservoirs through a number of branches. The most important of these reservoirs seems to have been at the Monastery of Vlatadon, which provided a large part of the city with water.[28] Another branch of the aqueduct, which sent water to the eastern part of the city, has been identified in the area of the Church of the Presentation of the Saviour (Ypapanti).[29] Other branches are also mentioned; some of them are probably earlier than the Vlatadon one; for example, the one that passed along the eastern wall, ended up at the Gate of Kalamaria and brought water in a reservoir behind the Church of Haghios Nikolaos Orfanos,[30] or the one passing outside the eastern walls, in the area of the Aristotle University of Thessaloniki.[31]

According to a view, the water of the Hortiatis aqueduct passed also from the Gate of Anna Palaiologina, to serve other reservoirs and fountains of the city, such as those at the Church of Haghios Georgios, and the fountain to the north of the Church of Haghia Sofia.[32]

The qanat

Description

The qanat is a very simple and effective way to tap the underground aquiferous stratum and transport water with the help of the natural inclination to the level of the settlement. The term *qanat* that has prevailed in the international literature is Arabian (in Persian it is called *kariz).* The word comes from the Assyrian word *qanu,* which means reed – from which comes the synonymous ancient Greek word *κάννα* (and through it the Latin *canna)* – and consequently whatever has the shape of a reed, i.e. pipe, duct.[33]

The qanat is a system of underground tunnels and pipelines, with the help of which the aquiferous stratum, whose water is transported to the surface with the help of the slope gradient, is tapped.[34] Consequently, the settlement should be located close to a natural slope. In the upper part of the slope, the mother (or searching) shaft – as it is called by Polybius (10. 28. 2) and Xenophon (*Hell.* 3. 1. 7) – is opened, in order to find the spring. From this highest point of the qanat the tunnels and (in greater depths) the closed pipes begin to carry water to the settlement, which has to be located on a lower level than that of the spring. Their gradient – necessarily smaller than that of the slope so that they gradually reach the surface – usually does not exceed 5‰, or 0.5m per km, the height of the tunnels ranges from 1.00 to 2.5m, their width from 0.75 to 1.5m, and their length from a few meters up to 50 km. The water often flows in a channel, while all the remaining space serves the need to be accessible by people for inspection, repairs, cleaning and generally maintenance.

In the international literature the qanat is considered a Persian (according to the prevailing view) or an Armenian invention, which, based on the existent archaeological evidence, is dated no later than the 8th

[26] In fact this is the southeastern part.

[27] Cf. Letsas 1961: 145; Dimitriadis 1983: 424; Spieser 1984: 13; Tamiolakis 2001: 24. Cf. also Cousinéry 1831: 106-107.

[28] Dimitriadis 1983: 423, 433; Stogioglou 1971: 52-53, who argues that this specific water supply belongs to an era earlier than the founding of the monastery. For other branches of the aqueduct in Thessaloniki, cf. also Tafrali 1913: 117-118; Letsas 1961: 145.

[29] Dimitriadis 1983, 61: 433.

[30] Hatzi Ioannou 1880: 66. For the 26 branches of the water pipes of the Hortiatis aqueduct in Thessaloniki in the Ottoman period see Dimitriadis 1983: 432-442. Cf. Tamiolakis 2001: 41-51.

[31] Hatzi Ioannou 1880: 66; Tamiolakis 2001: 46-49.

[32] Tafrali 1913: 117; Letsas 1961: 145; Tamiolakis 2001:41. This view is not accepted by Dimitriadis (1983: 433).

[33] One can also meet the terms *foggara* (in North Africa) and *madjira*, after which Madrid was named (the city of qanats). Cf. Hodge 1992: 392 note 8; Vavliakis (1989: 1) and Wulf (1968: 94) etymologize, not convincingly, the term *qanat* from the ancient Semitic word *gano* (dig), as they do also with the word *channel*.

[34] The most important study of the qanat system is that of Goblot (1979), with exhaustive bibliography. For additional bibliography, see Hodge 1992: 391 note 4. Also recently, Tölle-Kastenbein 1990: 39-42; Hodge 1992: 20-24; Grewe 1998: 33-40 and passim; Hodge 2000: 35-38; Schacht 2003: 411-423. In the Greek bibliography the work of Vavliakis (1989; 1997; 2000) almost monopolizes. In the 11th century the Arab mathematician Mohamed Al Karagi wrote a book in Persia, recording details about the qanat construction techniques. See, Grewe 1998: 33-40, esp. 33 notes 46-49.

century BC.[35] Recently, however, the view that the technique was probably known in Greece earlier, even if used for other purposes, has been supported. For example, the drainage construction of the Minyans in Lake Kopais in the 15th century BC is considered to essentially follow the qanat technique,[36] while mythological references to the labyrinth of Daedalus at Knossos are also connected with this particular water supply system.[37] In Greece, some scholars believe that we can talk about the existence of water works based on the logic of qanat (regardless of use) from the 6th century BC onwards. Some of the cases are the 6th century BC pipeline in Athens,[38] the Tunnel of Eupalinos in Samos (540-525 BC)[39] and the galleries of Stryme (earlier than the 4th century BC),[40] Olynthus (*c.* 420 BC),[41] Perachora (*c.* 300 BC), Aegina, and Skyros,[42] although there is no written testimony that refers to the presence, or the technique, of the qanat construction. Many relevant constructions are dated to the Ottoman period. It is worth noting that, already since antiquity, the qanat experienced a rapid spread from the Middle East to Arabia, India, the Iberian Peninsula and northwest Africa, especially in the Roman period, while -in later times- it reached Japan and the New World through the Spanish conquistadors (attested as far as the city of Los Angeles).[43]

However, not all of the above-mentioned underground aqueducts in Greece can certainly be called qanats. Today, there are some reservations concerning the use of the term *qanat* itself, and the wider term *underground aqueduct* is preferred by some scholars. From a point of view, this is not unjustified, since not all the underground aqueducts are qanats. Nevertheless, the Hortiatis underground aqueduct is indeed a qanat, and that is why we shall keep using this term here for the specific construction.

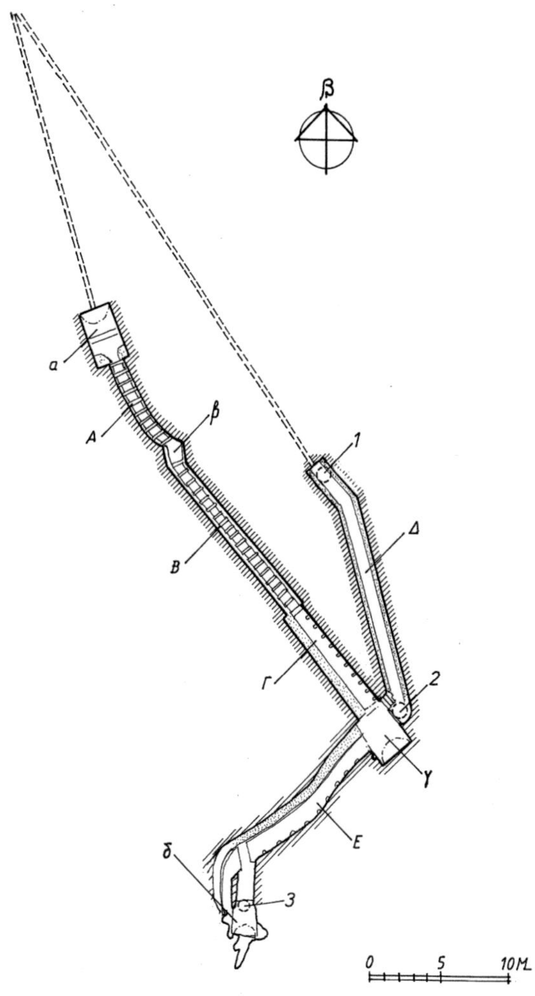

Figure 2. Plan of the water abstraction zone of the Haghia Paraskevi qanat.

The qanat of Haghia Paraskevi in Hortiatis (Figures 2-5) is the largest known in Greece.[44] The water abstraction zone[45] covers an area of about 600m², consists of underground tunnels (of a total length of 74m, average height of 1.50-1.60m and width of 0.56-0.75m), three shafts[46] and four water collection areas (Figure 2). The central part consists of three consecutive tunnels (A-Γ), with NNW-SSE direction. In the SE part

[35] The Persian origin of qanat is supported by the majority of scholars, such as Hodge 1992: 20; 2000: 35; Tölle-Kastenbein 1990: 39; Grewe 1998: 33. In contrast, Forbes (1956: 665-666; 1964: 157-158) considers more likely that the qanat was invented in Armenia. However, in Iran, the qanat is still the primary method of water supply, and it is estimated that in 1961 22,000 qanats were operating, with a total length of 170,000 miles, and a daily output of 48,384,000m³, equivalent of a large river (Wulf 1968: 94, 105; Hodge 2000: 37 note 4). Indeed, it is estimated that in Persia, in antiquity, there were about 40-50,000 qanats operating simultaneously, having a total length larger than the distance from the earth to the moon (Garbrecht 1988: 18).

[36] Knauss *et al.* 1984:186 ff. Cf. Strabo 9.2.40.

[37] Vavliakis 1997: 585-586; 2000: 47-48.

[38] Perhaps of the Peisistratid period. See Thompson 1956: 49-52. Cf. Vavliakis 1997: 586; Grewe 1998: 55-56. A first description was given by Curtius and Kaupert 1878:15,16.

[39] Seitanidis 2003: 56 and 63, indicating, however, a difference between the Tunnel of Eupalinos and the qanat. Cf. Grewe 1998: 58-69. A bit later, probably under Darius I, the Persians constructed the first qanat in Egypt, in the oasis of Kharga. See, e.g. Forbes 1964: 158. For the Kharga qanats, which were probably constructed in the Roman period, according to recent research, see Schacht 2003.

[40] Bakalakis 1967: 38-45.

[41] Robinson 1946: 103 ff.

[42] Cf. Vavliakis 1997, 586; 2000: 50.

[43] Hodge 1992: 21-22; 2000: 35.

[44] For the Haghia Paraskevi qanat, see Manoledakis 2007: 96 and in detail Manoledakis and Androudis 2007.

[45] By 'water abstraction zone' we mean the section with the tunnels. Often, the term qanat is used to define only this part, although qanat can be considered the entire construction, including the underground pipes that follow.

[46] These shafts are cut for the cleaning and ventilation of the qanat while it is being constructed, and its maintenance afterwards.

Figure 3. Roman bricks on the furrow walls of the Haghia Paraskevi qanat.

there is the largest water collection area, a rectangular arched room (γ), from which two tunnels diverge to different directions. One of them (E) is going SW, and ends up at an area (δ) near the cave with the spring. The other (Δ) is going NNW. It is located at the lower level, accessible by a masonry scale and has a shaft at each of its two ends (1 and 2). The depth of shaft 2 (7.90m) corresponds also to the maximum depth of the whole qanat. From the NW end of this tunnel, as well as from the corresponding end of tunnel A in area α, water continues its route in underground closed rectangular cross-section pipes, to its final destination.[47]

The tunnels are constructed in Quaternary sediments. The rocks in the qanat construction area consist mainly of crystallized limestones of the Triadic Period and clay states of the Jurasic Period.[48] The karst origin of the water is indicated by stalactites 2-3cm long on the roof of tunnel E. The average water production of the qanat is 22m^3/h.[49]

The Haghia Paraskevi qanat is not just excavated in the soil, but has also masonry in many places, consisting of bricks and semi-chipped stones. In the lower parts, bricklayers with crushed tile mortar domain. Higher, we find stones with lime mortar, while in some parts there are remnants of brick masonry. The entrance to the tunnels is located at the ground surface, at the north end of room α, while the water comes from the spring near the area δ.[50]

Construction phases and chronology

The dating of the construction of the Haghia Paraskevi qanat remains a difficult issue, as it applies to every qanat in the absence of written or other evidence. According to the only view published so far, it is considered as having been constructed in the Ottoman period, under Sultan Murat II.[51] However, this view clearly needs revision. It is mainly the preserved masonry parts in the qanat tunnels that can lead to such a revision.

In the lower parts, where the water flows on the furrow (groove) walls, bricks are preserved at several points, as, for example, in spaces β and γ, and in tunnels Γ and E, the thickness of which reaches at least

[47] Manoledakis and Androudis 2007: 289.
[48] The Haghia Paraskevi qanat has been geologically studied by Vavliakis *et al.* 1998: 73-87.
[49] Vavliakis *et al.* 1998: 76, 81-87.
[50] Manoledakis and Androudis 2007: 290.
[51] Vavliakis 1997: 587; Vavliakis *et al.* 1998: 78. Indeed, Vavliakis, misinterpreting a passage from Tamiolakis' book (Tamiolakis 2001: 18-20), mentions that the qanat was constructed in the 18th century by Murat II, who, of course, lived in the 15th century.

Figure 4. Part of Byzantine masonry of the Haghia Paraskevi qanat.

4cm (Figure 3). In some parts lime mortar with grey-white and yellowish colour is preserved. Noteworthy is the presence of a clay cylindrical pipe in tunnel A, 11cm thick and with a diameter of 26cm, of unknown function, which could lead water from another nearby source to the groove, or perhaps serve the ventilation of the tunnel. All these seem to belong to the first construction phase of the qanat, or at least of its masonry. This phase certainly belongs to Late Antiquity. Indeed, it is likely to be Roman, judging by certain elements: one is the similarity of the section of qanat to those of known Roman qanats in Europe, such as the so-called Raschpëtzer-Qanat in Luxembourg, dated from the 2nd century AD.[52] Another could be the fact that the dimensions of the aforementioned bricks are absolutely identical to those of the bricks in the lower parts of the buildings in the Galerius Palace in Thessaloniki.[53] Essential for the dating of the construction of the qanat in the Roman period is certainly the dating of the earliest part of the Hortiatis aqueduct bridge in this period, as will be presented in detail below.

Based on the above, we can say that the Haghia Paraskevi qanat, which as a masonry construction existed in Late Antiquity, may have been initially created during the Roman period. Thus, the view that there are no known qanats in Greece constructed during the Roman and Byzantine periods must be refuted.[54] This first construction primarily may have been just excavated in the soil, with a simple masonry on the furrow walls. Later, when the need to strengthen the structure, perhaps also to increase its production, appeared, most of the parts of the brick masonry that we see today were created. Besides, the vaulted roofing of, for instance, spaces γ and δ, with thinner and in many cases irregular bricks, presents similarities with the Middle Byzantine period's techniques (Figure 4).[55]

Certainly, similar needs for improvement and expansion of the construction appeared also later, for example, during the Ottoman period, and these works could indeed be connected to Murat II, for whom we know that he cared for the repair and maintenance of the directly associated with the qanat Hortiatis aqueduct.[56] The extended parts that are built only with stones, such as the entire tunnel Δ, seem to belong to this period (Figure 5).[57]

[52] Grewe 1998: 176 ff.
[53] Manoledakis and Androudis 2007: 290.
[54] Vavliakis 1997: 586.
[55] Manoledakis and Androudis 2007: 291.
[56] See *infra*.
[57] Manoledakis and Androudis 2007: 291.

Figure 5. Part of Tunnel A of the Haghia Paraskevi qanat, constructed during the Ottoman period.

Finally, we know that the last major repair of the qanat took place in 1918, and this is visible in several parts all over the monument, and indeed it is often difficult to distinguish it from the Ottoman ones. The restoration was made by the Municipality of Thessaloniki, to the ownership of which the whole water supply system passed after the Liberation of the city (1912) and until 1939.[58] In 1975 the qanat stopped watering Thessaloniki and today it supplies with water a hospital, a monastery and a military camp in the area between Panorama and Asvestochori.[59]

The aqueduct bridge of Hortiatis

Description

The aqueduct bridge (today known as 'the aqueduct of Hortiatis') is preserved in a total length of approximately 223m and heads northeast towards the village (Figures 1 and 6). To the opposite direction it continues almost parallel to the Hortiatis-Asvestochori road until it is lost under the territory of a modern military camp. It practically connects two hills, in order to keep the required inclination of the water route to the town steady.[60]

Figure 6. The south face of the Hortiatis aqueduct bridge.

[58] Tamiolakis 2001: 20, 57.

[59] Manoledakis 2007: 96-98; Manoledakis and Androudis 2007: 291.

[60] See in detail Manoledakis and Marki 2008.

The highest point of the monument (499.07m above sea level – 498.12m above ground level) is to be found at its southeast end, from where the aqueduct heads northwest for 118m (hereafter Section 1, Figure 7). From that point it changes direction and presenting an angle of 140° it heads west for another 101.5m, until it meets the ground level (hereafter Section 2, Figure 6). The maximum height of the aqueduct in the central part of Section 2 reaches 20.1m, and its maximum thickness 5m. In the centre of Section 2 the monument is divided into three horizontal zones, from which the lowest is slightly thicker by about 1.00m than the middle one and the latter slightly thicker than the upper one (Figure 6). On the upper zone there is a water conduit of rectangular section, measuring 0.5m across (Figure 8).The clay water pipes were placed within, however only few remains of them have survived. The conduit was covered with large slates, many of which remain in situ.[61]

Figure 7. The southwest face of the aqueduct (Section 1) from the west.

In the centre of Section 2 there are two large open arches (Figure 6). The largest, in the lower zone, has a height of 8.5m, a maximum width of 5.3m, and gives the impression of an underpass on the dirt road. Above this there is the second open arch, with another one to its west.

Figure 8. Section 2. Slates covering the conduit on the aqueduct.

The north face of the aqueduct (Section 2) is reinforced by buttresses on either side of the large central arch, of which the two that are located to the east of the arch (e.g. Figure 13, on the left) are preserved to a large extent – the western one, which is the largest, is

[61] Manoledakis and Marki 2008: 361.

Figure 9. South side. Arched room at the western part.

wholly preserved – while to the west of the arch the remains of other buttresses are visible, damaged to a large degree, so that it is difficult to determine their exact number.

On the south face, at a distance of 28.5m west of the western open arch, an arched room is opened into the aqueduct masonry, which probably served the guard of aqueduct and its water (Figure 9). To the east of the central arch there is a small brick-framed arch. At the western end of the aqueduct there is a water collection area.

Section 1 is a straight stonewall, with a few rows of bricks only at its lower part (Figure 7). Only the southwest face is actually visible, while the largest part of the northeast face rises just 0.5-1.5m from the natural slope.[62]

The aqueduct is founded directly on the natural limestone rock. It is built with worked stones and freestones of various sizes and bricks, while lime mortar has been used as a binder.[63]

Construction phases and chronology

The exact dating of the aqueduct, as well as the discrimination between its construction phases is not an easy task. A careful observation leads to the conclusion that the monument received many repairs and interventions throughout its history and that its image changed significantly several times. The view that has been expressed and many times repeated in the past that in the three horizontal zones of the aqueduct we can distinguish the three successive layers of its life phases, the Roman, the Byzantine and the Ottoman ones (Figure 6), is simplistic and erroneous.[64] Much of what we see today is the product of the Byzantine and the Ottoman periods. To the initial construction phase undoubtedly belong the thick bricks (up to 8cm thick) in the lower part of the north face, which suggest that the aqueduct was

[62] Manoledakis and Marki 2008: 361-362.

[63] Manoledakis and Marki 2008: 363.

[64] Tamiolakis 2001: 16, 24, who actually repeats the view of P. Theodoridis.

initially built in the Roman period (Figure 10). It is likely that this construction was part of a general activity of public buildings construction, particularly water supply ones, by the Romans during the Imperial period. G. Bakalakis, reported in the mid-20th century the incorporation of a Roman altar into the aqueduct,[65] which, however, is nowhere visible today, nor was it identified during the recent archaeological works on the monument.[66]

Figure 10. The thick bricks in the lower part of the north face. From the initial construction phase of the aqueduct (Roman period).

J.-M. Spieser, based on the description from O. Tafrali regarding a reservoir column with a capital in the Citadel of Thessaloniki, [67] which received water from the Hortiatis aqueduct, ventured a connection of this reservoir to that of the Pergamon Acropolis, which also had a column with capital; he consequently ventured a dating of the earliest water supply of Thessaloniki from Hortiatis - and hence the construction of the aqueduct- in the Hellenistic period, during which, as he adds, the city had already enough population.[68] This view, although interesting, is certainly difficult to be proven, since it is based just on the description of a structure that no longer exists.

Until a few years ago, the dominant view on the dating of the monument was that of G. Tamiolakis - the only one systematically involved with the aqueduct until the 2000s - namely that the monument was built at the beginning of the 4th century, during the years of Galerius' administration.[69] In those years, as he claimed, not only the necessary financial resources were available, but also many other indications of the intention of the Romans to construct large water supply systems or bath facilities, in Greece generally and in Thessaloniki particularly.

However, the thick bricks in the lower part of the north face, to the west of the large open arched, always created to me the impression that they were earlier than the 4th century (Figure 10).[70] Therefore, within the frame of a study for the consolidation and conservation of the aqueduct in 2008, after the initiative of the Municipality of Hortiatis and the former 9th Ephorate of Byzantine Antiquities,[71] I sent, after having taken the necessary license, parts of the aforementioned thick bricks to the Cultural and Educational Technology Institute of the 'Athena' Research Centre in Xanthi, where they were dated with the Thermoluminescence method. The dating AD 46 was given as the year of the firing of the bricks, with an error deviation of about one century. Since dating becomes less likely as we move away from the year

[65] Bakalakis 1953-1955: 361.

[66] For this works see *infra*.

[67] Tafrali 1913: 117.

[68] Spieser 1984: 14.

[69] Tamiolakis 2001:16. Cf. also Nigdelis 2006: 67.

[70] Manoledakis 2007: 100.

[71] This study (Marki *et al.* 2008) constituted practically the first archaeological and architectural study of the aqueduct in its long history. Its results were presented in 2008 (Manoledakis and Marki 2008), while the results of the restoration works of the monument, which started in 2011, based on the above-mentioned study, finished in 2014 and revealed some new data on the aqueduct, soon to be published by S. Akrivopoulou, the archaeologist responsible for the works, and the author of this paper.

Figure 11. Parts of arches that have later been walled up, on the north side (Section 2) of the aqueduct.

AD 46, we can generally say that the aqueduct must have been built in the 1st century AD.[72] This makes the dating of the first construction phase of the qanat in the Roman period even more likely, as was mentioned above.

The history of the aqueduct is reflected in both its today's masonry and in the written sources. The monument we see today is characterized by the existence of different phases and repairs.[73] It is characteristic that we can distinguish traces of arches of the Roman and Early Christian periods that have been later walled up (Figure 11). The today existing arches must belong to the Middle Byzantine period,

[72] Manoledakis and Marki 2008: 364.

[73] A detailed description of the monument and its phases, as they clearly appear after the recent archaeological research, will be soon released by S. Akrivopoulou and M. Manoledakis (see above, note 71).

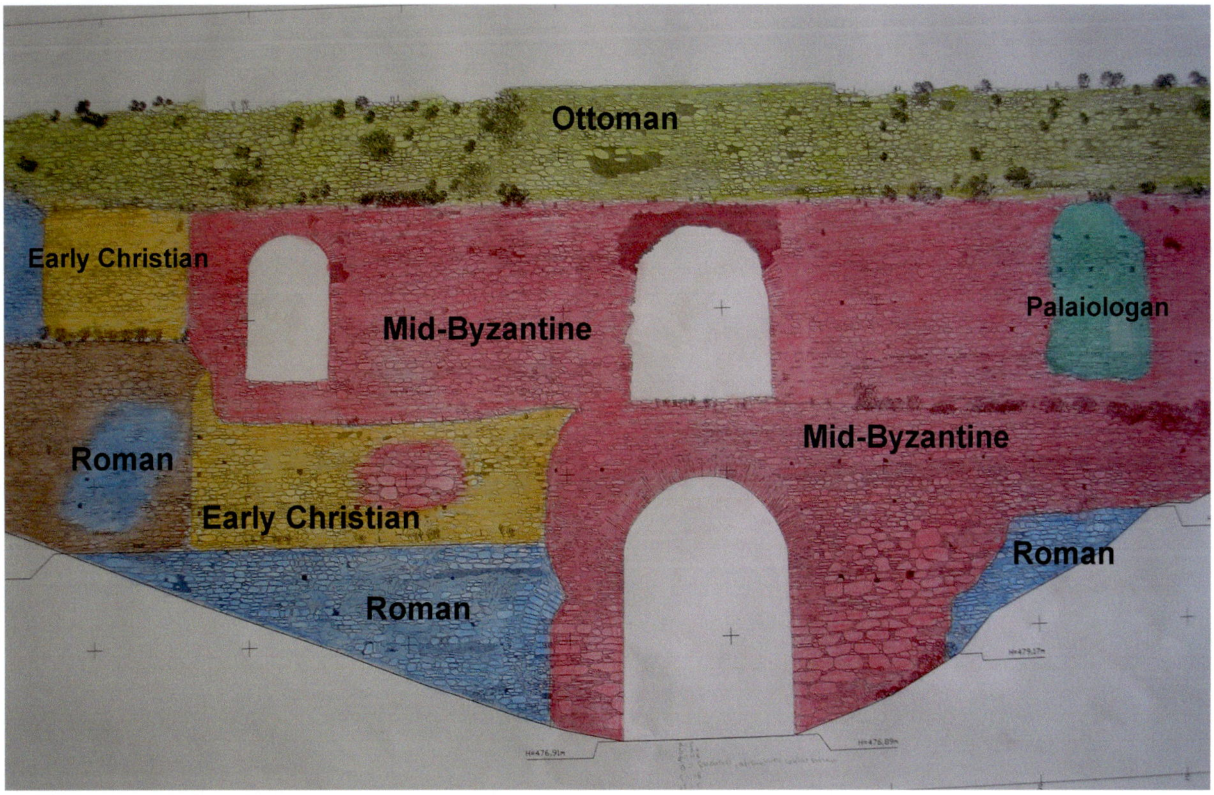

Figure 12. The south face (Section 2) of the aqueduct. To the left of the large central arch, as well as in other places, parts of arches are visible, which were later walled up.

which means that the current image of the aqueduct significantly differs from the initial one.[74] Some of the construction phases that are distinguished can more easily be linked to specific historical periods and events than others.

The surviving traces of arches indicate that in its initial phase the aqueduct probably had at least two successive zones of close to each other arches, according to other examples of Roman aqueducts. Parts in a lower zone on the south face, which clearly juts out from its overlying, later zones, belong to this initial, Roman phase (Figure 12). In the western part of the south face a part of an arch is visible, which must have been walled up no later than the mid-Byzantine period. Some relics of the same phase, with small stones, thick bricks and parts of arches, are also preserved elsewhere.

A second construction phase, which appears to belong to the Early Christian period, is visible on the north face, above the Roman part with the thick bricks to its west (Figure 13), as well as on the south face, above and near the aforementioned Roman parts (Figure 12). Here, irregularly placed freestones are interrupted by horizontal rows of bricks, slightly thinner than the Roman ones, while on the north face several parts of arches are again visible, in the areas of the buttresses.[75] This Early Christian phase could be related to the water supply constructions that Constantine the Great is known to have carried out in the area of Thessaloniki.[76]

The central part of Section 2, namely the part with the arches, belongs to the Middle Byzantine period (Figures 12-13). It is the period during which the small arches of the earlier phases have been walled up, the lower part of the monument has been reinforced with large stones, and the three arches we see today have been opened, at least one of which received later interventions, judging by its pointed

[74] Manoledakis and Marki 2008: 365.
[75] Manoledakis and Marki 2008: 365.
[76] Marki and Hatziioannidis 2004: 279-287.

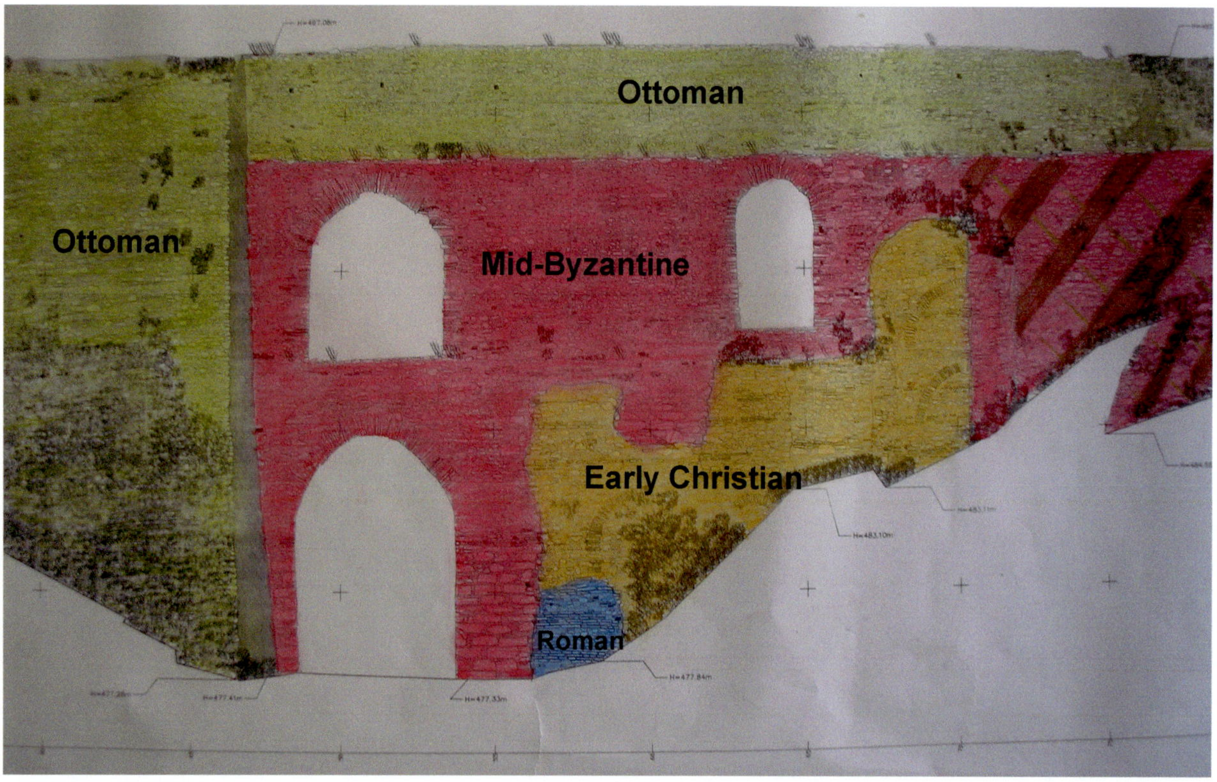

Figure 13. The north face (Section 2) of the aqueduct.

ending. Fieldstone and thin bricks dominate in the masonry. On the upper zone of the arches, to the east of the central one and at the same distance from that as the western one, there was another arch, of the same size as the western one, which gave an impression of symmetry. This arch was walled up with small stones and thin bricks during the Late Byzantine (Palaiologan) period, when the small open arch that exists today was created (Figure 12 and Figure 14). It is the only clearly visible Late Byzantine part of the aqueduct.[77]

These Byzantine period repairs may be associated with the general development of the region of Hortiatis, due to the building and the flourishing of the Monastery of Hortaitis.[78] It is from this period that we have the first indirect reference to the aqueduct from the *Typikon* for the Monastery of Christ Pantokrator of 1137;[79] it is there implied that in the 12th century Thessaloniki was supplied with water from Hortiatis, that the Monastery of Christ Pantokrator had rights over the income received from the water supply of the city, and that the Hortiatis water was used in industrial activities.[80] A few decades later, Archbishop Eustathius of Thessaloniki, when describing the siege of the city by the Normans in 1185, refers to the 'water from Hortiatis' that watered Thessaloniki, to a leak in the reservoir of Eptapyrgion that the city inhabitants suffered from, as well as to the efforts of the city's governor to repair it.[81] Similar references are found in golden bulls of 1316 and 1317.[82]

Other parts of the monument are characterized by greater uncertainty as to their construction phase (Figure 13, with stripes), which, however, dates from the Byzantine period and later. One or more construction phases may belong to the Ottoman period, during which bricks are completely absent

[77] Manoledakis and Marki 2008: 365-366.
[78] For the Hortaitis Monastery, see just recently Manoledakis 2007:109-127, with the previous bibliography.
[79] Gautier 1974: 121.
[80] Cf. Letsas 1961: 146; Spieser 1984: 12; Manoledakis 2007: 102-104.
[81] *De Thessalonica capta*, Bonn, 435-436 (S. Kyriakidis, *Eustazio di Thessalonica, La espugnazione di Thessalonica*, Palermo 1961: 78).
[82] *Actes de Chilandar I* (ed. by M. Zivojinovic, V. Kravari, Ch. Giros), Paris 1998: 48, 61, 135-139, 230 n. 33, 6-10, 237 n. 34, 240 n. 35, 27-30; Dölger 1960: 66, 69; Manoledakis 2007: 104.

(Figure 13).The entire upper zone of Section 2 (Figure 6), the arched room of the south side (Figure 9), and the large eastern buttresses of the north side (Figure 13, on the left), definitely belong to this period.

Figure 14. South face. Small open arch to the east of the central arches.

As regards Section 1, except for a small zone in the lower part of the southwest face showing the features of the Early Christian masonry, all the rest seems to belong mainly to the Ottoman period, perhaps with some traces of earlier, Byzantine interventions, confined on the southwest face (Figure 7).[83]

During all these years, the great significance of the water of the Hortiatis aqueduct for Thessaloniki is indicated as undoubted through the references of the written sources. This significance becomes most evident in a testimony delivered by Hierax, the Grand Logothete of Istanbul (16th c.), which refers to the era of the siege of Thessaloniki by the Turks in 1430.[84] According to this testimony, when Murat II was besieging the city, he encountered such a strong resistance that he was ready to abandon his attempt. But then he was approached by the monks of the Vlatadon Monastery, who advised him to cut off the water supply of the city from Hortiatis, so that the city residents would suffer from water scarcity and surrender. The sultan, always according to Hierax, followed the advice of the monks and put a soldier to guard the traitor monks, so that no one hurts them. According to a view, this testimony of Hierax is based on a tradition, which corresponds to reality and explains the privileges of the Vlatadon Monastery during the Turkish occupation, due to this very betrayal.[85] However, the majority of scholars consider the accusation of Hierax unfounded, even malicious, since no other relevant source refers to such an event.[86] It is indeed hard to believe that the defence of Thessaloniki was so strong in this time to drive Murat to abandon the siege,[87] or that the reservoir of the Vlatadon Monastery was the only source of water supply of Thessaloniki, so that its occupation would result in the surrender of the city.[88]

However, Murat II is definitely connected to the aqueduct of Hortiatis and specifically to the first of its Ottoman phases, since shortly after the fall of Thessaloniki in 1430 he arranged a repair of the aqueduct and built twenty new fountains in Thessaloniki, as well as the famous 'Bey Hamam', which received water from it.[89] Moreover, for the storage, cleaning and maintenance of the repaired aqueduct, Murat settled some landless peasants in the area of the modern village of Asvestochori, in return for certain privileges.

[83] Manoledakis and Marki 2008: 366.
[84] Sathas 1872:256-257. Cf. also Tafrali 1913: 116; Spieser 1984: 12.
[85] See the relevant views in Tamiolakis 2001: 27.
[86] Dimitriadis 1983: 423; Letsas 1961: 147-148; Tamiolakis 2001: 26-27; Vakalopoulos 1969: 92; 1997, 194.
[87] See e.g. Manoledakis 2007: 92-95.
[88] Cf. also Tamiolakis 2001: 27.
[89] Vasdravelis 1952: 167, text 129.

Thus Yeni Köy (i.e. New Village) was founded, which, when its inhabitants began to produce and sell lime in the 17th century, was renamed as Kireç Köy (i.e. Asvestochori).[90]

The operation of the aqueduct is also testified in the 16th century, in the diary of Gabriele Cavazza, secretary of a Venetian ambassador who passed from Thessaloniki going to Istanbul,[91] as well in the 17th century, in the *Seyahatnâme* of Evliya Çelebi (1667), which refers twice to the Thessaloniki water supply facilities from Hortiatis.[92]

Specific events concerning the aqueduct are mentioned in the 18th century: apart from the destruction of the reservoir of the Vlatadon Monastery by rioters in 1714, which was accompanied by attacks on monks,[93] another event is mentioned in a firman of 10 September 1722: 'Some residents of Thessaloniki opened up wells on Mount Hortiatis and managed to find large quantities of water, which they introduced to the Hortiatis aqueduct. Then they sold part of it to the poor and kept the rest. In this way, however, water was shut off in most of the city's fountains, so that foreigners and the poor suffered from water shortage'.[94]

In the 19th century Michael Hatzi Ioannou called the aqueduct of Hortiatis 'the most noteworthy aqueduct of the present city (Thessaloniki)'.[95] In 1927, nine years after the renovation of the Haghia Paraskevi qanat, the Municipality of Thessaloniki stopped the water supply of the Vlatadon Monastery from Hortiatis and channelled the latter's water to new reservoirs.[96] Since 1945 the aqueduct bridge was not in use, but the water continued passing through the same point, just in an underground pipeline. Thessaloniki continued to be supplied with water from Hortiatis until 1975.[97]

The preserved part of the Hortiatis aqueduct bridge is also connected to the most tragic event in the history of the mountain and of the village in particular, since in its area the events that led to the Hortiatis' Holocaust of 2 September 1944 where the slaughter of 149 people took place.[98]

Conclusion

Thessaloniki experienced a population increase and an economic boom in the Roman period, testified from both written sources and the appearance of the first large public buildings in the city, among them bath complexes. Thus, the underground waters of the city were no longer sufficient to meet its increasing needs in water. The water of Mt Hortiatis, located very close to Thessaloniki, was an ideal, but also an expensive solution. Therefore, an aqueduct was constructed, which, using the qanat tapping technique, as well as several bridges – where required because of the natural relief – carried water from the Haghia Paraskevi region of Hortiatis to the city, covering a distance of about 20km.

The dating of the building material of the first construction phase of the Hortiatis aqueduct showed that the monument was built in the 1st century AD. This dating perfectly fits the reference by the Roman authors to Thessaloniki as 'the metropolis of Macedonia' (Strabo 7 frg. 21) and as 'the largest city of Macedonia' (Lucian, *Lucius* or *the Ass* 46). It also perfectly fits with the building development of the city that is archaeologically documented at that time. Therefore, the Hortiatis aqueduct can be included among the Roman aqueducts of Greece.[99]

---

[90] Dimitriadis 1983: 217, 424; Tamiolakis 2001: 20. For Neochorion, see Hatzi Ioannou 1880: 49. Cf. also Tsekos 1957; Manoledakis 2007:105-106.

[91] Mertzios 1947: 125.

[92] *'The most delicious water in Thessaloniki is the one coming from the sources of Hortiatis, sweet, clean and very digestive. With this water the khans, mosques, baths and public stores are supplied, as well as the mansions of prominent citizens'* (*Seyahatnâme*, book 8, ch. 'Water supply')...... *'The freshwater of Hortiatis comes to Thessaloniki in underground pipes and is distributed to all baths, mosques and the city's imarets'* (*Seyahatnâme*, book 8, ch. 'On the sources and rivers of Mount Hortatzis'). For the water supply of Thessaloniki with the water of Hortiatis in the 17th century see also Vakalopoulos 1997: 231-233.

[93] Stogioglou 1971: 54. In the same year the Hortiatis aqueduct was also seen by P. Lucas (1719: 56).

[94] Vasdravelis 1952: 167, text 129. Cf. Spieser 1984: 12. For these texts, see also Manoledakis 2007: 106-108.

[95] Hatzi Ioannou 1880: 66.

[96] Stogioglou 1971: 54; Tamiolakis 2001: 26, 57, who also mentions the renovation of the aqueduct in 1918.

[97] Tamiolakis 2001: 23.

[98] Manoledakis 2007: 167-168, with the previous bibliography. All the details of the event with relevant documents can be found in Valachas and Theochari 2008.

[99] For the Roman aqueducts in Greece see Lolos 1997: esp. 302-312.

This is certainly not the case with the other two aqueducts that supplied the city with water, the aqueducts of Retziki and Lembet, as the data available on these (considerably less than on the aqueduct of Hortiatis) begin from the Byzantine period onwards. In this period the two aqueducts supplied the western part of the city with water, while the aqueduct of Hortiatis watered the eastern part. In this case, it is highly probable that the Roman fountains, baths and nymphaea of Thessaloniki, in both its eastern and western parts, such those identified in the Dioiketerion Square,[100] in the Agora,[101] under the churches of Haghios Demetrios, Panayia Acheiropoietos and Haghia Sofia, on Olympou, Phillipou, Profitou Ilia and Galileou Streets, at the junctions of Egnatia and Antigonidon Str., Baltadorou and Venizelou Str., Armenopoulou and Apostolou Pavlou Str., on Platonos, Sofokleous, and Haghiou Nikolaou Streets,[102] in the Galerian palace,[103] as well as the 'hydreion' in the sanctuary of Serapis[104], received water exclusively from the aqueduct of Hortiatis, the only one confirmed in the Roman period.

In the same case, the inscription of the 2nd century referring to the water from the Altia region[105] is also likely to be associated with the Hortiatis' aqueduct and this could mean the location of the settlement called Altos somewhere around the route of this aqueduct. This is of course completely hypothetical, and provided that all proposals made both to complete the inscription and to date the three aqueducts are correct, and that Altos had not a different, yet unknown, aqueduct. Similar reservations about a connection with Hortiatis exist also for the Roman and Early Christian water pipes found in the suburbs of Thessaloniki, such as Thermi and Toumba.

Through the examination of the preserved parts of the aqueduct of Hortiatis, both in the qanat of Haghia Paraskevi and in the aqueduct bridge at the entrance of the village, as well as through the references of the texts from the 12th until the 20th century, it becomes clear that the aqueduct was an extremely critical for the city of Thessaloniki public work, and for this reason it was repaired or even reconstructed several times, during the Early Christian, Middle Byzantine, Late Byzantine and Ottoman periods.

Bibliography

Abbreviations used

BCH	*Bulletin de correspondance hellénique*
IG X, 2 1	*Inscriptiones* Graecae, X: Inscriptiones Epiri, Macedoniae, Thraciae, Scythiae. Pars II, fasc. 1: Inscriptiones Thessalonicae et viciniae, ed. Charles Edson. Berlin 1972.

Adam-Veleni, P. 2003. Θεσσαλονίκη: Ιστορία και πολεοδομία. In: D. B. Grammenos (ed.) *Ρωμαϊκή Θεσσαλονίκη* (Αρχαιολογικό Μουσείο Θεσσαλονίκης 1): 121-176. Thessaloniki.
Adam-Veleni, P. 2004. Κρήνη στην αρχαία αγορά της Θεσσαλονίκης. Ένα 'χαμένο' κτίσμα από την εποχή του Φ. Πέτσα. *To Archaeologiko Ergo stē Macedonia kai stē Thrakē (= AEMTh)* 18: 223-238.
Allamani-Souri, V. 2003. Η θέση της Επαρχίας Μακεδονίας στο Ρωμαϊκό Imperium. In: D. B. Grammenos (ed.) *Ρωμαϊκή Θεσσαλονίκη* (Αρχαιολογικό Μουσείο Θεσσαλονίκης 1): 67-120. Thessaloniki.
Athanasiou, F., Malama, V., Miza, M. and Sarantidou, M. 1999. Τα λουτρά των ανακτόρων του Γαλερίου. *To Archaeologiko Ergo stē Macedonia kai stē Thrakē* (= *AEMTh*) 13: 191-206.
Bakalakis, G. 1953-1955. Κισσός. *Makedonika* 3: 353-362.
Bakalakis, G. 1967, Ανασκαφή Στρύμης. Thessaloniki: AUTh.
Cousinéry M. E. M. 1831. *Voyage dans la Macédoine* II. Paris.

[100] Tassia 1993: 332-334.
[101] Adam-Veleni 2004, who mentions that this fountain must have been constructed in the 2nd century BC, using the underground waters, which the Agora area was rich in.
[102] For these constructions, see Marki 1983: 14; Vitti 1996: 96, 174 nr. 45, 179 nr. 56, 191 nrs. 69 and 71, 194 nr. 81, 202 nr. 89, 204 nr. 90, 228 nr. 112, 229 nr. 115, 234 nr. 121, 239 nr. 126, 241, nrs. 127 and 129, 243 nr. 132. Cf. also the reservoirs and pipes in Vitti 1996: 192-193 nrs. 77-78, 197 nr. 82, 226 nr. 110, 243 nr. 130.
[103] Vitti 1996: 110, 150-151; Athanasiou *et al.* 1999, with the previous bibliography.
[104] *IG* X 2 1, 83. Cf. Vitti 1996: 175 nr. 47, with previous bibliography.
[105] See above, in the first chapter.

Cramer, J. A. 1828. *A Geographical and Historical Description of Ancient Greece* vol. I. Oxford.
Curtius, E. and Kaupert, J. A. 1878. *Atlas von Athen*: 15.16. Berlin.
Dimitriadis, B. 1983. *Τοπογραφία της Θεσσαλονίκης κατά την εποχή της τουρκοκρατίας 1430-1912*. Thessaloniki: Kyriakides Bros. Editions.
Dimitsas, M. 1874. *Αρχαία Γεωγραφία της Μακεδονίας*. Athens: Institute of Balkan Studies.
Dölger, F. 1960. *Regesten der Kaiserurkunden des Oströmischen Reiches von 565-1453, Teil 4 (1282-1341)*. Munich/ Berlin.
Doulgeri, E. 1998. Σωστικές ανασκαφές υδραγωγών στη Θέρμη και Τούμπα Θεσσαλονίκης. *To Archaeologiko Ergo stē Macedonia kai stē Thrakē* (= *AEMTh*) 12: 217-222.
Eleftheriadou, K. 1992. Κάτω Τούμπα. *Archaiologikon Deltion* 47: 424.
Forbes, R. J. 1956. Hydraulic Engineering and Sanitation. In: Ch. Singer, E. J. Holmyard and A. R. Hall (eds) *A History of Technology* vol. II: 663-694. Oxford: Clarendon Press.
Forbes, R. J. 1964. *Studies in Ancient Technology* vol. I. Leiden: Brill.
Garbrecht, G. 1988. Mensch und Wasser im Altertum. In: Frontinus-Gesellschaft e.V. (eds) *Geschichte der Wasserversorgung III, Die Wasserversorgung antiker Städte*: 13-42. Mainz am Rhein.
Gautier, P. 1974. Le typikon du Christ Sauveur Pantocrator. *Revue des études byzantines* 32: 1-145.
Goblot, H. 1979. *Les qanats: une technique d'acquisition de l'eau*. Paris: Mouton Éditeur.
Grewe, K. 1998. *Licht am Ende des Tunnels*. Mainz am Rhein: Philipp von Zabern Verlag.
Hatzi Ioannou, M. 1880. *Αστυγραφία Θεσσαλονίκης, ήτοι, τοπογραφική περιγραφή της Θεσσαλονίκης: περιγραφή μιας εποχής*. Thessaloniki: Makedonia editions.
Hodge, A. T. 1992. *Roman Aqueducts and Water Supply*. London: Duckworth Archaeology.
Hodge, A. T. 2000. Qanats. In: Ö. Wikander (ed.) *Handbook of Ancient Water Technology*: 35-38. Leiden: Brill.
Knauss, J., Heinrich, B. and Kalcyk, H. 1984. *Die Wasserbauten der Minyer in der Kopais - die älteste Flussregulierung Europas*. München: Technische Universität.
Lazaridis, S. 2001. Το υδραγωγείο του Λεμπέτ. *Thessalonikeon Polis* 4: 217-224.
Letsas, A. M. 1961. *Ιστορία της Θεσσαλονίκης 1961. Εταιρεία Μακεδονικών Σπουδών*.Thessaloniki: Society for Macedonian Studies.
Lioutas, A. 2000. Υδροδοτικά Θεσσαλονίκης. In: K. Σουέρεφ (επιμ.) *Υδάτινες σχέσεις, Το νερό ως πηγή ζωής κατά την αρχαιότητα*: 39-46. Thessaloniki: University Studio Press.
Lolos, Y. A. 1997. The Hadrianic Aqueduct of Corinth (With an Appendix on the Roman Aqueducts in Greece). *Hesperia* 66.2: 271-314.
Lucas, P. 1719. *Troisième Voyage du sieur Paul Lucas, fait en M.DCCXIV par ordre de Louis XIV* vol I. Rouen.
Makropoulou, D. 1992. Άνω Πόλη (οδός Αιόλου 3). *Archaiologikon Deltion* 47: 424-425.
Manoledakis, M. 2007. *Από τον Κισσό στον Χορτιάτη*. Thessaloniki: Kornelia Sfakianiakis Editions.
Manoledakis, M. and Androudis, P. 2007. Το σύστημα υδρομάστευσης (qanat) της Αγ. Παρασκευής Χορτιάτη. *To Archaeologiko Ergo stē Macedonia kai stē Thrakē* (= *AEMTh*) 21: 285-292.
Manoledakis, M. and Marki, E. 2008. Το υδραγωγείο του Χορτιάτη. *To Archaeologiko Ergo stē Macedonia kai stē Thrakē* (= *AEMTh*)22: 361-368.
Marki, E. 1983. Η Θεσσαλονίκη στην Αρχαιότητα. *Αρχαιολογία* 7: 11-16.
Marki, E. and Hatziioannidis, A. 2004. Υδρευτικά έργα του Μ. Κωνσταντίνου στη Θεσσαλονίκη. *To Archaeologiko Ergo stē Macedonia kai stē Thrakē* (= *AEMTh*) 18: 279-287.
Marki, E., Siaxabani, Ch., Manoledakis, M., Passia, A., Michailidis, G. and Vlysidou, Z. 2008. *Study for the consolidation and conservation of the Hortiatis aqueduct in Thessaloniki*. Thessaloniki.
Mertzios, K. 1947. *Μνημεία Μακεδονικής ιστορίας. Εταιρεία Μακεδονικών Σπουδών*.Thessaloniki: Society for Macedonian Studies.
Nalpandis, D. 1991. Θεσσαλονίκης Υδρευτικά. *To Archaeologiko Ergo stē Macedonia kai stē Thrakē* (= *AEMTh*)5: 271-287.
Nigdelis, P.M. 2006. *Επιγραφικά Θεσσαλονίκεια*. Thessaloniki: University Studio Press.
Orlandos, A. 1940. Η Κινστέρνα της εν Θεσσαλονίκη Μονής των Δώδεκα Αποστόλων. *Makedonika* 1: 377-383.
Papageorgiou, P. N. 1901. Θεσσαλονίκης βυζαντινοί ναοί και επιγράμματα. I. Ο Ναός των Δώδεκα Αποστόλων. *Byzantinische Zeitschrift* 10: 23-39.
Papakonstantinou-Diamantourou, D. 1990. Χώρα Θεσσαλονίκης. Μια προσπάθεια οριοθέτησης. In: Μνήμη Δ. Λαζαρίδη: *πόλις και χώρα στην αρχαία Μακεδονία και Θράκη. Πρακτικά Αρχαιολογικού Συνεδρίου, Καβάλα 9-11 Μαΐου 1986*: 99-107. Thessaloniki: Ministry of Culture.
Papakonstantinou-Diamantourou, D. 1997. Φίληρος η... προβληματική!. In: Αφιέρωμα στον *N. G. L. Hammond*: 349-356. Thessaloniki: Εταιρεία Μακεδονικών Σπουδών (Society for Macedonian Studies).

Papazoglou, F. 1988. *Les villes de Macédoine a l'époque romaine*. Bulletin de Correspondance Hellénique, Suppl. XVI. Paris.

Robinson, M.D. 1946. *Excavations at Olynthus. Part XII, Domestic and Public architecture* (The Johns Hopkins University Studies in Archaeology 36). Baltimore: The Johns Hopkins Press.

Sathas, K. N. 1872. *Μιχαήλ Ατταλειάτης-Νικήτας Χωνιάτης-Θεόδωρος Μετοχίτης-Θεόδωρος Ποτάκιος-Χρυσόβουλοι λόγοι-Ιέρακος Χρονικόν-Κατάλογοι των εν ταις Βιβλιοθήκαις του Αθώνος και του εν Κωνσταντινουπόλει Μετοχίου του Παν. Τάφου χειρογράφων*. Vol. II. Venice: Typois tou Chronou editions.

Schacht, I. 2003. A Preliminary Survey of the Ancient Qanat Systems of the Northern Kharga Oasis. *Mitteilungen des Deutschen Archäologischen Instituts, Abteilung Kairo* 59: 411-423.

Seitanidis, G. C. 2003. *Μορφολογικά-υδρολογικά και περιβαλλοντικά στοιχεία που συνδέονται με το Ευπαλίνειο Όρυγμα της Σάμου*. Thessaloniki.

Soueref, K. 2000. Σχέσεις με το νερό. Η Τούμπα Θεσσαλονίκης ως αφορμή. In: K. Σουέρεφ (επιμ.) *Υδάτινες σχέσεις, Το νερό ως πηγή ζωής κατά την αρχαιότητα*: 17-37. Thessaloniki: University Studio Press.

Spieser, J.- M. 1984. *Thessalonique et ses monuments du IVe au VIe siècle*. Athens/Paris: Diffusion De Boccard.

Stogioglou, G. A. 1971. *Η εν Θεσσαλονίκη πατριαρχική μονή των Βλατάδων*. Thessaloniki: Patriarchal Institute of Patristic Studies.

Tafrali, O. 1913. *Topographie de Thessalonique*. Paris: P. Geuthner.

Tamiolakis, G. 2001. *Η ιστορία της ύδρευσης της Θεσσαλονίκης*. Thessaloniki: University Studio Press.

Tassia, A. 1993. Η σωστική ανασκαφή της ΙΣΤ΄ Εφορείας στην Πλατεία Διοικητηρίου. *To Archaeologiko Ergo stē Macedonia kai stē Thrakē* (=AEMTh) 7: 329-341.

Thompson, H. A. 1956. Activities in the Athenian Agora. *Hesperia* 25: 46-68.

Tölle-Kastenbein, R. 1990. *Antike Wasserkultur*. München: H. Beck.

Tsekos, Ch. 1957. *Ιστορία του Ασβεστοχωρίου*. Thessaloniki (own edition).

Vakalopoulos, A. 1969. *Ιστορία της Μακεδονίας 1354-1833*. Thessaloniki: Vanias Editions.

Vakalopoulos, A. E. 1997. *Ιστορία της Θεσσαλονίκης 316 π.Χ.-1983*. Thessaloniki: Kyriakides Bros. Editions.

Vasdravellis, I. K. 1952. *Ιστορικά Αρχεία Μακεδονίας. Α' Αρχείον Θεσσαλονίκης 1695-1912* (Μακεδονική Βιβλιοθήκη 13). Thessaloniki: Εταιρεία Μακεδονικών Σπουδών (Society for Macedonian Studies).

Vavliakis, E. 1989. *Τα συστήματα qanat στην Ελλάδα. Μελέτη συστημάτων qanat στην Επαρχία Φυλλίδας Σερρών, από μορφολογική, υδρογραφική και κοινωνικο-οικονομική άποψη*. Επιστημονική Επετηρίδα Τμήματος Γεωλογίας Α.Π.Θ. 9. Thessaloniki.

Vavliakis, E. 1997. Τα συστήματα υδρομάστευσης/υδραγωγής και άρδευσης ως βασικός παράγων υποδομής προϊστορικών και ιστορικών πολιτισμών του ελλαδικού χώρου. In: *Πρακτικά 1ου Διεθνούς Συνεδρίου για την Αρχαία Ελληνική Τεχνολογία*: 583-591. Thessaloniki: Piraeus Bank Group Cultural Foundation.

Vavliakis, E. 2000. Υδρομάστευση και Υδραγωγή: το Κανάτ και τα Αρχαιοελληνικά Συστήματα Υδροδοτήσεως Υψηλής Τεχνολογίας. In: K. Σουέρεφ (επιμ.) *Υδάτινες σχέσεις, Το νερό ως πηγή ζωής κατά την αρχαιότητα* 1: 47-50. Thessaloniki: University Studio Press.

Valachas, Th. and Theochari, D. 2008. *2 Σεπτεμβρίου 1944, Χορτιάτης. Ό,τι απόμεινε από τη μέρα εκείνη, η μνήμη είναι*. Thessaloniki.

Vavliakis, E., Staninos, E. and Stefos, N. 1998. Μελέτη του υδρομαστευτικού συστήματος τύπου Qanat (Κανάτ) Αγ. Παρασκευής Χορτιάτη Θεσσαλονίκης (Β. Ελλάδα). In: *Πρακτικά 4ου Πανελληνίου Γεωγραφικού Συνεδρίου της Ελληνικής Γεωγραφικής Εταιρείας, 12-14 Οκτωβρίου 1995*: 73-87. Athens: Greek Geographical Society.

Vitti, M. 1996. *Η πολεοδομική εξέλιξη της Θεσσαλονίκης από την ίδρυσή της έως τον Γαλέριο*. Athens: En Athinais Archaeologike Etaireia.

Wulf, H. E. 1968. The Qanats of Iran. *Scientific American*: 94-105.

Dr Manolis Manoledakis
Associate Professor of Classical Archaeology
School of Humanities,
International Hellenic University,
Thessaloniki, Greece
m.manoledakis@ihu.edu.gr

The Hadrianic aqueduct of Athens and the underlying tradition of hydraulic engineering

Eustathios D. Chiotis

Αρχή παιδεύσεως ονομάτων επίσκεψις[1]
(The investigation of the meaning of terms is the beginning of education)

Antisthenes (445-360 BC)
Αντίδωρον[2] to Theodosios Tassios

Abstract

The Hadrianic aqueduct of Athens is a monumental work of the Roman Emperors Hadrian and Antoninus Pius, constructed during the period AD 125 – 140, with unique hydraulic features. The backbone of the aqueduct is an array of deep wells along a 17.8km long continuous tunnel, whereas the typical Roman aqueducts of Greece (Nikopolis, Corinth, Samos, Lesbos, Knossos, Gortyn) are mainly constructed on the ground and are characterized by water bridges upon arcades across valleys, although occasionally comprise tunnels in order to pass through topographic highs.

In the dry land of Attica (of poor surface water resources), the Hadrianic aqueduct mainly captured groundwater along the tunnel, whereas the typical Roman aqueducts of Greece transferred water from distant springs or rivers. The route and the constructional details of the tunnel are adequately known, given that a section of the aqueduct was recognized and located in 1840 by Friedrich Stauffert[3] and remained the main hydraulic work supplying Athens with water till the 1930s. The sustainability of the aqueduct is due to its design for underground water capturing, a concept already applied to the archaic aqueducts of Aegina and Megara seven centuries earlier.

Despite the fragmentary nature of archaeological data, a preliminary effort is undertaken to trace the remains of the water distribution network in the city of Athens and correlate conduits, reservoirs, fountains and baths associated with the Hadrianic aqueduct. Furthermore, its period of operation is examined, as well as the reasons for the construction of a Late Roman aqueduct with water bridges, still preserved in Athens.

Keywords: aqueduct, groundwater, Hadrian, Athens, tunnel, water bridge

Introduction

The Roman Emperor Hadrian was a dedicated philhellene, especially fond of Athens and among his numerous benefactions to the city was the aqueduct. Hadrian visited Greece three times when he was Emperor (AD 125, 129 and 132) and, as commonly stated, the aqueduct was founded most likely during his first Imperial visit (AD 125).

The main structure of the Hadrianic aqueduct is a 17.8km long tunnel, the northern end of which lies in the center of the modern Olympic Village of the 2004 Olympic Games, at the foothills of the Parnes Mountain, at a depth of 30.5m (Figure 1).

The tunnel ends near the Haghios Demetrios Church, at Ampelokipoi (AD in Figure 1), at a bottom depth of about 6m.[4] From this point it continued as a covered channel constructed in a few meters deep trench for another 2.2km up to the Lycabettus Hill near the city of Athens. The water was conveyed by gravity into a reservoir, in front of which a monumental façade was built. Thanks to early travelers, descriptions

[1] As cited in Arrian (*Discourses of Epictetus*, i, 17).
[2] (EN: *Gift in return*).
[3] Papageorgiou-Venetas 2010: 109
[4] Some critical elevations of the tunnel bottom are: at the well 299 in the northern end 212m, at Haghios Demetrios *c.* 140m and at the bottom of the Lycabettus reservoir *c.* 133m. The elevation at the NE boundary of the National Garden is *c.*100m, high enough for the natural flow of water to the ancient city of Athens.

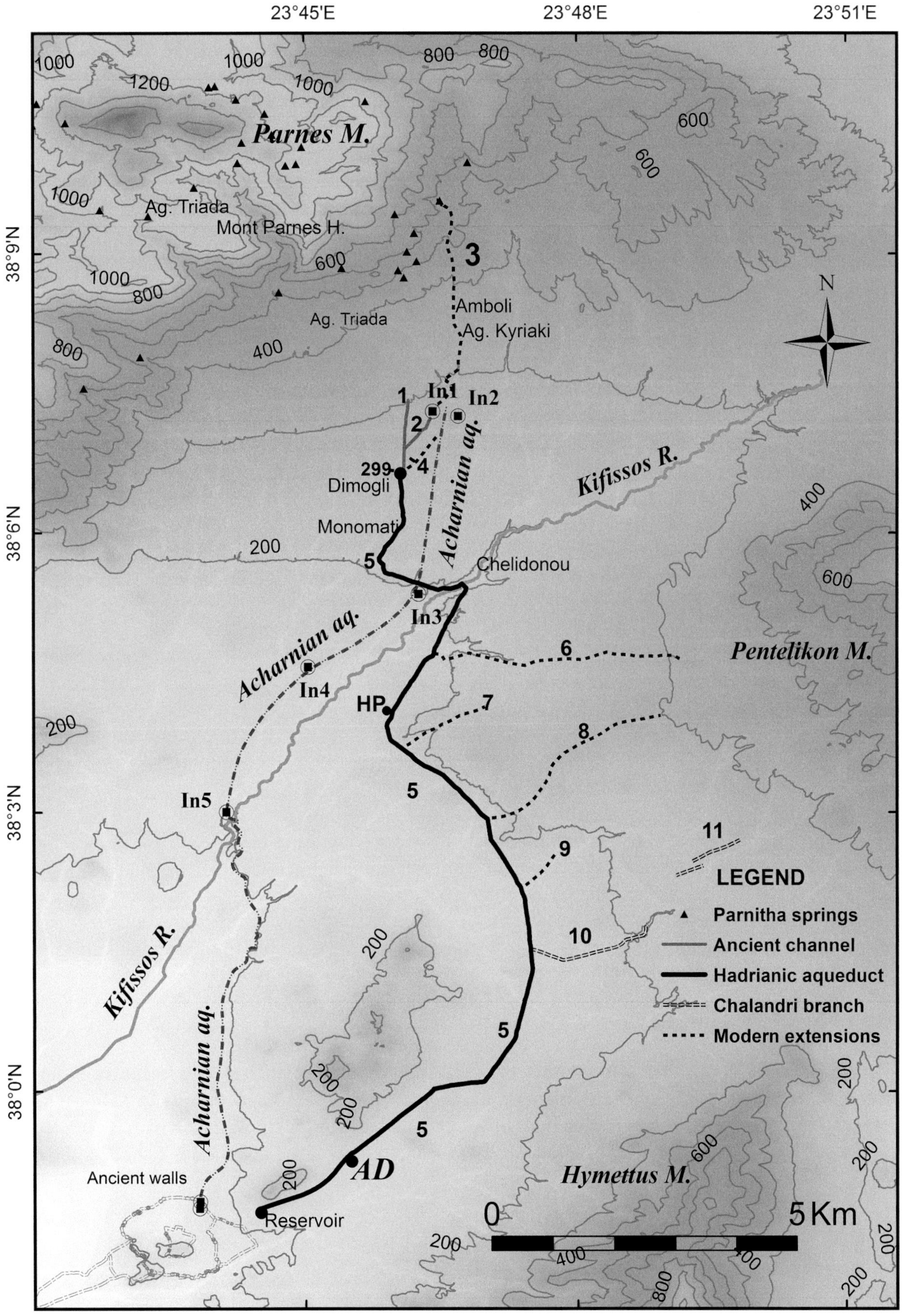

Figure 1. The course of the Hadrianic aqueduct from the North (well 299 at Demogli) up to the resevoir on the Lycabettus Hill; the surface channels at Demogli (1 and 2); the modern branches (3 and 4); the Acharnian aqueduct (from the Parnes Mt to Athens); lateral modern extensions 6 to 9 and the ancient Chalandri branch 10 and 11.

and drawings of the façade and its inscription (according to which the aqueduct was completed in AD 140) are available to us. A fragment of the inscription can still be seen in the National Garden of Athens.

Every 35-40m, wells up to 41m deep were dug along the tunnel using the shafts-and-gallery technique.[5] The aqueduct fell into obscurity during Late Antiquity, apparently due to tunnel collapse near the Haghios Demetrios church at Ampelokipoi. The water continued to overflow and was therefore considered as a spring. In the Ottoman period this 'spring' was exploited for irrigation of the local fields and the left over was conveyed to Athens via a stone-built channel. Experts knew that the Haghios Demetrios 'spring' was associated with the Hadrianic aqueduct, the aligned wells of which were still visibly full of water. Friedrich Stauffert, an architect of the Municipality of Athens during the period 1835-1843, when Athens was still a small town of 25,000 people, tracked the aqueduct for about 5km up to Chalandri, where he lost the continuation.[6] In 1840, he counted 109 shafts of an average section of 1.5m^2 and was the first to recommend the cleaning of the aqueduct for the water supply of Athens.

The process of i) tracing the tunnel ii) cleaning the aqueduct from mud and collapsed blocks of rock and iii) supporting the walls of the tunnel was an expensive and long project, which started in the middle of the 19th century and took a few decades to complete; 465 shafts had been reported cleaned by 1879.[7] The aqueduct remained the main water supply system of Athens until the Marathon Reservoir was completed in 1935 – which for the first time supplied water from areas outside of the hydrologic basin of Athens, surrounded by the Mountains Hymettus, Pentelikon and Parnes (Figure 1). During the summer period, the Hadrianic aqueduct, at its maximum capacity, supplied modern Athens daily with ca. 10,000m^3 (116lit/sec) of water and was in operation until the middle 20th century. After a chemical analysis conducted in 1893, the quality of the water proved to be very good, due perhaps to $CaCO_3$ deposition along its route.[8]

In addition to the detailed description of the aqueduct by Kordellas, Ziller (1877) also wrote short remarks; Pappas (1999) provided plans and data, but the first systematic study was accomplished by Susan Leigh (1998), whose excellent treatise constitutes a dependable source, although important new evidence was brought to light from recent excavations. Impressive underground observations and images have been recently uploaded from a continuing research program conducted by the engineer and speleologist P. Defteraios, in preparation for his Ph.D. dissertation.[9]

The sustainability and some constructional details of the aqueduct have been recently described;[10] this paper shall emphasize on other aspects and mainly on the tunneling technique many meters below the water table, the similarities with the Raschpëtzer aqueduct in Luxemburg and the water distribution network in Athens. In addition, the evolution of pre-Roman hydraulic technology in Greece and its application to the Hadrianic aqueduct will also be examined and correlated to the Iranian qanats in the Appendix.

Description of the aqueduct

The aqueduct tunnel collected surface and groundwater of various origins. Two recently excavated surface channels at the foot of the Parnes Mountain conveyed water from springs in modern times: Channel 1 was part of the so named Karydhia aqueduct,[11] the northern extension of which is only loosely known, and Channel 2, part of the so called Varybopi[12] aqueduct which continues as a shaft-and-gallery

[5] The manholes were sunk at regular intervals of 35-37m, according to Travlos (1971: 242).
[6] Cited in
[7] Kordellas 1879: 85.
[8] *Praktika tēs en Athēnais Archaiologikēs Hetaireias* 48, Χρονικά (Περί της κοίτης του Ιλισού Ανασκαφής) 1893: 118; the quality of the Kallirhoe spring water at the Ilissos riverbed is compared to that of the Hadrianic aqueduct; chemical analysis indicates that the water sample from the Lycabettus Hill reservoir was of much better quality than from Kallirhoe (E. Ghintoni, IGME, oral communication).
[9] http://urbanspeleology.blogspot.gr/p/blog-page_2754.httm
[10] Chiotis and Marinos 2012: 23, 36.
[11] Pappas 1999: 190.
[12] Pappas 1999: 190.

aqueduct up to Haghia Kyriaki and further northwards (modern branch 2). At point A, Channels 1 and 2 are joined together into a single one which pours into the inclined gallery BC (Figure 2).

Pappas supported that the well 299 at C was the northernmost end of the ancient aqueduct and that all hydraulic works northwards were modern extensions.[13] This view was also adopted by the excavators of the site.[14] Asimakou and Kalliga, suggested in particular that the surface Channels 1 and 2 are works of the last quarter of the 19th century.

Although the Karydhia and Varybopi aqueducts are modern extensions at the end of the 19th century, it is supported here that the Channels 1 and 2 have replaced ancient ones, transferring water from springs into the aqueduct, not directly to the well 299, but through the typical inclined Roman gallery BC, which was constructed exclusively for this purpose (Figure 2).

Along the section AB it is possible to distinguish ancient remains: whereas the channels 1 and 2 were in general covered in modern times with plates of Pentelikon marble, ancient stone plates of well cemented breccia are maintained along the section AB, quarried locally in the ancient technique of opening a narrow groove around the block to be excavated. Thus, the rock dug from the excavation of the trench was used in antiquity to cover the trench. During the modern operation of the aqueduct, the ancient trenches were extended and plastered.

The Hadrianic aqueduct was constructed almost 5 centuries after the so called Acharnian aqueduct (dated in the 3rd quarter of the 4th century BC), the course of which is approximately traced in Figures 1 and 2, based on the ancient inscriptions found along its course.[15] The northernmost inscriptions (Figure 1, points In1 and In2), were found 1500m north of the well 299. It would be therefore unwise for the engineers of the Hadrianic aqueduct to construct the inclined gallery and let the adjacent springs (the main resources of the Acharnian aqueduct) unexploited. It is therefore supported that the ancient channel 2 collected water from these springs.

The inclined gallery, which connects the surface Channels 1 and 2 to the Hadrianic tunnel is 80m long, dips ca. 20°. It is a typical Roman construction: the walls are made in *opus latericium* and the roof is composed of three terracotta bricks, which are slightly rotated forming a pattern known as 'pitched-brick' vaulting (Figure 3).

According to Lancaster, the earliest known Roman examples of 'pitched-brick' vaulting with curved bricks appear in hydraulic systems in the 1st half of the 2nd century AD at Athens, Eleusis, and Argos, while others, at Dion and Gortyn, may also date from the same period.[16] It is furthermore argued by Lancaster, that the technique was introduced from Parthia by military engineers, specialized in hydraulics, who participated in Trajan's Parthian War and, upon returning home, they brought along new construction techniques.

This view should probably be reconsidered in the light of opposing evidence. Similar bricks have been used in Eleusis at a grave dated in the middle 1st century BC.[17] Furthermore, vaulting technique was used in much earlier tunnels in Assyria, in the Palace of Sargon II at Khorsabad.[18]

An additional control point (In3) of the Acharnian aqueduct is provided by an inscription recently excavated near the Hadrianic aqueduct reported at a depth of 8m and identified via the characteristic feature of the elliptical section of the excavated conduit.[19] However, the fact that it was found at the

[13] The well 299 in Figure 1 corresponds to the point C in Figure 2.

[14] Platonos-Yiota 2004: 61; Asimakou and Kalliga 2013: 167.

[15] Vanderpool 1965: 171.

[16] Lancaster 2010: 464.

[17] Mylonas 1932: 45.

[18] Mylonas 1932: 48.

[19] Platonos-Yiota 2004: 440; Georgousopoulou 2013: 156.

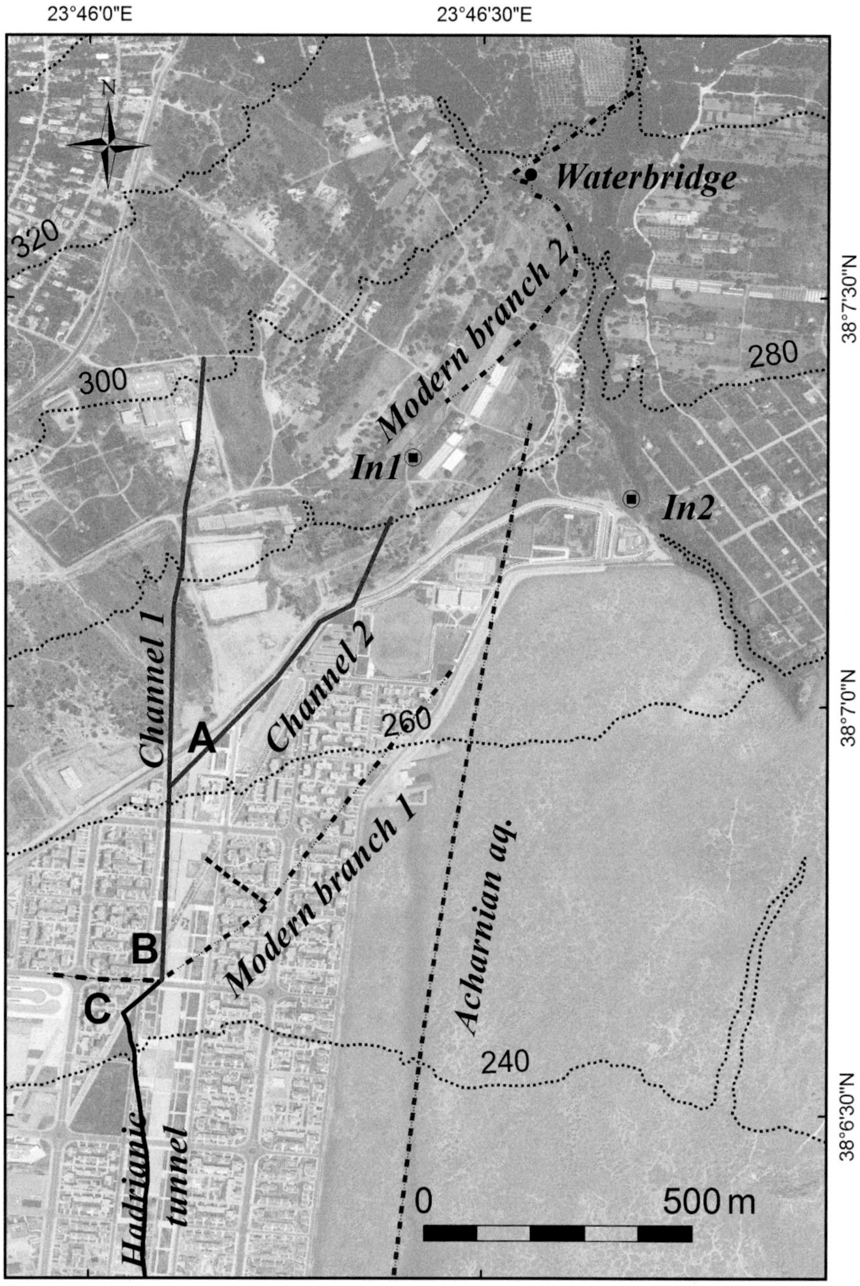

Figure 2. Projection on Ktimatologio air photos: The Olympic Village and the surface channels 1 and 2; the inclined gallery BC; the northernmost end of the Hadrianic tunnel (C), and the Acharnian aqueduct; the locations of the Inscriptions 1 and 2 (In1 and In2).

depth of 8m (the Acharnian aqueduct was usually placed in a cut-and-covered trench, only a couple of meters deep), raises the question whether the conduit belonged to a system supplying water from the Hadrianic aqueduct to the Roman farms in the area.

A boundary stone (*HOPOC*) found at Metamorphosis was considered as a proof that the Acharnian aqueduct reached this area.[20] However, according to the excavation report, it was found at a distance from the course of the Acharnian aqueduct and exactly on the route of the Hadrianic aqueduct (point HP in Figure 1). Thus, a '*HOPOC*' stone without a more specific indication does not constitute a solid evidence for its association with the Acharnian aqueduct.

Another section of the Acharnian aqueduct is suggested here for the first time, at the point where the aqueduct crossed the ancient city walls (Figure 1). A water duct was excavated in 1959 by Threpsiades, passing through an opening of the ancient walls, of similar structure to the Acharnian aqueduct.[21] The terracotta U-shaped half-drums, each of internal dimensions 0.42m height by 0.40m width, bear a combination of Greek letters, such as XP, OA, OY for matching the proper half-drums. Based on the shape of the letters, the duct is dated by the excavator to the beginning of the 4th century BC and its abandonment period is dated by a late Roman

[20] Platonos-Yiota 2004: 59 and 70.

[21] *Archaiologikon Deltion* 16, Χρονικά, B'1, 1960: 29-32

Figure 3a. Top. The entrance to the inclined gallery at the Olympic Village. The walls are made of opus latericium, and the roof is composed of three curved, slightly rotated bricks. (Photo: Courtesy of P. Defteraios).

Figure 3b. Bottom. The lower part of the inclined gallery at the Olympic Village. The roof is composed of corbeled rectangular bricks; further down, by horizontal stone plates (Photo: Courtesy of P. Defteraios).

lamp found in the sediment filling the aqueduct. However, it is reasonable to assume that the Acharnian aqueduct was abandoned earlier, when the Hadrianic aqueduct was put in operation.

The extension of the above described aqueduct was most likely revealed and retained in place in a recent excavation, during the foundation of the new building of the Technical Chamber of Greece in Athens (3-5, Palaion Patron Germanou Str.); however, the excavation report is not published yet. It is suggested that the water of the Acharnian aqueduct from this point (at elevation of *c.* 72m) could possibly reach the Agora of Athens by gravity flow. It could certainly supply Plato's Academy as well, along the Demosion Sema, as inferred from the Hellenistic pipeline at the site C8 in Figure 7 (described below) – which could represent a branch of the Acharnian aqueduct at the site of the Kimonian branch of the Peisistratean aqueduct.[22]

[22] Chiotis and Chioti 2012: 429.

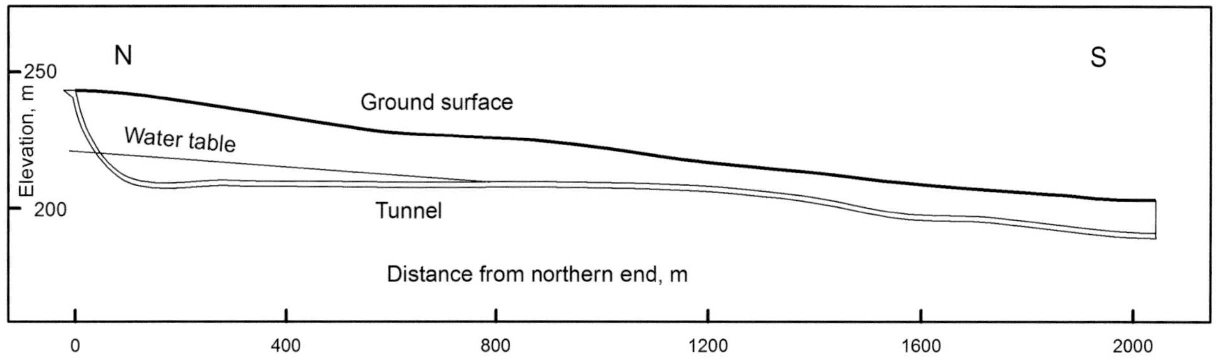

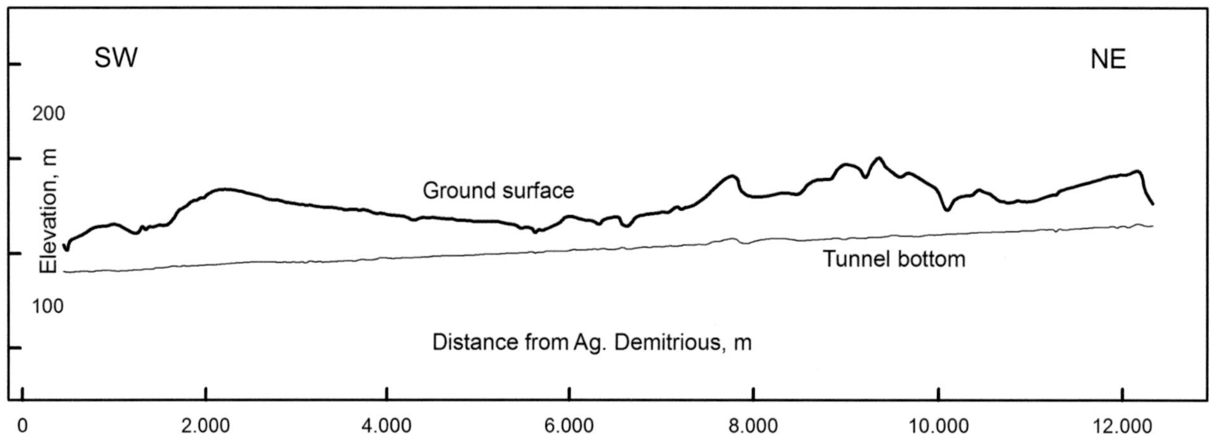

Figure 4a. Top. Section along the northern part of the Hadrianic aqueduct depicting: the ground surface, the water table, the inclined gallery and the aqueduct (after Chiotis and Chioti 2012: 432).
Figure 4b. Bottom. Surface elevation vs tunnel mileage measured from Haghios Demetrios to the Kalyftaki stream (section restricted by the available data. After Chiotis and Chioti 2012: 433).

Lateral underground branches of the Hadrianic aqueduct, approximately parallel to the course of the main streams, supplied additional water. The better known and most important is the ancient Chalandri branch which brought water from the Pentelikon Mountain.[23] However, the most significant source of water was and remains the groundwater along the wells and the main tunnel excavated below the water table.

The tunnel bottom of the 30.5m deep well 299 lies 10m below the undisturbed water table,[24] and remains below the water table over a distance of approximately 800m (Figure 4a).[25] Downstream, where the older cone of compacted sediments occurs below a thin alluvial mantle, the wells are rather shallow, slightly deeper than 10m and stop in the upper zone of the hard rock. The tunnel depth in general varies mostly between 15 to 35m (Figure 4b) – except at the final section (Haghios Demetrios church), where it gradually becomes shallower until it surfaces. Longitudinal sections of the Figure 4 were compiled from original data kindly provided by the EYDAP Engineer E. Nestorides.[26]

The Hadrianic tunnel was intentionally designed unusually deep in order to tap groundwater and thanks to that the aqueduct is sustainable and still functional. The water is infiltrated through both the wells and the tunnel itself. The average gradient of the tunnel is ca. 0.4% although it is locally higher near the Kifissos River, where the tunnel has to pass below the riverbed and the adjacent streams at a safe distance.

[23] Pappas 1999: 187. Sections 10 and 11 in Figure 1 come from Kifissia Sheet (Curtius and Kaupert 1881).
[24] This numbering starts from Haghios Demetrios. Earlier numbering started from the Lycabettus Reservoir and the northernmost well was numbered 367, considerably fewer than the 465 wells reported by Kordellas, due to loss of wells. Assuming a 37.5m average distance between wells, 475 wells are estimated along the proper tunnel, 17.8km long.
[25] Chiotis and Marinos 2012: 27.
[26] Chiotis and Marinos 2012: 31.

Hydrogeological conditions

The Hadrianic aqueduct was built completely underground in order to benefit from the local hydrogeological conditions. This deep underground design intended to capture groundwater through an area of lower precipitation, lacking sufficient surface water resources.[27]

There is, however, a common view that the Perissos and Philothei water bridges belong to the Hadrianic aqueduct (Figure 5). Leigh has convincingly shown that at both water bridges their elevations are lower than the adjacent Hadrianic tunnel and could not therefore supply it with water.[28] Most likely, the Perissos and Philothei water bridges belong to a Late Roman aqueduct which collected water from springs at Kifissia and Heracleion.[29]

The geological formations along the course of the aqueduct, shown in Figure 5, are:

- Quaternary talus sediments
- Older foothill cones
- Red Pikermian sediments
- Lignite-bearing Neogene marls and clays and
- The metamorphic basement represented by Athenian schists.

Thick Quaternary talus sediments outcrop near the Parnes Mountains and constitute the best aquifer in the area. The older formations: Older Cones, Pikermian, Neogene and Schists are impermeable, covered by an alluvial aquiferous mantle. The aqueduct taps water mainly from the Quaternary cone over a length of *c.* 800m, by means of wells 20-30m deep and the tunnel itself (Figure 4a). Further downstream, in the stratigraphically underlying older cone of compacted sediments, the wells are less deep. The whole design of the aqueduct indicates a good knowledge of the hydrogeological conditions in advance which could only be obtained from previously dug wells. The stratigraphy and water yield of such wells would be decisive for planning the course and the depth of the aqueduct.

Copying the ancient engineers, the groundwater of the Quaternary breccias near the northern end of the Hadrianic tunnel was also captured in 1901-1903 along a modern shafts-and-gallery extension (Modern branch 1 in Figure 2) with 7 shafts 33m deep, connected through a tunnel 320m long. The project was undertaken by Petros Protopapadakis, Chief Engineer of the Municipality of Athens (and later, Prime Minister of Greece), and yielded an additional summer rate of about 750m^3 daily.[30]

Along the aqueduct's route additional water is collected from the perched aquifers on top of the above described impermeable formations. If occasionally good aquifers are crossed by the wells, lateral galleries are opened above the tunnel floor to tap them.

Remarks on tunneling

Tunneling faced two significant problems: rock stability and digging at rather significant depths below the water table. The first one was solved by minimizing the tunnel dimensions and by constructing artificial support. The tunnel dimensions were marginally ergonomic for a small in size person, i.e. 0.4-0.8m wide and 1.2-1.8m high. The walls were supported by stone masonry, whereas the roof by either a brick vault (Figure 3a) or a corbelled roof (made of stone or terracotta plates) (Figure 3b).

According to Ziller, the wells and the tunnel north of Ampelokipoi were dug in solid rock.[31] The wells are 9-15m deep and their diameter varies from 1.20m -1.50m. The tunnel measures 0.70m in width and

[27] The annual precipitation is *c.* 400mm in Athens and almost doubles on the Parnes and Pentelikon Mountains.
[28] Leigh 1998: 61-65.
[29] Chiotis and Chioti 2011: 182, 188.
[30] Paraskevopoulos 1907: 446.
[31] Ziller 1877: 121-122.

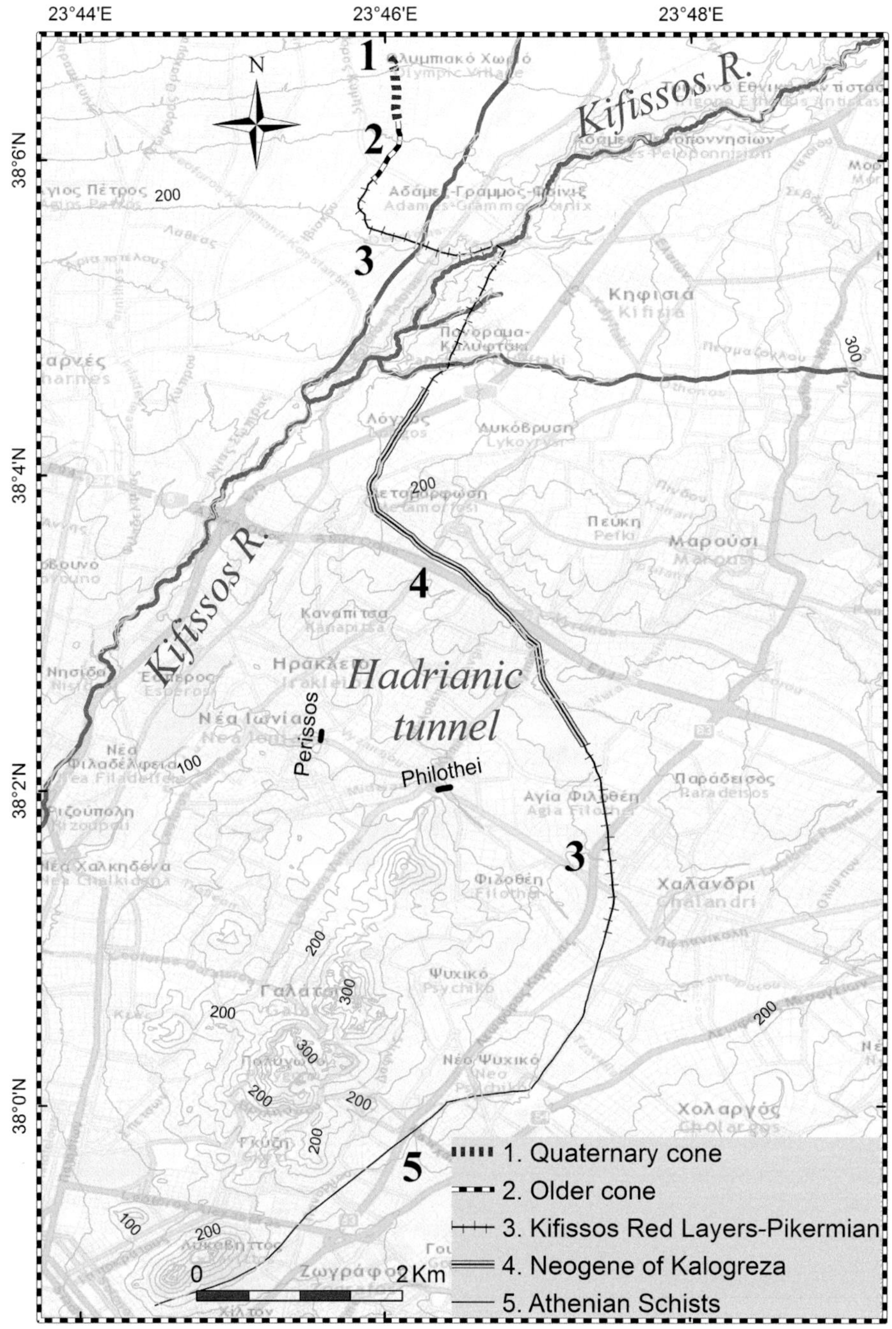

Figure 5. Projection on ESRI' world digital map: geological formations along the tunnel of the Hadrianic aqueduct; Late Roman water bridges at Philothei and Perissos.

1.60m in height and the walls are covered with waterproofing plaster (perhaps to avoid water loss at tunnel elevations above the water table). In case of fractured rock walls or soft formations the walls are supported with masonry and the roof by a vault of three terracotta bricks.[32] Beyond Koukouvaounes (modern Metamorphosis), the tunnel crosses thick clay layers and the wells are lined with horizontal layers of 6 curved bricks.[33] Wall masonry and roof support by curved bricks, stone or terracotta plates are also described in a recent publication.[34]

In addition to Haghios Demetrios, another location where the tunnel almost surfaces is at the southern bank of the Kifissos River; this location is ideal for initiating part of the aqueduct by digging upstream.[35] Uphill tunneling and water drainage at the excavation front would cause a local depression of the water

[32] Ziller 1877: table VIII-28.
[33] Ziller 1877: table VIII-6.
[34] Angelakis *et al.* 2014: 100
[35] At this region the Kifissos River could also supply the aqueduct, but there is no such evidence.

table, enabling a simultaneous excavation of wells at more than one locations, ahead of the tunnel face. Thus, the wells could be dug in advance by independent teams and the course of the tunnel would be defined by the array of the wells on the ground; the tunnel, which would have to meet the well bottoms, was excavated step by step from one well to the next one uphill.

The basic components of the aqueduct, i.e. the tunnel and the wells, require craftsmen of different qualities. The tunnel was probably excavated by one person at a time, assisted by a worker for removing the rock cuttings. It is suggested that the wells were undertaken by a team of specialized craftsmen, as in the Laurion mines.[36]

An intriguing similarity with the Hadrianic aqueduct can be recognized at the ancient Raschpëtzer aqueduct in Luxemburg which was also designed within the aquifer, which is a sandstone layer interbedded within marls. The total length is 600m and the wells reach the depth of 36m, contrary to other aqueducts of the region which are shorter in length, with wells only 6 -10m deep. The sandstone aquifer yields 180m^3 daily. The aqueduct is dated via dendrochronology in AD 130, i.e. during the Hadrianic period (however with a rather wide age error margin).

Following the false trend of considering any shafts-and-gallery aqueduct as qanat, the Raschpëtzer aqueduct is reported as such, although there is neither a single mother well nor an indication of Persian influence and the climate is far from arid.[37] I would suggest that it has similarities with the Roman aqueduct at the Colony Caesarea Mauretaniae (modern Cherchel) in Algeria and the Hellenistic aqueduct in Alexandria of Egypt. Indeed, Philip Leveau has underlined the importance of Roman hydraulic works in Algeria, in the area of Cherchel, and the remarkable analogies with the Hellenistic galleries in sandstone layers in Alexandria, a center of important innovations in hydraulic engineering.[38] He also connects the construction of the hydraulic installations in Cherchel with the migration of Greek engineers from Egypt.

Thanks to accurate underground surveying and geological mapping at the Raschpëtzer aqueduct, it became clear that the tunnel was excavated stepwise, starting from the bottom of a well near the tunnel entrance and heading upstream to the bottom of the next well. Errors in the heading of the tunnel between two wells were adjusted by joining the tunnel section in progress to the next well, through an S-like path to correct the error in the X-Y plan and via a 'stairway' to correct the error in the Z axis. Similar S-like tunnel sections and gaps in elevation on both sides of a well are also reported by Defteraios, regarding the northernmost section of the tunnel of the Hadrianic aqueduct.[39]

The S-shaped sections of the Hadrianic tunnel resulted similarly from the correction of the tunnel course in order to meet the next well upstream. In parallel with tunneling, a separate team excavated the wells ahead of the tunnel face, at locations and depths accurately determined by the surveyors' team. In case of tunnel deviation from the correct heading an S-shaped curved paths was necessary to correct for the deviation and meet the next shaft. When the lateral deviation was small and the tunnel section passed higher or lower than the shaft bottom, then the tunnel was enlarged vertically to pass at the scheduled elevation; in this case, there was only an elevation difference on both sides of the wells. The larger section of the tunnel occurred downstream when the tunnel crossed the shaft too high and vice versa.[40]

Hadrianic distribution network in Athens

Recent excavations at center of modern Athens, particularly in the National Garden (NG in Figure 6a) and west of the Parliament, near the point H1, unearthed numerous remains of ancient hydraulic works.

[36] Kakavoyiannis and Koursoumis 2013: 83. Since the end of the 6th century BC, craftsmen of the Phrearrians Demos of Laurion, Attica were specialized in the construction of mining shafts by contact.

[37] Kayser and Waringo 2003: 278.

[38] Leveau 2012: 157. In the Roman colony Caesarea Mauretaniae, an area with an average annual precipitation of 633mm, the Roman galleries capture water from a layer of sandstone and remained in use in modern times.

[39] http://urbanspeleology.blogspot.gr/p/blog-page_2754.html

[40] Defteraios *et al.* 2017: 52.

This is not unexpected since the area lies at a higher altitude in relation to the ancient city, in the route of ancient aqueducts; actually, all ancient aqueducts cross this area – except of the Acharnian aqueduct.

Although the course of the Hadrianic tunnel is accurately known, its continuation beyond the Lycabettus reservoir up to the National Garden has not been identified yet. Ziller supported that the water was conveyed from the reservoir to the city through lead pipes, one of which was found in situ (diameter 18cm, thickness 3cm).[41] No more lead pipes have been reported and this single lead pipe could be simply part of a fountain.

Branches of the Hadrianic aqueduct in Athens

Leigh described in detail the remains of the Hadrianic aqueduct in the Agora, by expertly combining the ancient structures next to the City Eleusinion (H5, Figure 6a), with petrographic study of the mortars.[42] The easternmost section consists of a channel covered by a vault, formed by 3 curved bricks held in place by mortar. The channel of the aqueduct is built in *opus mixtum* and measures 0.60m wide in the interior and 0.80m to the top of the side walls.[43]

The western extension of the channel is of different masonry and rests on a wall of poros blocks founded on concrete bedding. Because of the steep slope of the ground, the northern continuation of the aqueduct channel was carried on an arcade. The arcade was abandoned and replaced by a continuous foundation wall that carried a new channel. The abandonment of the earlier channel was probably caused by the Herulian sack of AD 267, and the reconstruction was probably connected with the revival of the Agora area, around AD 400.

The continuation of the channel beyond the SW fountain at the Athenian Agora is unknown, but a similar vaulted conduit with walls in *opum mixtum* seems to belong to the water distribution network[44] (C6, Figure 6a). It was abandoned in the 4th century AD (as dated from the numerous lamps found inside and around, from an adjacent workshop).

Another section of the Hadrianic aqueduct (H3, Figure 6a) was discovered in 1926, during excavations at the former Royal Stables. It was first reported as a Roman aqueduct lined with clay plates and contained many shafts.[45] It is briefly described by Semni Karouzou as a probable part of the aqueduct constructed by Hadrian.[46] It was discovered south of the ancient city wall and was casually repaired in Late Antiquity. The body of a Hellenistic statue and pieces of a 4th century AD sarcophagus were used to replace bricks and plug a hole at the aqueduct.

Sections of the Hadrianic Aqueduct have been encountered in various excavations, but not recognized as such yet. The conduit excavated by G. Mylonas west of the Parliament (1.7m below the solid rock and close to a hypocaust), constitutes in my opinion part of a primary branch of the Hadrianic aqueduct (H1).[47] The construction of the conduit is similar to that of the inclined gallery at the northern end of the Hadrianic tunnel, and the vaulted roof is composed of three curved bitched bricks. According to Mylonas an extension of the conduit was also found in the center of the square in front of the National Monument and at the NW end of the National Garden. Ventilation and cleaning shafts were also recorded at regular intervals.

---

41 Ziller 1877: 120.
42 Leigh 2000: 118.
43 Picture of the vaulted channel at the Agora excavation and plan of the aqueduct respectively are displayed at: http://agora.ascsa.net/id/agora/image/2004.01.0724?q=hadrianic%20aqueduct&t=&v=list&sort=&s=8, and http://agora.ascsa.net/id/agora/image/2002.01.2315?q=aqueduct&t=&v=list&sort=&s=68
44 *Archaiologikon Deltion* 40, Χρονικά (Οδ. Ηρακλειδών 18-42), Β'1, 1985: 20.
45 *Archaiologikon Deltion* 9, Παράρτημα (Αι ανασκαφαί των βασιλικών Στάβλων), 1924-25: 70.
46 Amandry 1947-1948: 386 fig.1, 391.
47 Mylonas 1932: 45.

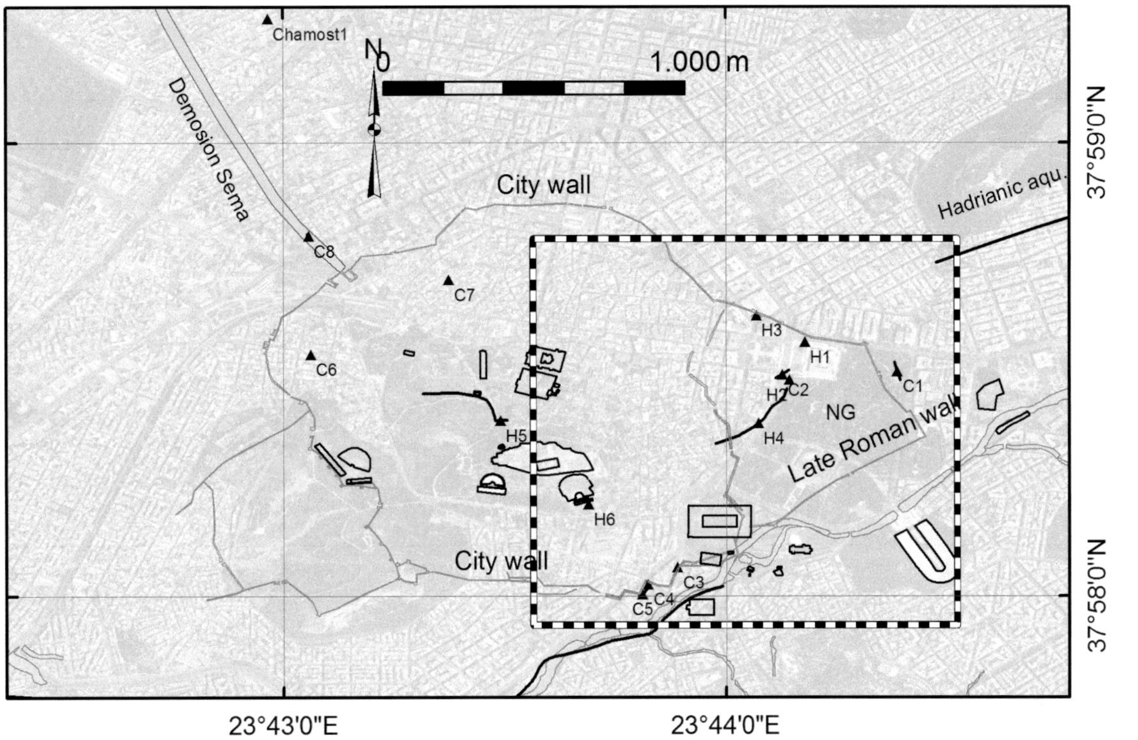

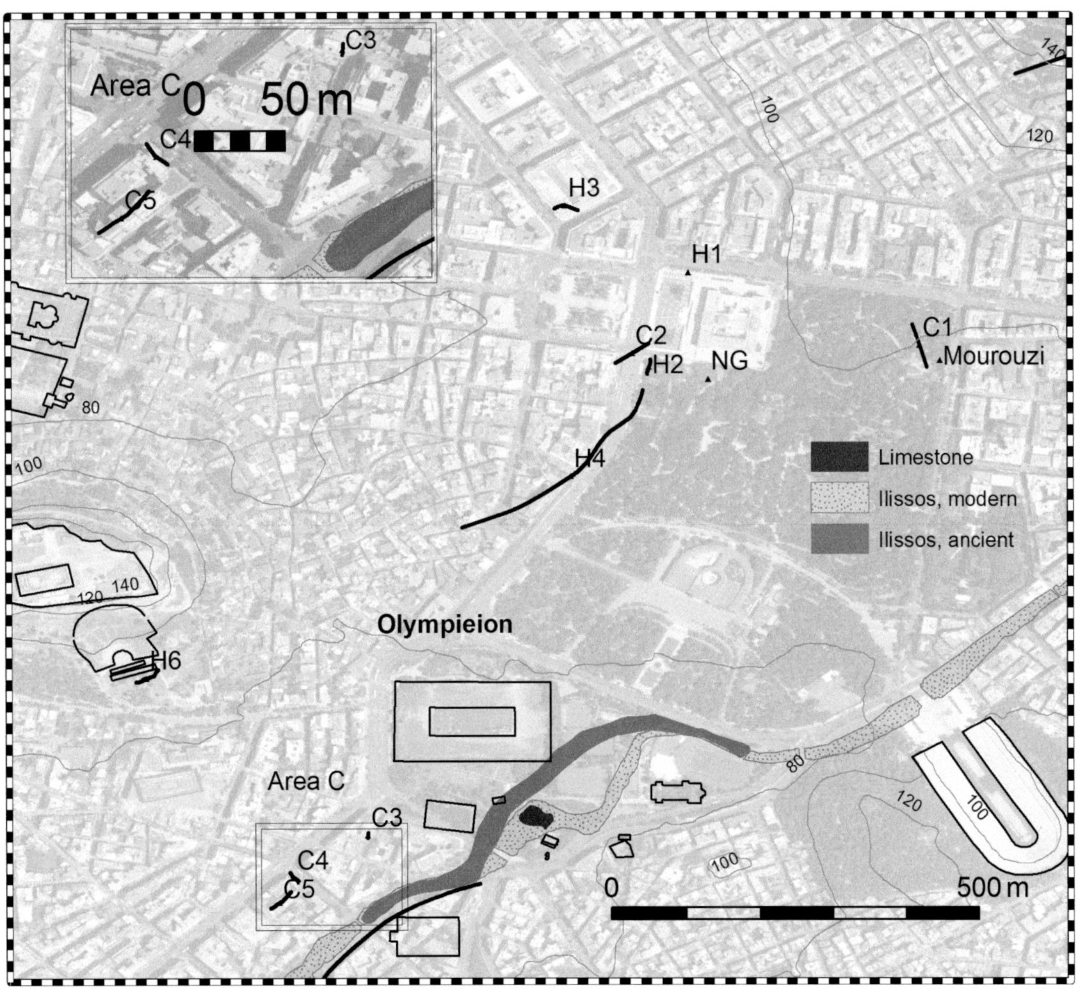

Figure 6. Projection on Ktimatologio air photos: (a) Hadrianic aqueduct remains (H1-H6), above and (b) heavy vaulted conduits of Roman concrete (C1- C8), below.

Recent excavations confirm Mylonas' description. A 27.3m long conduit built in *opus latericium,* along which two shafts were interpolated, was excavated at the entrance F of the Syntagma Metro Station.[48] The channel (H2, Figure 6a), is apparently the extension of the conduit described by Mylonas, directed indeed towards the NW corner of the National Garden; it was however found open, without any roof.

The section H4 in the Figure 6a is hypothetical, drawn on the basis of geomorphological criteria along the local crest line in combination with the distribution of Roman baths in the area.

Ziller described another conduit (H6, Figure 6a) which, I would suggest, could belong to a branch of the Hadrianic aqueduct.[49] He reported that during excavations searching for hydraulic works, conducted by the Director of the Royal Garden (and ordered by Queen Amalia), a conduit was discovered behind the scene of the Dionysius Theater, which was constructed in the system 12 and 18 depicted in Ziller's table VIII.[50] Ziller described the characteristic roof of three curved bricks, but he assumed that the conduit belonged to the aqueduct of the Royal Garden and that it was refurbished, when the theater was extended.

Secondary distribution network

The description of pipelines or conduits in excavation reports is in general either circumstantial or omitted. Furthermore, some confusion often occurs regarding their function -as in the case of certain heavy, vaulted conduits made of Roman concrete. A typical example is the vaulted conduit C1 (Figure 6b), *c.* 72.5m long (dipping from North to South), which was excavated in two periods at the higher NE area of the National Garden, in the yard of the Presidential Garrison.

As shown from the first excavation period, the conduit was founded in a trench on Athenian schists, 0.50m below the surface. This conduit, had walls 0.55 -0.60m thick and external width 2.05m; the excavators ascribed it to the Hadrianic aqueduct.[51]

Another section of the same vaulted conduit C1 was unearthed during the second excavation period;[52] it was interpreted as a sewer belonging to a wide network located at many places in ancient Athens, and was dated in the early Roman period. The argument was based firstly on the fact that six terracotta pipes emptied into the conduit and secondly on pottery findings dated in the 3rd or early 4th century AD, indicating the period of the conduit abandonment.

In my opinion, drainage from an area outside of the city walls (Figure 6a), would be more convenient in the opposite direction, outwards, along a path following the modern Queen Sophia's Avenue. In addition, a terracotta pipeline composed of U-shaped elements would be sufficient for sewerage instead of the heavy concrete conduit.[53] Furthermore, in a drainage network the size of the ducts increases progressively; in this specific case the conduit is oversized from the beginning and too heavy to belong to a sewer. Therefore, I would suggest, that this conduit should be associated with a secondary network of water distribution of the Hadrianic aqueduct.

A section of a similar vaulted conduit, C3, was excavated 0.30-0.55m below the surface (Figure 6b); it is made of stonewalls bound with hydraulic mortar and measures 4.40m in length, 2.10 in width and 2.15 -2.25m in height. It is attributed by the excavator to a network of the Imperial period (2nd-3rd century AD). [54] The numerous remains of conduits, similar in construction and contemporary, support the assumption that they belong to a common network. The relation of this particular conduit to a water supply network is indicated by an adjacent water reservoir (14.60 x 6m), excavated by the P. Bouyia

[48] *Archaiologikon Deltion* 52, Χρονικά, Β'1, 1997: 46-47.

[49] Ziller 1877: 115.

[50] Also, Dörpfeld and Reisch 1896: Table I.

[51] *Archaiologikon Deltion* 38, Χρονικά (Λεωφ. Βασ. Σοφίας και Ηρώδου Αττικού 2), Β'1, 1983: 24.

[52] *Archaiologikon Deltion* 50, Χρονικά (Φρέαρ Εθνικού Κήπου/Οδ. Ηρώδου Αττικού), Β'1, 1990: 32-35, plan 2.

[53] See also Chiotis and Chioti 2014: 316.

[54] *Archaiologikon Deltion* 60, Χρονικά, Β'1, 2005: 81-83.

downwards of the previous discussed location.[55] The reservoir, whose internal sides were coated with a hydraulic plaster, was abandoned in the 4th century AD.

A similar vaulted conduit, C5, with plaster coated walls, dated in the 2nd to the 3rd century AD and of a total known length of 30m, was unearthed close to the surface in three adjacent excavations.[56] The floor is covered by a layer of stones bound with plaster.

All the above massive and vaulted conduits share certain common characteristics: they have similar massive dimensions; their walls are *c.* 0.50m thick, made of Roman concrete; their internal surface is plastered and are dated mainly in the 2nd century AD. In addition, they are built near the surface, apparently in a trench founded on the Athenian schists and were abandoned in the 4th century AD. Therefore, we may conclude that they belong to a secondary network of water distribution fed by primary branches of the Hadrianic aqueduct.

Having established the typical features of this network, some more conduits can be attributed to it: the 47m long conduit C2, constructed close to the primary branch H2;[57] the 20m long conduit C4, built close to a significant water reservoir,[58] built on the fill of the moat (trench) of the classical precinct wall after Sulla's sack in 86 BC and on top of the Vallerian Wall constructed a few years prior to the Herulian raid in AD 267; the previously mentioned vaulted conduit C6.[59]

In addition, the vaulted conduit C8 (2nd century AD), excavated along the Demosion Sema, presents some common features.[60] It was built at a depth of 3.40m at the same location with a 5.60m long, late Hellenistic pipeline. This pipeline was composed of symmetrically coupled U-shaped terracotta parts at a depth of 4.90m, well protected below two layers of amphorae; it was refurbished and reinforced by curved terracotta parts commonly used for lining water wells. The pipeline was equipped with an access shaft 0.90m in diameter, and was considered as a sewer by the excavator. I believe that the pipeline of U-tubes and the access shaft were carefully designed as a branch of an older aqueduct directed towards the Academy, similar in design to the Acharnian aqueduct. It was most likely built to replace the Kimonian branch of the Peisistratean aqueduct mentioned earlier. The Acharnian type aqueduct was later replaced by the vaulted conduit of Roman concrete, with 0.85m thick walls and 1.50m internal height, which was again abandoned and plugged in the 4th century AD.

Another vaulted conduit, C7, which can possibly be attributed to the same network, is a heavy construction of monumental dimensions, found near a Roman hypocaust.[61]

The co-existence of hydraulic lines of various dates is a common observation at many locations and indicates that generally the courses of water supplying systems developed along the same routes throughout the centuries. The distribution of water to a network of lower level was probably regulated by reservoirs, as suggested by the excavation of some cisterns adjacent to the network. This holds at least for the reservoir in the NE corner of the Syntagma square, found 2.5m below the surface, and coated internally with a hydraulic plaster containing ceramic powder.[62]

Apart from the water distribution, another function of the secondary network would be managing the water overflow of the Hadrianic aqueduct. The study of the 'First Maps of the City of Athens' by Prof. Manolis Korres, that includes both ancient waterworks and the network of fountains of Athens in the

---

55 *Archaiologikon Deltion* 60, Χρονικά, B'1, 2005: 83-85.

56 *Archaiologikon Deltion* 23, Χρονικά (Οδ. Κορύζη 6), B'1, 1968: 68, plan 26; 43, Χρονικά (Οδ. Κορύζη 4), B'1, 1988: 29; 55, Χρονικά (Οδ. Κορύζη 8), B'1, 2000: 78-80 note 27, figs. 10-11.

57 *Archaiologikon Deltion* 49, Χρονικά (Λεωφόρος Αμαλίας), B'1, 1994: 28-32, plan 1.

58 *Archaeologike Ephemeris*, Χρονικά (Γωνία Λεωφ. Συγγρού και Βούρβαχη), 1971: 16-18, fig.5.

59 *Archaiologikon Deltion* 40, Χρονικά (Οδ. Ηρακλειδών 18-42), B'1, 1985: 20.

60 *Archaiologikon Deltion* 33, Χρονικά (Οδ. Πλαταιών 4 και Αγησιλάου), B'1, 1978: 19, plan 4.

61 *Archaiologikon Deltion* 30, Χρονικά (Οδ. Λεπενιώτου 5-7), B'1, 1975: 24-25, plan 4.

62 *Archaeologike Ephemeris*, Χρονικά (Πλατεία Συντάγματος), 1973: 69. Another reservoir was previously mentioned at the excavation conducted by P. Bouyia (see above, note 55: *Archaiologikon Deltion* 60, Χρονικά, B'1, 2005: 83-85).

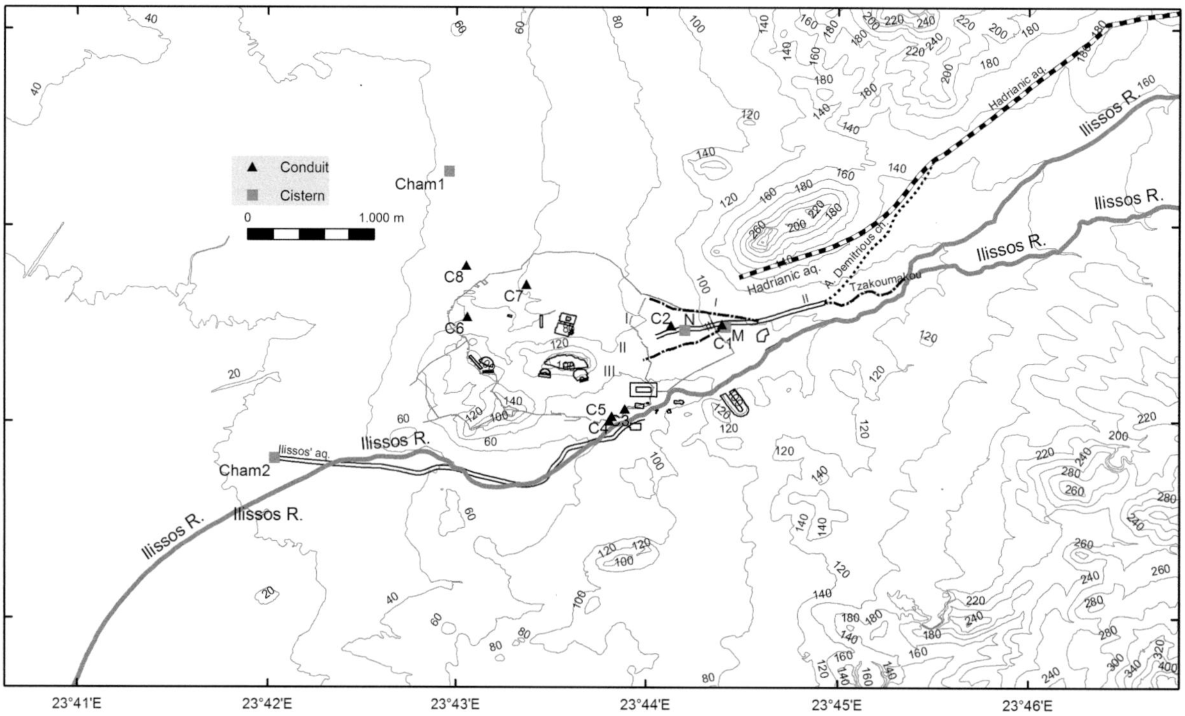

Figure 7. Ancient cisterns Chamosterna 1 (Cham1), Chamosterna 2 (Cham2), Mourouzi (M) and National Garden (N); vaulted conduits C1 to C8; the Hadrianic aqueduct; the stone-built channel from the Haghios Demetrios 'spring'; the Tzakoumakou channel; the Ilissos' aqueduct ('tou Vounou'); the distribution channels I, II, and III supplying fountains during the Ottoman period.

Ottoman period, has been a great contribution in many aspects. Korres was in position to locate on the maps the sites of two important Roman cisterns both known under the common name Chamosterna. They are distinguished in this study by the names Chamosterna 1 and Chamosterna 2. According to Kordellas, the Chamosterna 1 at Kolokynthou was huge;[63] it measured 22.5x28.0x3.0m and was similar in construction with the arcade water bridges at Philothei and Perissos. Because of that he assumed that the ancient aqueducts passing over these water bridges supplied water to the Chamosterna 1, at Kolokynthou. The second cistern, Chamosterna 2, was fed by the Ilissos' aqueduct (Figure 7) – also known as aqueduct 'tou Vounou'- which is described by Ziller and depicted on the Curtius and Kaupert map of Athens.[64]

I consider the above cisterns to be in proper locations to collect the redundant water overflowing from the vaulted conduits of the described secondary network (Figure 7). In particular, the conduits near the Ilissos riverbed could feed the Chamosterna 2; the later could at the same time collect water from the Ilissos' aqueduct possibly transferring it to Piraeus.

As Leigh pointed out, according to Vitruvius, conduits brought water from Athens to Piraeus; however this water was not drinkable: 'No one drinks from it, but they use it for baths and so forth. They drink from wells and thus avoid its ill effect'. [65]

In addition to the Chamosterna cisterns shown in Figure 7,[66] another couple of ancient cisterns, Mourouzi, at the western extension of the Mourouzi Street, and NG, in the National Garden, are also marked. The

[63] Kordellas 1879: 91.
[64] Ziller 1877: 108; Curtius and Kaupert 1881; Korres 2010: 99.
[65] Leigh 2001: 65.
[66] Curtius and Kaupert 1881.

Mourouzi cistern is depicted on the Kleanthis-Schaubert 1:2000 map, according to which Korres estimated its size at 8.0 x 5.5m.[67]

According to F. Stauffert, the water supply system of Athens in the 1830s was fed by two aqueducts;[68] namely the stone built channel from Haghios Demetrios and the aqueduct of Tzakoumakou, which are marked upon the 'First Maps of Athens' (Figure 7). The water was collected in the Mourouzi cistern near the Royal Palace and was distributed in Athens through 3 branches, also marked in Figure 7 (I, II and III). The Mourouzi cistern is also depicted on the Sommer's map of 1841, along with the supplying aqueduct directed from east to west.[69]

Based on the above, it is suggested that even prior to the cleaning of the Hadrianic aqueduct in the 19th century, the water supply Athens in the Ottoman period was based on ancient hydraulic works.

It is inferred that the secondary network of vaulted conduits was abandoned in the 4 century AD. This does not necessarily imply that all the primary branches of the Hadrianic aqueduct ceased supplying water as well. For example the primary branch supplying the Agora might be longer in operation. The construction of the late Roman aqueduct, which supplied the water mill in the Agora, strongly indicates abandonment of the Hadrianic aqueduct in the 5th century AD. Coins, pottery, and lamps found in the mill provide a date for the mill construction at the third quarter of the 5th century AD.[70]

Tertiary network of water distribution

The tertiary distribution network, supplied by the secondary one via cisterns, is generally composed of terracotta pipelines, the commonest of which are made of coupled U-shaped parts, forming an elliptical duct. Actually, similar ducts are used both for water distribution and for waste water removal. A typical example of tertiary water distribution is the combination of various duct types excavated along an ancient road, including an elliptical pipeline, inversed- Π terracotta duct and a stone-built channel.[71]

The Roman baths below the church of Haghios Nikodemos were apparently supplied by two alternative systems. There are actually two baths built at different levels and perhaps at different periods. The northern one, which is better preserved and dated by Travlos probably in the mid-second century AD, was supplied by an elliptical pipeline, presumably fed by the Hadrianic aqueduct.[72] The second alternative, associated with both baths is a well in possible connection with the underground National Garden-Hymettos aqueduct.

It is suggested that traditional wells remained in use, particularly for drinking water, concurrently with the Hadrianic aqueduct.

Tradition of hydraulic technology in Greece before the Hadrianic aqueduct

As perceptively underlined by Leigh, in regard with the ancient hydraulic technology, it is necessary to think in terms of a continuous engineering tradition, extending from Assyria through the works of the Nabataeans, to the Greeks, the Carthaginians, and the Etruscans, then to the Romans and beyond.[73] It is therefore interesting to track the possible contribution of the Greek tradition, already well developed since the Mycenaean period, to the design of the Hadrianic aqueduct and hydraulics engineering in general.

---

67 Korres 2010: 99.
68 Stauffert, as cited in Papageorgiou-Venetas 2010: 109.
69 Sommer 1841, as accessed from http://www.nhmuseum.gr/el/fakelos-syllogon/antikeimena/16042_el/.
70 Franz *et al.* 1988: 81.
71 *Archaeologike Ephemeris*, Χρονικά, 1971: 20.
72 *Archaeologike Ephemeris*, Χρονικά, 1856-57: 1450; Travlos 1971: 181.
73 Leigh 1998: 118.

The military engineers of the Mycenaean period confronted the demand to supply fortified cities with potable water, which could not be cut off or contaminated by the enemy during sieges. The Klepshydra fountain in the Athenian Acropolis, the secret underground cistern at Mycenae and the underground spring at Tiryns are representative examples of the engineers' ingenuity for the water supply of a citadel.

Numerous remains of Mycenaean hydraulic works have been unearthed in the Boeotian Thebes, an area with abundant springs in antiquity. An interesting hydraulic structure was excavated by Keramopoullos in the citadel in the so called 'Kadmos House', the most important archaeological site at Thebes.[74] A well was found near a Mycenaean pottery kiln; at its bottom (at a depth of 3.60m), a rock-cut duct 1.10m high and 0.50m wide was ramified in 5 directions. This structure probably constituted a groundwater collection work, a technique applied later in the Classical period, e.g. in the Corinthian wells.[75]

Early tunneling techniques in Greece were also partially affected by the underground burial structures (and vice versa). The similarity of the hydraulic galleries at Thebes to the 'dromos' of the Mycenaean chamber tombs has already been emphasized by the excavators.[76] In addition, peculiar chamber tombs on Aegina Island confirm a long tradition in underground works since the 8th century BC; they were accessed by shafts and in some cases were more complicated in structure than the shafts-and-gallery technique.[77] The geological conditions were favorable on Aegina, where solid layers of limestone (the roof of the tombs) covered soft layers of marls, within which the chambers were easily carved.

At Corinth, underground waterworks capturing spring aquifers were cleverly adjusted to the local geology and were applied from the early 6th century BC onwards. Vertical shafts were dug at regular intervals to facilitate rocks removal and improve ventilation. The shafts had toe-holes for climbing and their long dimension was aligned with the tunnel's direction,[78] a pattern adopted also in the Hadrianic aqueduct.

The shafts-and-gallery tunneling technique has been applied rather early in Greece. The Eupalenian aqueduct on Samos Island is dated in 550 BC.[79] It is composed of four differently constructed sectors. The first one is a stone built surface channel carrying spring water to an underground gallery (built in the shafts-and-gallery technique) which in turn transfers the water to the famous Eupalinos tunnel. The final sector conveys the water to the Pythagoreion harbor through a shafts-and-gallery tunnel, *c.* 500m long.[80] It should be noted here Graeber's view that the Aegina and Megara aqueducts were the models for the Eupalenian aqueduct.[81]

Two aqueducts of the shafts-and-gallery type, dated in the period of Thebe's prosperity in the Classical period were also excavated.[82]

The National Garden-Hymettus aqueduct is another underground aqueduct tapping groundwater. It was constructed following the shafts-and-gallery tunneling technique, perhaps in the 4th century BC and, despite the absence of any springs, continues to irrigate the National Garden of Athens.[83]

The Acharnian aqueduct of the 4th century BC constitutes, in a sense, an improvement of the Archaic Peisistratean aqueduct; the latter was built in Athens in the late 6th century BC and conveyed spring water through a terracotta pipeline with openings, placed in a cut and covered trench.[84] In the Acharnian aqueduct mortared bricks along the straight joint of the U-shaped drums and plastering of the elliptical junction can compose a watertight terracotta pipeline. Such a structure was installed in the branch D

[74] Keramopoullos 1911: 149.
[75] Landon 2003: 44.
[76] Papadakis 1911: 132-142.
[77] Papastavrou 2006: 129.
[78] Hill 1964: 54.
[79] Kienast 1995: 186.
[80] Zambas *et al.* 2017: 64.
[81] Graeber 1905: 558.
[82] Symeonoglou 1985.
[83] Chiotis and Chioti 2012: 408; Chiotis and Marinos 2012: 23.
[84] Chiotis and Chioti 2012: 424.

of the Megara aqueduct, to avoid water loss in the gallery and keep the water clean.[85] The waterline of the Late Roman aqueduct in Athens, to which the Perissos and Philothei water bridges belong, was also similar.

Proper siphons are not known in Attica, but they are known in Crete – in the Hellenistic aqueduct of Lyttos and the Roman aqueduct of Gortyn.[86]

Inclined wells along the aqueduct gallery were introduced to the Hellenistic aqueduct on Rhodos Island.[87] A unique and ingenious combination of an inclined gallery with three vertical wells for water uplift is well preserved at Perachora in Corinthia.[88]

The different attitude and scale of hydraulic works of Greeks and Romans is best exemplified at ancient Gadara (modern Jordan), a city of the *Dekapolis*. Until the Hellenistic period water supply of Gadara was based on cisterns and a few springs. Economic prosperity and population increase during the late 3rd century BC necessitated the construction of a shafts-and-gallery aqueduct, 30km long with 700 shallow wells, for the transportation (in a terracotta pipeline) of water from a spring, yielding about 4l/sec. To pass the tunnel through a ridge up to 72m below the surface, over a distance of 600m, an auxiliary gallery was built, a few meters apart, connected with the aqueduct gallery with short inclined shafts dipping about 45°.[89] As I consider, this was an ingenious solution for creating a ventilation circuit at the gallery faces, an approach resembling mining technology.

The Roman aqueduct known as Qanāt Fir'aun, or Dekapolis aqueduct, was constructed during the 1st and 2nd centuries AD, in the present day Syria and Jordan. It is 207km long (106km of which as tunnel), with about 2900 inclined shafts. Undoubtedly, it constitutes the most complex, ambitious and impressive ancient aqueduct. It is noted that the inclined shafts were dug out initially from 20 to 200m intervals and at a depth from 5 to 70m. Then, from the shaft bottom, tunneling started with pilot tunnels in opposite directions, which were enlarged to the final dimension after the meeting of the tunnel sections, a step which permitted the correction of any deviations from the planned geometry.

Based on the above examples, one can discern that groundwater hydrogeology and tunneling in the shafts-and-gallery technique were mastered in Greece in a much earlier period, already in the 6th century BC. However, there is another common parameter for the completion of the above discussed aqueducts of such complex geometry (i.e. the Eupalenian tunnel, the underground section of the Hellenistic aqueduct of Gadara, the Hadrianic aqueduct Athens and the Roman aqueduct of Dekapolis), and that is the application of mathematics: as already emphasized by Theodosios Tassios, building the Eupalinos tunnel would not be possible without applying the Pythagorian mathematics. Similarly, the route of the Roman Dekapolis aqueduct was fixed using triangulation, as known from Pythagoras, Hipparchos of Nicaea, Menelaus of Alexandria and Heron.[90] In both the Eupalinos and the Decapolis aqueduct, measurements were written on the rock wall in Greek letters – an indication of the Greek origin of their surveyors.

Summarizing the above discussion, the Hadrianic aqueduct in Athens is a hydraulic work built in the Roman period, but technically based on the long hydraulic tradition in Greece and Asia Minor – in combination with Hellenistic mathematics and the mining technology of deep shafts (Laurion). A transfer of hydraulic models from the Middle East, or elsewhere, cannot be excluded in the 7th or early in the 6th century BC, especially in Aegina and Megara, but cannot be traced. From the 6th century BC onwards, up to the Roman revolution in civil engineering works (e.g. the introduction of Roman concrete, exemplified by the impressive water bridges), hydraulic engineering developments in Greece must be considered indigenous.

---

[85] Avgerinou 2017: 47.
[86] *Lyttos*: Oikonomakis 1984: 66-99. Also, Amanda Kelly's paper (in present volume). *Gortyn*: Giorgi 2007: 287–320.
[87] Voudouris *et al.* 2013.
[88] Tomlinson 1969.
[89] Döring 2017: 178.
[90] Döring 2017: 189.

Concluding remarks

Water supply of modern Athens was based on ancient aqueducts until the Marathon reservoir was constructed in 1935. Even before the repair of the Hadrianic aqueduct in the 19th century, the small city of Athens benefited for centuries from the ruined ancient aqueducts. The impressive network of fountains of the Ottoman period – recently revealed thanks to the meticulous study of the 'First Maps of Athens' by Korres[91] – was fed by overflows both from the Hadrianic aqueduct at the Haghios Demetrios Church (Figures 1 and 7) and from the National Garden-Hymettus aqueduct, a section of which is known near the Haghios Thomas Church, at Goudi. In addition, the National Garden continues to be irrigated from the National Garden-Hymettus aqueduct.

During the Ottoman period the stone-built channel from Haghios Demetrios followed a smooth course towards Athens (Figure 7), thus indicating that the construction of the reservoir on the Lycabettus Hill was not a necessity; it was however in accordance with the Roman concept of *castellum aquae*, located in Athens at an imposing location, selected also for the demonstration of its monumental façade.

The tunneling techniques made significant progress in Greece in areas rich in springs, such as Thebes and Corinth, where the shafts-and-gallery aqueducts gradually developed as an extension of the simpler previous techniques (that of a single well with galleries at the bottom, and spring-tunneling).[92] The gradual development of a short spring-tunnel into a complicated network of tunnels and shafts in only a couple of centuries can be better observed at the Peirene fountain in Corinth. The tunnels followed the sinuous aquifer in a mining-like technique, without a predefined plan. The excavation of shafts and galleries was adjusted to the local geological conditions and the winding course of the aquifer.

The sustainability of the ancient aqueducts in Athens until today is due to their adequately deep gallery and shafts which enter into the water table. The archaic aqueducts at Aegina and Megara (6th century BC), were the earliest of this type in Greece, even earlier than the outstanding Eupalinos aqueduct (*c.* 550 BC), the end sectors of which were also constructed in the shaft-and-gallery technique.

An important particularity of the Hadrianic aqueduct is the significantly greater depth of the shafts, up to *c.* 41m. Over significant part of its course, the tunnel and the wells were dug below the water table (for improved groundwater capturing). This required an innovation in tunneling for the drainage of the groundwater from the excavation front. The Hadrianic aqueduct was actually a development of the shafts-and-gallery technique (as used in the Aegina, Megara and National Garden- Hymettus aqueducts), applied to greater depths.

Neither the above aqueducts, nor the Hadrianic one, can be considered as qanats (cf. the definition and diffusion of which is examined in the Appendix). One reason is because the shafts-and-gallery technique was applied in Greece earlier than the Achaemenid Empire (538/532-332 BC) – which is credited with the development and diffusion of the Iranian qanats. Secondly, the Greek shafts-and-gallery aqueducts capture groundwater almost along their entire course, whereas Iranian qanats are typically fed from a mother well and their transportation course crosses dry rocks in arid areas. And, lastly, because aqueducts such as, the Eupalenian in Samos, the Hellenistic at Gadara, the Hadrianic at Athens and the Roman Dekapolis aqueduct were built according to the advanced applied mathematics of their time. By contrast, the surveying of the Iranian qanats is a simple empirical exercise.

In conclusion, as demonstrated in the examination of the hydraulic tradition in Greece, the Hadrianic aqueduct is the evolutionary product of mathematics and indigenous hydraulic and mining technology. It could be argued that the prefabricated terracotta pieces for roof support constitute a Roman influence, not without a counterpoint, but this is an issue of minor importance.

---

91 Korres 2010: 93.

92 The final constructions varied in form due to the different local conditions: straight aqueducts with sometimes overlapping tunnels at Thebes and meandering network of galleries at Corinth.

There are indications that similarly to the Roman Raschpëtzer aqueduct in Luxemburg, the Hadrianic tunnel was excavated at intervals upstream, starting from a well to the next one uphill. Furthermore, it is inferred that the construction of the Hadrianic aqueduct was undertaken simultaneously in more sectors starting from lateral deviations from the tunnel at the surface and by digging upstream. Thus, the water was drained by gravity flow and the water table was depressed and the excavation of more adjacent wells below the water table was possible. The wells were sunk in advance defining the course of the aqueduct and the tunneling followed at close distance.

The water distribution network from the Hadrianic aqueduct to ancient Athens can be traced from patchy - but structurally consistent - remains. The primary distribution netwok consisted of shallow tunnels, a couple of meters below the surface, exemplified in construction details by the Agora branch, near the City Eleusinion. Two main branches can be traced, one passing west of the parliament along the Amalia's Avenue (H1, H2, H4 in Figure 6a), which was subdivided in two sections at least, one directed to the Agora (H5) and the other one SW of the Dionysus Theater (H6). The second branch of the primary network was traced only at one location (H3), but it is guessed that it was directed towards the Plato's Academy (based on the remains of the secondary distribution netwok).

The latter one is represented by remains of vaulted tunnels of Roman concrete, with thick walls, constructed in shallow covered trenches. The secondary distribution netwok was constructed in the 2nd century AD and abandoned in the 4th century. It also served for collecting the water overflow to large cisterns, the largest of which was the Chamosterna at Kolokynthou, in the middle of a fertile land. The second cistern, near the Ilissos River, could collect the redundant water of the Hadrianic aqueduct in this region and the water from the Ilissos' aqueduct and transfer it, possibly, to Piraeus. The tertiary distribution netwok consisted mainly of terracotta pipelines (or built channels), fed from cisterns. The co-existence of hydraulic lines of various periods indicates that the courses of the main hydraulic works were developed along the same routes throughout the centuries. The operation of a new Late Roman aqueduct in the 3rd quarter of the 5th century AD for the supply of the Agora water mill indicates the abandonment of the Hadrianic aqueduct before that.

Acknowledgements

Professor Theodosios Tassios, President of the *Association of Ancient Greek Technology Studies* (EMAET), who has promoted for many decades the study of ancient Greek technology, has constantly inspired me to study the ancient hydraulics. He also contributed through his expert advice on the improvement of this study. The Chief Editor of this volume, Dr Georgia Aristodemou, improved my original text and kindly tolerated my corrections. My collaboration with Panayiotis Defteraios contributed significantly to my understanding of the underground structure of the Hadrianic aqueduct, thanks to his speleological investigations and his kind cooperation. Effie Nestorides (EYDAP), third generation Engineer, of a family with great contribution to the study and promotion of the Hadrianic aqueduct, has been constantly supportive and encouraging for many years.

My long involvement with the study of the ancient hydraulic works would not be possible without my family's appreciative and unceasing encouragement.

Addressing my thanks to all of them is a great pleasure.

Appendix

Comments on the nature and definition of qanats and comparison to the Hadrianic aqueduct

Following Antisthenes, the definition of the various types of underground aqueducts is here investigated, with emphasis on qanats. The distinction between qanats and underground aqueducts is often misunderstood and because of that there is an increasing tendency of naming qanat any shafts-and-gallery aqueduct. Thus, qanats are incorrectly defined as 'gently sloping, artificially built subsurface channels, which lead groundwater from the margins of the mountains to the area where water is needed, sometimes many kilometers away'.[93] Obviously, this broad definition includes practically almost all the underground aqueducts.

The same definition was recently applied to the ancient Greek underground aqueducts, which were characterized as *aqueduct-like qanats* – a neologism that intensified the confusion. Strangely enough, it was further argued that 'when comparing the etymology/definition of the terms aqueduct and qanat it should be effortlessly concluded that *Qanat is identical to Aqueduct, or at least it is included therein*'.[94]

May combined technical description and rather doubtful historical assumptions to define the qanat as 'a collection and conveyance system for groundwater that was developed in Persia'.[95] Accordingly, a qanat consists of an underground tunnel which uses gravity to convey water from the water table (or springs) at higher elevations, to the surface of lower lands. Qanats also have a series of vertical shafts used for tunnel excavation, maintenance, ventilation and lighting. The oldest qanats have been found in the northern part of Iran and date back to *c.* 3000 years ago, when the Arians (Aryans) settled in present day Iran. From 550-331 BC Persian rule extended from the Indus to the Nile, during which time qanat technology spread. [96]

An essential and distinguishing technical feature of qanats omitted in this definition-but stated by Lightfoot-is that 'a qanat is a subterranean aqueduct engineered to collect groundwater from a mother well'.[97] To emphasize the importance of this comment, it is noted that during the 20th century the 'mother wells' of the Iranian qanats were up to 150 m deep (in Tehran); however, the deepest 'mother wells' in Iran could reach depths of 300 or even 320m.[98]

It is self-evident that qanats were expensive in construction and maintenance, and it might take 20 to 30 years to dig a new qanat.[99] It is also significant that the term qanat is used in Iran in a particular cultural, technical and geographical context. It seems, therefore, that 'it cannot be used to designate all the types of water capturing tunnels, in various periods, in diverse geographical and climatic zones'.[100]

Boucharlat proposed the introduction of a more general term when describing tunnels capturing underground water, which would also include the qanats. He preferred in French the term 'galerie de captage'.[101]

In this paper, following Wilson, the general term shafts-and-gallery aqueduct is used when describing all hydraulic structures where an inclined tunnel along a 'chain' of wells uses the gravity based water flow.[102] This definition applies to all underground aqueducts irrespectively of the origin of water source. It includes Assyrian tunnels transferring water from a river, the Hadrianic aqueduct in Athens tapping groundwater almost all along its course, the *falaj* in the United Arab Emirates and Oman tapping shallow

[93] Weingartner 2007: A1556.
[94] Voudouris et al. 2013: 1341
[95] Mays 2010: 3.
[96] The period of Achaemenid Empire is defined from 538 to 332 BC by Magee (2007: 87) or from 532 to 332 BC by T. J. Wilkinson *et al.* (2012: 167).
[97] Lightfoot 2000: 215.
[98] Wilkinson 2012:16.
[99] Esfandiari 2007: 60.
[100] Salesse 2001: 713.
[101] Boucharlat 2000: 158.
[102] Wilson 2008: 290.

aquifers along valleys of seasonal streams and the typical Iranian qanats which tap water from strongly dipping aquifers in alluvial fans.[103] In this way, the confusion of naming qanat any shafts-and-gallery structure is avoided. Further subdivision of the shafts-and-gallery aqueducts should be based on the origin of water, the depth and dip of aquifer, and historical, geographical and climatic data.

According to Wilson, the shafts-and-gallery tunneling technique -that many authors associate with the Persian qanats – was firstly developed in the Assyrian Empire for the construction of long-distance artificial canals supplying water usually from distant rivers. They crossed intervening ridges in the form of tunnels several hundred meters (or even several kilometers) long. The shafts-and-gallery tunneling technique solved the problem of underground surveying; the shafts were set at sufficiently close intervals and thus, short, manageable sections were created. The Assyrian king Assurnasipal II (884-859 BC) dug a canal 19.5km long from the Upper Zab River to supply Nimrud; it passed under a rock ridge at Negoub via a tunnel 7km long, dug between pairs of vertical shafts. The system supplied the city of Nimrud and also irrigated the fields around it.[104]

Physical geography and climatic change at the beginning of the 1st millennium BC are indispensable factors both for understanding the causes that led to the development of *falaj* in the Peninsula of Oman, and for defining *falaj*.

Boucharlat demonstrated that the groundwater draining galleries named *falaj* firstly appeared in the present-day United Arab Emirates and Oman early in the 1st millennium BC; he associated the introduction of the draining galleries with a dramatic growth in population.[105] He also investigated the type of aquifers and concluded that galleries in the United Arab Emirates were shallow, and tapped water from valley aquifers or dry *wadi*. In contrast, the Iranian qanats appeared in the Partho-Sasanian period or even later and collected water from deeper aquifers.

There was a rapid settlement during the Iron Age II (*c.*1000-600 BC) throughout the southeastern Arabia. The *falaj* irrigation system is in close association with a number of piedmont settlements in the southeastern Arabia dated in Iron Age II (based on calibrated radiocarbon dates reported by Magee). The *falaj* irrigation permitted intensive and year-round agriculture in the small 20-30km strip of alluvial piedmont that flanks the Hajar Mountains to the East and West. This dating also confirms that the *falaj* system was not an Achaemenid-period technological adaptation from Iran.[106]

More than a dozen underground shafts-and-gallery aqueducts have been dated to the Iron Age or have been directly associated with Iron Age settlements. Where the water source could be determined, it was associated with *wadi* flow or superficial aquifers in small depressions receiving local runoff. It is also noted that Iron Age galleries differ from later Islamic qanats, which are deeper galleries catching water from deep aquifers.[107]

The shallow depth of the aquifer was clearly confirmed at the excavation of the Al Madam *falaj* system (a water catchment gallery). The gallery was located 1.3m below the surface and originally 1.5m high; subsequently the floor was lowered by 4.80m in order to reach the declining water table in a period of drought, due to which both the *falaj* and the adjacent village were finally abandoned.[108]

The generally accepted view that the qanat was invented in the well-watered regions of northwestern Iran around the 8th century BC and then diffused across the Middle East by the Achaemenid Empire was

[103] In strongly dipping aquifers, as quite often happens in alluvial fans in Iran in the piedmont, along normal fault lines of significant vertical displacement, the possible zone of water capturing with a slightly inclined gallery is very narrow and this explains why in the typical Iranian qanats the aquifer is tapped by a single mother well only. In addition, deeper aquifers are most likely artesian. Iranian qanats combining these features are distinguished here as typical ones.

[104] Wilson 2008: 290.

[105] Boucharlat 2003: 169.

[106] Magee 2007: 90.

[107] Mouton *et al.* 2011: 16.

[108] Cerro 2012: 134.

recently challenged via archaeological evidence. It is presently considered that the emergence of the qanat is more likely to have occurred in the southern regions of Iran, Pakistan and Arabia.[109] Palaeoclimatic evidence suggests that at the beginning of the 1st millennium BC these regions suffered from declining precipitation; this pattern does not apply in southeastern Arabia (United Arab Emirates and Sultanate of Oman). Extensive fieldwork recorded a rapid and intense establishment of new settlements around 1000 BC. Archaeological research at several of these settlements revealed the existence of 'qanat' systems dated at the first few centuries of the 1st millennium BC.[110] These 'qanat' systems constitute the earliest evidence of this irrigation system until today.[111]

However, the identification of *falaj* with *qanat* is a precarious simplification (given that *falaj* appeared earlier than *qanats* and taps shallow open aquifers). Furthermore, three types of *falaj* are distinguished in Oman, in relation to the water sources: *ghaili falaj* taps water from the gravel of a river or *wadi*; *aini falaj* draws water from one or more natural springs and *daudi falaj* taps an underground aquifer.[112]

It is further emphasized that the conventional *falaj* in Oman (either *ghaili* or *aini*) constitutes the response of small communities to the changing climatic conditions, by which surface channels were gradually transformed into shallow underground galleries. By contrast, it seems that the typical Iranian *qanat* was planned in advance and executed as a major project for tapping deeper water resources from confined aquifers via a single mother well and this is their distinguishing feature.

Table 1 summarizes the basic features of the shafts-and-gallery aqueducts discussed previously. Particular emphasis is given on the origin of water in order to indicate the significant existing variability. Among the aqueducts listed, only the Xinjiang aqueduct in China can be considered as qanat of the Iranian type.

The impressive variability in both the period of first installation and the diversified water capturing methods, results from the technological adjustment to local historical and environmental conditions over a long period starting from the Bronze and the Iron Age. As expressed by Wilkinson,[113] water-supply systems are sensitively engineered to suit local hydrogeological conditions.

It is therefore a misrepresentation of historical evidence to merge all these types of aqueducts under the common model of Iranian *qanats*, particularly because as Boucharlat concludes: 'none of the Iranian qanat can be directly dated from the pre-Islamic period and even from the following centuries, despite some assumptions put forward from time to time, especially for the Khorasan province, corresponding to the NE quarter of the country'.[114] Boucharlat additionally supports 'that the first generation of shafts-and-gallery aqueducts was very likely *polycentric* during varied periods of the 1st millennium BC. Much later, the second generation might have been actually implemented in Iran around the middle of the 1st millennium BC and was soon spread elsewhere'.

Finally, Chiotis (2007) associates the development of early underground aqueducts with world-wide climatic change and concludes that 'early invention of underground aqueducts has been recorded in the process of adaptation of communities to natural decline of precipitation and springs, particularly intensive in the turn of the second to the first millennium BC. Thus, areas rich in water in the first half of the Holocene were gradually transformed into deserts in Libya and Western Egypt, with few oases near springs. Subsequently, the decline of the springs forced the invention of underground aqueducts, the foggaras in Libya, where outcrops of fossil aquifers were the last refugia for settlements; in the Peninsula of Oman, the Hajar Mountains were similarly the last refugia, thanks to their higher precipitation, which

[109] Wilkinson *et al.* 2012.
[110] The application by Wilkinson *et al.* (2012) of the term qanat instead of *falaj* is rather confusing, because of the existence of several different types of *falaj*; Magee (2007), however, avoided this misconception.
[111] Wilkinson *et al.* 2012: 168.
[112] Al-Marshudi 2007: 34.
[113] Wilkinson 2003: 45.
[114] Boucharlat 2017: 296.

could support irrigation in narrow zones at the foothills of the mountain belt by means of underground aqueducts'.[115]

Table 1
Indicative types of the origin of water in shafts-and-gallery (SAG) aqueducts*

Origin of water	**Typical SAG aqueducts**	**Bibliographical Reference**	**Comments**
Shallow unconfined aquifers below dry valleys (wadi)	Ghaili falaj at Al Madam, UAE	Cero 2012: 134	Iron Age II (*c.*1000-600 BC)
River	Assyrian tunnels	Wilson 2008: 290	early 9th century BC (first half)
Spring	1. Eupalenian tunnel (Samos, Greece)	Zambas *et al.* 2017	Horizontal blind tunnel with **SAG** end sections, 550 BC
	2. Qanāt Fir'aun, or Dekapolis aqueduct	Döring 2017	Tunnel and inclined shafts, 1st and 2nd centuries AD
Confined aquifers	1.Peirene (Corinth, Greece)	Hill 1964	Twisting spring tunnels and shafts, 6th c. BC
	2. Raschpëtzer aqueduct (Luxemburg)	Kayser and Waringo 2003	2. early 2nd century AD (first half) ,**SAG**
	3. Perachora aqueduct (Corinthia, Greece)	Tomlinson 1969	Hellenistic. Inclined gallery-vertical shafts
	4. Rhodos (Greece)	Voudouris *et al.* 2013	Hellenistic. Horizontal tunnel-inclined shafts
Unconfined aquifer (SAG at the level of the water table)	Aegina, Greece Megara, Greece	Graeber 1905: 558	Earlier than the Eupalenian tunnel
Tunneling below the water table	1. Hadrianic aqueduct of Athens	Chiotis, present paper	AD 125-140 shafts up to 41m
	2. Hadrianic aqueduct of Athens, modern branches 1 and 2	Chiotis, present paper	19th -20th century
Springs	Aini falaj (UAE)	UAE 2006: 5	Modern
Groundwater from mother wells	Daudi falaj (UAE)	UAE 2006: 13	Modern, shafts 6.5 to 30m.
Mother wells in steep confined aquifers	Qanat at Xinjiang (China)	Hu *et al.* 2012: 216	Modern, shafts: 40-70m, 100m max

* Abbreviations used
SAG = shafts-and-gallery aqueducts
UAE= United Arab Emirates

[115] Chiotis 2017: 4.

Bibliography

Al-Marshudi, A.S. 2007. Institutional arrangement in the Sultanate of Oman. In: *Proceedings of the International History Seminar on Irrigation and Drainage*: 31-42. Tehran: Iranian National Committee on Irrigation and Drainage.

Amandry, P. 1947-1948. Chronique des Fouilles en 1946. *Bulletin de Correspondance Hellénique* 71-72: 382-422.

Angelakis, A. N., Mamassis, N., Dialynas, E. G. and Defteraios, P. 2014. Urban Water Supply, Wastewater, and Stormwater Considerations in Ancient Hellas: Lessons Learned. *Environment and Natural Resources Research* 4.3: 95-102.

Aravantinos, V. 2014. Το αρχαιολογικό έργο στη Θήβα: οι άνθρωποι και τα αρχαία. In: Β. Αραβαντινός – Ε. Κουντούρη (επιμ.) *100 χρόνια Αρχαιολογικού Έργου στη Θήβα. Οι πρωτεργάτες των ερευνών και οι συνεχιστές τους*. Συνεδριακό κέντρο Θήβας, *15-17 Νοεμβρίου 2002, Αθήνα 2014:* 1-57. Athens: Ministry of Culture.

Asimakou, E. and Kalliga, K. 2013. Το Αδριάνειο Υδραγωγείο της Αθήνας: ένα τεχνικό έργο της αυτοκρατορικής Ρώμης στην υπηρεσία της νεότερης Αθήνας. In: Μ. Δόγκα-Τόλη – Στ. Οικονόμου (επιμ.) *Αρχαιολογικές Συμβολές* vol. A (Attica): 161-171. Athens: Museum of Cycladic Art.

Avgerinou, P. 2017. The Archaic underground aqueduct in Megara, in the Chapter 4: Updated Appraisal of Ancient Underground Aqueducts in Greece. In: A. N. Angelakis, E. D. Chiotis, S. Eslamian and H. Weingartner (eds) *Underground Aqueducts Handbook*: 44-48. Boca Raton: CRC Press.

Boucharlat, R. 2000. Les galeries de captage dans la péninsule d'Oman au premier millénaire avant J.-C.: questions sur leurs relations avec les galeries du plateau iranien. In: P. Briant (ed.) *Irrigation et drainage dans l'Antiquité, qanats et canalisations souterraines en Iran, en Égypte et en Grèce*: 157-183. Paris: Thotm Éditions.

Boucharlat, R. 2003. Iron Age water-draining galleries and the Iranian 'Qanat'. In: D. T. Potts, H. Al Naboodah and P. Hellyer (eds) *Archaeology of the United Arab Emirates. Proceedings of the First International Conference on the Archaeology of United Arab Emirates*: 159-172. London: Trident Press Ltd.

Boucharlat, R. 2017. Qanāt and Falaj: Polycentric and Multi-Period Innovations: Iran and the United Arab Emirates as Case Studies. In: A. N. Angelakis, E. D. Chiotis, S. Eslamian and H. Weingartner (eds) *Underground Aqueducts Handbook*: 279-301. Boca Raton: CRC Press.

Cerro, C. del 2012. Some evidence of crisis and abandonment at the end of the Iron Age in Al Madam 1-Thuqaibah, Sharjah. In: D.T. Potts and P. Hellyer (eds) *Fifty Years of Emirates Archaeology. Proceedings of the Second International Conference on the Archaeology of the United Arab Emirates*: 132-139., London/Dubai: Motivate Publishing.

Chiotis, E. 2017. Underground Aqueducts: Types, Definitions and General Conclusions. In: A. N. Angelakis, E. D. Chiotis, S. Eslamian and H. Weingartner (eds) *Underground Aqueducts Handbook*: 1-16. Boca Raton: CRC Press.

Chiotis E. D. and Chioti L. E. 2011. Παραγωγικές δραστηριότητες στην αρχαία Αγορά της Αθήνας κατά τους υστερορρωμαϊκούς χρόνους. In: Α. Γιαννικουρή (ed.) *Η Αγορά στη Μεσόγειο από τους Ομηρικούς έως τους Ρωμαϊκούς χρόνους, Διεθνές Επιστημονικό Συνέδριο*//The Agora in the Mediterranean from Homeric to Roman times, International Conference, Κως 14-17 Απριλίου 2011: 181-196. Athens: Ministry of Culture.

Chiotis E. D. and Chioti L. E. 2012. Water supply of Athens in antiquity, In: A. N. Angelakis, L. W. Mays, D. Koutsoyiannis and N. Mamassis (eds) *Evolution of Water Supply Through the Millennia*: 407-445. London: IWA Publishing.

Chiotis, E. D. and Chioti, L. E. 2014. Drainage and Sewerage Systems at Ancient Athens, Greece. In: A. N. Angelakis and J. B. Rose (eds) *Evolution of Sanitation and Wastewater Technologies through the Centuries*: 313-329. London: IWA Publishing.

Chiotis, E. D. and Marinos, P. G. 2012. Geological aspects on the sustainability of ancient aqueducts of Athens, *Bulletin of the Geological Society of Greece*, 46: 16-38. (http://www.geosociety.gr/images/news_files/EGE_XLVI/2_CHIOTIS_p16-38.pdf).

Curtius, E. and Kaupert, J. A., 1881. *Karten von Attika*, Berlin: D. Reimer.

Defteraios, P., Chiotis, E. and Mamassis, N. 2017. Underground evidence on the tunneling technique of the Hadrianic aqueduct of Athens, in the Chapter 4: Updated Appraisal of Ancient Underground Aqueducts in Greece. In: A. N. Angelakis, E. D. Chiotis, S. Eslamian and H. Weingartner (eds) *Underground Aqueducts Handbook*: 50-54. Boca Raton: CRC Press.

Döring, M. 2017. Qanāt Fir'aun: n underground Roman water System in Syria and Jordan. In: A. N. Angelakis, E. D. Chiotis, S. Eslamian and H. Weingartner (eds) *Underground Aqueducts Handbook*: 173-196. Boca Raton: CRC Press.

Dörpfeld, W. and Reisch, E. 1896. *Das Griechische Theater: Beiträge zur Geschichte des Dionysos-theaters in Athen und anderer griechischer Theater.* Athen: Barth & Von Hirst.

Esfandiari, O. 2007. Qanat: Iranian's most remarkable ancient irrigation system. In *Proceedings of the International History Seminar on Irrigation and Drainage*: 55-61 Tehran: Iranian National Committee on Irrigation and Drainage.

Fahlbusch, H., 2014. The Roman Water Supply System of Pergamum. In: Ch. Ohlig and Ts. Tsuk (eds) *Cura Aquarum in Israel II: Water in Antiquity. Proceedings of the 15th International Conference on the History of Water Management and Hydraulic Engineering in the Mediterranean Region Israel 14-20 October 2012* (Schriften der Deutschen Wasserhistorischen Gesellschaft 21): 165-184. Siegburg.

Frantz, A., Thompson, H. A and Travlos, J. 1988. *The Athenian Agora, XXIV. Late Antiquity: AD 267-700*: 1-156: Princeton: American School of Classical Studies at Athens.

Georgousopoulou, Th. 2013. Το υδρευτικό σύστημα της ΒΔ Αττικής κατά τους κλασικούς και ελληνιστικούς χρόνους. In: M. Δόγκα-Τόλη – Στ. Οικονόμου (επιμ.), *Αρχαιολογικές Συμβολές*, τ. A (Attica): 153-160. Athens: Museum of Cycladic Art.

Giorgi, E. 2007. Water technology at Gortyn in the 4th -7th c. AD: transport, storage, distribution. In: L. Lavan, E. Zanini, and A. Sarantis (eds) *Technology in Transition A.D. 300–650*: 287-320. Leiden: Brill.

Graeber, F. 1905. Die Wasserleitung des Peisistratos und die Wasserversorgung des alten Athen. *Zentralblatt der Bauverwaltung* 25: 557-560.

Hill, B. H, 1964. *Corinth, I. 6. The Springs: Peirene, Sacred Spring, Glauke*. Princeton: American School of Classical Studies at Athens.

Hu, Wen Jun, Zang, Jie Bin and Liu, Yong Qiang. 2012. The qanats of Xinjiang: historical development, characteristics and modern implications for environmental protection. *Journal of Arid Land* 4: 211–220.

Kakavoyiannis E. and Koursoumis, S. 2013. Εντοπισμός, καταγραφή και χαρτογράφηση των αρχαίων μεταλλευτικών φρεάτων της Λαυρεωτικής. *Archaeologike Ephemeris* 152: 77-102.

Karouzos, Ch., 1926. Thebes. *Archaelogikon Deltion* 10. Μελέται: 7-10.

Kayser, P. and Waringo G. 2003. Die unterirdische Wasserleitung der Raschpëtzer. Ein Monument antiker Ingenieurbaukunst aus Luxemburg. In: *Wasserversorgung aus Qanaten – Qanate als Vorbilder im Tunnelbau, Internationales Frontinus-Symposium, Walferdange, Luxemburg* (Schriftenreihe der Frontinus-Gesellschaft vol. 26): 277-292.

Keramopoullos, A. D. 1911. Ανασκαφή του ανακτόρου του Κάδμου εν Θήβαις. *Praktika tēs en Athēnais Archaiologikēs Hetaireias:* 143-152.

Keramopoullos, A. D. 1917. Θηβαϊκά. *Archaelogikon Deltion* 3. Μελέται: 1-503.

Kienast, H.J., 1995. *Samos XIX. Die Wasserleitung des Eupalinos auf Samos.* Bonn: R. Habelt Vlg.

Kordellas, A. 1879. *Αι Αθήναι Εξεταζόμεναι υπο Υδραυλικήν Έποψιν*. Athens: Filokalia's Publications.

Korres, M. (ed) 2010. *Οι πρώτοι χάρτες της πόλεως των Αθηνών*. Athens: Melissa Publications.

KTIMATOLOGIO S. A., http://gis.ktimanet.gr/wms/ktbasemap/default.aspx, Air photos of Greece available online.

Lancaster, L. 2010. Parthian Influence on Vaulting in Roman Greece? An Inquiry into Technological Exchange under Hadrian. *American Journal of Archaeology* 114: 447–72.

Landon, M. E. 2003. Beyond Peirene: Toward a Broader View of Corinthian Water Supply. In: C. K. Williams II and N. Bookidis (eds) *Corinth* XX. *Corinth, the Centenary, 1896–1996*: 43-62. Princeton: American School of Classical Studies at Athens.

Leigh, S. 1998. The Aqueduct of Hadrian and the water supply of Roman Athens. Unpublished Ph.D. dissertation, University of Pennsylvania.

Leigh, S. 2000. Interdisciplinary research on the aqueduct of Hadrian in the Athenian Agora. In: G. C.M. Jansen (ed.) *Cura Aquarum in Sicilia. Proceedings of the Tenth International congress on the History of Water management and Hydraulic Engineering in the Mediterranean Region (Syracuse, May 16-22, 1998)*: 117-124. (BaBesch Supplement 6) Leiden: Peeters.

Leigh, S. 2001. A survey of the early Roman hydraulics in Athens, 65-82. In: A.O. Koloski-Osrtow (ed.) *Water use and hydraulics in the Roman city* (Archaeological Institute of America, Boston, Colloquia and Conf. Papers 3): 65-82. Dubuque (Iowa): Kendall/Hunt Publications.

Leveau Ph. 2012. D'Alexandrie d'Égypte à Caesarea de Maurétanie: transfert de technologie hydraulique et diffusion d'un nouveau modèle urbain. In: Stéphanie Guédon (ed.) *Entre Afrique et Égypte: relations et échanges entre les espaces au sud de la Méditerranée à l'époque romaine* (Scripta Antiqua 49). Paris: Ausonius Éditions.

Lightfoot, D. R. 2000. The Origin and Diffusion of Qanats in Arabia: New Evidence from the northern and southern Peninsula. *The Geographical* Journal 166: 215-226.

Magee, P. 2007. Beyond the Desert and the Sown: Settlement Intensification in Late Prehistoric Southeastern Arabia. *Bulletin of the American Schools of Oriental Research* 347: 83-105.

Mouton, M., Benoist, A. and Cordoba, J. M. 2011. The Snake figuration in Iron Age Society. *Liwa, Journal of the National Center for Documentation and Research* 3: 3-25.

Mylonas, G. 1932. Ελευσινιακά Πάρεργα. *Archeologikon Deltion* 14, Παράρτημα: 41-48.

Oikonomakis, N. E. 1984. Το Υδραγωγείο της Λύκτου. *Lyktos* 1: 66-99.

Papadaki, A. 2000. Παρατηρήσεις στην τοπογραφία των Αρχαίων Ποτνιών (Τάχι Θηβών). In: B. Αραβαντινός (επιμ.), *Γ' Διεθνές Συνέδριο Βοιωτικών Μελετών (Θήβα 4-8 Σεπτεμβρίου 1996)* (Επετηρίς της Εταιρείας Βοιωτικών Μελετών τ. 3): 357-369. Athens: Society of Beoteian Studies.

Papadakis, N. G., 1911. Ανασκαφαί τάφων εν Βοιωτία. *Praktika tēs en Athēnais Archaiologikēs Hetaireias* 66: 132-142.

Mays, L.W. 2010. A Brief History of Water Technology during Antiquity: Before the Romans. In: L. W. Mays (ed.) *Ancient Water Technologies:* 1-28. Arizona State University: Springer.

Papageorgiou-Venetas, A. 2010. *Πόλεις και μνημεία στην Ελλάδα του 'Οθωνος*. Athens: En Athinais Archaeologike Etaireia.

Pappas, A. 1999. Η *'Υδρευσις των Αρχαίων Αθηνών*. Athens: Eleutheri Skepsis Editions.

Paraskevopoulos, G. P. 1907. *Οι Δήμαρχοι των Αθηνών (1835-1907)*. Reprinted in 2001. Athens: Municipality of Athens.

Papastavrou, E., 2006. *Υπόγειοι λαξευτοί τάφοι της Αίγινας* (Archaeologike Ephemeris 145). Athens.

Platonos-Yiota, M. 2004. *Αχαρναί. Ιστορική και Τοπογραφική Επισκόπηση των Αρχαίων Αχαρνών, των γειτονικών Δήμων και των οχυρώσεων της Πάρνηθας*. Athens: Municipality of Acharnai.

Robinson, B. A. 2011. *Histories of Peirene. A Corinthian fountain in three millennia*. Athens: American School of Classical Studies.

Salesse, E. 2001. Du nouveau à propos des galeries de captage émergentes. Quelques réflexions sur les actes du séminaire du Collège de France de mars 2000. In: P. Briant (ed.) *Irrigation et drainage dans l'Antiquité. Qanâts et canalisations souterraines en Iran, en Egypte et en Grèce* (Séminaire du Collège de France, mars 2000). *Topoi* 11: 711-736.

Sommer, J. A. 1841. *Athenes et ses Environs*. In: A. F. Stademann (ed.) *Panorama von Athen*. Munich.

Symeonoglou, S. 1985. *The Topography of Thebes from the Bronze Age to Modern Times*. Princeton: Princeton University Press.

Tomlinson, R. A. 1969. Perachora: the remains outside the two sanctuaries. *The Annual of the British School at Athens* 64: 155–258.

Travlos, J. 1971. *Pictorial Dictionary of Ancient Athens*. Washington/ NY: Praeger Pub.

UAE 2006. Current status of aflaj in the Al Ain area. http://www.enhg.org/resources/articles/al_ain_falaj/Al_%20Ain_Falaj_Report.pdf.

Vanderpool, E. 1965. The Acharnian Aqueduct. In: Χαριστήριον εις Α. Κ. *Ορλάνδον*, τ. 1: 166-174. Athens: En Athinais Archaeologike Etaireia.

Voudouris, K. S., Christodoulakos, Y., Steiakakis, E. and Angelakis, A. N. 2013. Hydrogeological Characteristics of Hellenic Aqueducts-Like Qanats. *Water* 5: 1326-1345.

Weingartner, H. 2007. Water supply by qanats. A contribution to water shortage in Mediterranean? In: *Proceedings of the 10th International Conference on Environmental Science and Technology. Kos, Greece 5-7 September 2007*: 1555-1561. Syros: University of the Aegean Department of Environmental Studies Global Network for Environmental Science and Technology (Global-NEST).

Wilkinson, T. J. 2003. *Archaeological Landscapes of the Near East*. University of Arizona Press.

Wilkinson, T. J. 2012. Earth and Water: The Land of Iran, 12-27. In: S. David and A. P Mousavi (eds) *Ancient Iran from the Air*: 12-27. Darmstadt: Zabern.

Wilkinson, T. J. Boucharlat, R., Maurits, W., Ertsen, M. W., Gavin Gillmore, G., Kennet, D., Magee, P., Rezakhani, K., De Schacht, T. 2012. From human niche construction to imperial power: long-term trends in ancient Iranian water systems. *Water History* 4: 155-176.

Wilson, A.I. 2008. Hydraulics Engineering and Water Supply. In: J. P. Oleson (ed.) *Oxford Handbook of Engineering and Technology in the Classical World*: 285-318. New York: Oxford University Press.

Zambas, C., Dounias, D. and Angistalis, G. 2017. The Aqueduct of Eupalinos on Samos, Greece, and its Restoration. In: A. N. Angelakis, E. D. Chiotis, S. Eslamian and H. Weingartner (eds) *Underground Aqueducts Handbook*: 63-80. Boca Raton: CRC Press.

Ziller, E. 1877. Untersuchungen über die antiken Wasserleitungen Athens, *Mitteilungen des Deutschen Archäologischen Institutes, Athenische Abteilung* 2: 107-131.

Dr Eustathios D. Chiotis
Mining and Petroleum Engineer
Former Director at the Public Petroleum Corporation &
The Institute of Geology and Mineral Exploration of Greece
echiotis@otenet.gr

The Hadrianic aqueduct in Corinth

Yannis Lolos

Abstract

The Hadrianic aqueduct of Corinth, bringing water from Stymphalos, was one of the largest works of engineering conducted in Greece during Roman times. Its course as well as its construction and other technical details had been studied by the undersigned in the past. That investigation was able to follow the line of the aqueduct as far as the western outskirts of Corinth but not within the city proper. This paper precisely discusses the course and final destination of the aqueduct within the city. Thanks to recent fieldwork we were able to map traces of the aqueduct on the slopes of Acrocorinth, and consequently narrow down the area where the main water reservoir should be sought. In addition, we discuss some other water channels known from this area and whether they could belong to the Hadrianic line or not. Finally we address the subject of the final recipients of the Stymphalian water, namely the fountains and the public baths.

Keywords: Aqueducts, Corinth, Stymphalos, Peirene, Hadrian

Twenty-seven years ago I attempted to trace the Roman aqueduct that once brought water to Corinth from Stymphalos.[1] The mid-2nd century AD traveler Pausanias mentions this major hydraulic work on two occasions in his *Periegesis*, and attributes it to the emperor Hadrian.[2] In the course of my survey, I observed traces of the aqueduct in 69 locations, enough to allow me to reconstruct the line of the aqueduct over an approximate length of ca. 85 km.[3] In accordance with the terrain, the water followed first a southeasterly course as far south as Mount Megalovouni of Nemea before it turns to the northeast in the direction of Corinth (Figure 1). The last trace of the aqueduct in the direction of Corinth that I was able to map at the time was the one preserved on the western flank of the Anapnoga ravine, some 1,400m to the southwest of the Roman forum but just within the line of the Classical city-walls of Corinth (no. 1 on Figure 2).[4] It is an almost 7m-long section of the *specus*, oriented roughly north-south and preserving part of its base and of the inner wall, the rest having collapsed down the precipitous slope (Figure 3). Typical for this aqueduct, the built channel features the rubble and mortar masonry with opus signinum applied to its bottom and side-walls (Figure 4). Its coordinates (close to the southern end and in the EGSA87 coordinate system) are 400198.4 (Easting) by 4194680.7 (Northing) and its exact elevation at the level of the base of the channel equals 190.912 masl. Based on topography, the aqueduct would have continued further south in order to cross the ravine via a bridge no traces of which are visible.

The continuation of the aqueduct eastwards, i.e. within the city, and its final destination had remained unknown despite my efforts as well as the efforts of other scholars who looked into the water supply of Corinth, notably Henry Robinson's and Mark Landon's. In August 2016, I revisited the area with Dimitris Bartzis, an architect engineer specializing in ancient architecture and native of Archaia Korinthos, equipped with a differential GPS of high accuracy in an attempt to follow the course of the aqueduct within Corinth. Our starting point was precisely the last certain segment of the Hadrianic aqueduct, high on the western bank of the Anapnoga ravine. From there we walked eastwards until past the eastern ravine of the hill of Haghios Georgios, trying to keep at an elevation of 190-170 m. Overall we covered a distance

[1] Lolos 1990-1991;1997; 2010.
[2] Pausanias, 2.3.5; 8.22.3.
[3] In this pre-GPS era, I calculated the length by tracing the line of the aqueduct on the 1 to 50,000 military map using an opisometer.
[4] Lolos 1991-1992: no 69; Lolos 1997: 279 and plates 71, b and c.

Figure 1. The course of the Hadrianic aqueduct of Corinth (with dotted line; source: Lolos 2010: 12–13).

Figure 2. Aerial view of the north slope of Acrocorinth, with the reconstructed route of the Hadrianic aqueduct and the points (1-4) where it was found. White asterisks indicate possible traces of the aqueduct. Points 100-102 indicate traces of an aqueduct around the hill of Haghios Georgios. The hatched area indicates a possible location for the *castellum divisorium* of the Hadrianic aqueduct. The black dotted line represents the Classical city-walls, and the blue lines the supply tunnels of Peirene (both taken from http://www.ascsa.edu.gr/index.php/excavationcorinth/maps-and-gis-data-for-corinth-and-greece

Figure 3. The channel of the Hadrianic aqueduct above the western bank of the Anapnoga ravine. (no. 1 on Fig. 1)

Figure 4. The base and inner side wall of the channel above the western bank of the Anapnoga ravine.

of *ca.* 2 km but not in a straight line, walking mostly through maquis which hindered ground visibility, with patches of olive groves laid out in terraces. In fact, terracing characterizes the northern slopes of Acrocorinth, and may not be irrelevant to the line of the Hadrianic aqueduct as we came to realize.

The line of the aqueduct above the eastern bank of the ravine is not obvious but more than 300 meters northeast of the ravine we observed a vaulted tunnel opened into the conglomerate (no. 2 on Figure 2: Easting: 400511.64, Northing: 4194838.63). It is 1.90 to 2m wide, and its visible length reaches ca. 13m (Figures 5-6). The location of this tunnel and its elevation (191.27 at the level of the opening - the floor of the tunnel would have been lower) as well as the semi-circular profile of its vault all point to it being part of the Hadrianic line. This would make it the fourth tunnel of the aqueduct, but only the second one surviving seemingly without later modifications.[5] In comparison to the tunnel of Spathovouni, which is the third tunnel mapped along the course of the aqueduct due east of the village of Spathovouni (to the southwest of

Figure 5. The opening of the tunnel above Anapnoga (no. 2 on Fig. 1).

[5] The tunnels of Siouri and of Alonaki were probably widened in modern times, Lolos 1997: 289-290.

Figure 6. The interior of the tunnel above Anapnoga.

ancient Corinth), this one is almost twice as wide, which forces us to accept its inclusion to the Hadrianic line with reservations.

In the same campaign we found a section of the underground *specus* on the north slopes of Acrocorinth, above the excavated sanctuary of Demeter, some 760 m east of the tunnel just described (no. 3 on Figure 2; Easting: 401281.44, Northing: 4194818.02, Elevation: 187.22 masl). We feel more confident that this is part of the Hadrianic system. Its presence was betrayed by illicit digging (Figure 7). The channel shows the characteristic vaulted roof consisting of rubble and mortar radially laid out on a wooden scaffolding (Figure 8). The imprint of the wooden boards is clearly visible on the underside of the roof. The diameter of the vault is *c.* 1.50m, i.e. similar to the diameter found in other sections of the aqueduct. The width of the channel below is certainly smaller because its side walls typically project towards the interior by 0.10 to 0.22m.[6] The channel must continue its underground course towards the east for *c.* 87 m where one more clandestine digging exposed part of the vault (no. 4 on Figure 2; Figure 9; Easting: 401364.95, Northing: 4194793.15, Elevation: 187.15 masl). The presence of the aqueduct underground is suggested on the surface by a linear bump, which we noticed also in other areas along the projected course of the aqueduct. Between Anapnoga (no. 2) and the northern slopes of Acrocorinth (nos. 3 and 4), we mapped possible traces of the channel in three places (marked on the map as white asterisks), betrayed mostly by manholes over its roof. It is clear that the aqueduct passed a few meters higher than the excavated sanctuary of Demeter, and that it must have been cut by the modern road leading to Acrocorinth.

It is reasonable to assume that an aqueduct of this size and water discharge (estimated to almost 80,000 m^3 per day) would have terminated at a central reservoir acting as a storage and distribution center. No traces of such *castellum divisorium* have been reported so far from the area east of our last documented

[6] Lolos 1997: 281.

trace of the aqueduct (no. 4), namely from the environs of the hill of Haghios Georgios. Instead, we mapped traces of an aqueduct running at an elevation of 176-178 masl to the southeast of the church (nos. 100-102 on Figure 2).[7] The first section (no. 100: Easting: 401661.49, Northing: 4194681.27, Elev. 177.98 masl), visible over a total length of *c.* 8.5 m and a maximum height of 0.95m, may correspond to the *substructio* of the channel (Figure 10). The second section (no. 101: Easting: 401727.58, Northing: 4194620.76, Elev. 178.51 masl), above the western bank of the ravine, can be traced over a length of *c.* 4.15 m and is preserved to maximum visible height of 0.45m (Figure 11). The longest section of this aqueduct is the one visible above the eastern bank of the ravine and follows a roughly north-south direction (no. 102; *southern end*: Easting: 401746.84, Northing: 4194601.81; Elev. 177.72 masl; *northern end:* Easting: 401752.65, Northing: 4194638.34, Elev. 176.24 masl). It can be traced over a length of *c.* 37m, and shows a maximum preserved height of 1.20m (Figure 12). In this case we are likely dealing with the inner side wall of the aqueduct rather than with its foundation. The continuation and

Figure 7. The *specus* of the Hadrianic aqueduct above the sanctuary of Demeter (no. 3 on Fig. 1).

[7] Two of these sections are briefly discussed in Lolos 1997: 279 (note 19) where I miscalculated the elevation to *c.* 162 instead of 177m.

Figure 8. View of the interior of the *specus*.

Figure 9. The last documented section of the *specus* of the Hadrianic aqueduct (no 4 on Fig. 1) showing part of its vaulted roof.

Figure 10. The masonry of an aqueduct (probably its base) near Haghios Georgios (no. 100 on Fig. 1)

Figure 11. The masonry of an aqueduct to the southeast of Haghios Georgios (no. 101 on Fig. 1).

Figure 12. The line of an aqueduct above the eastern bank of the rema of Haghios Georgios (no 102 on Fig. 1).

destination of this channel are unclear, as the dense maquis did not allow us to trace it further. All three sections feature a characteristic rubble, cement and tile masonry (Figure 13), which is markedly different to the masonry of the Hadrianic channel. This difference as well as the substantial lower elevation of this aqueduct in relation to the one found on the northern slopes of Acrocorinth (the section no. 100 lies almost 10 m lower than the section no. 4) indicate that we are dealing with two separate channels. It is also conceivable that the Haghios Georgios aqueduct is a later extension of the Hadrianic line, but we can be fairly certain that the two were not built together from the start. One more argument against this section being part of the original, Stymphalian line is its very location, way too east in relation to the center of the Roman city, i.e. its forum. On the contrary it would fit well the heart of the late Roman city – itself located to the east of the Roman forum.

On the lower slopes of Acrocorinth, at an elevation of *c.* 140 masl, an underground conduit was discovered in 1965 during a road widening operation.[8] The remains are no longer visible but the description of the

[8] This channel is discussed by M. Landon (2003: 56), and in more detail in his doctoral dissertation (1994: 350-352).

Figure 13. Detail of the masonry of the aqueduct.

discovery helped me to place the site of that excavation on Figure 1, some 200 m northeast of the last point where the Hadrianic channel was mapped (point 4). A segment of an aqueduct of a north-south orientation, 32.50 m long, had come to light, on both sides of the modern road ascending to Acrocorinth. It is a brick and mortar construction with vaulted concrete ceiling laid over a frame of wooden planks. Its inner dimensions were found to be 1.30 m (H.) by 0.70 m (W.), but the width was reduced at the bottom of the channel by the presence of two 'curbs' built at the junction of the bottom with the walls to prevent water leaking. Similar 'quarter rounds' were observed at several places along the Hadrianic line.[9] Landon tentatively identified this conduit 'as a sub-branch of the Hadrianic system, designed to carry the water from a distribution reservoir somewhere higher up the slope, and probably further west, to the eastern part of the city's central district '.[10] Our discovery of what appears to be the channel of the main Hadrianic line further up the slope, and the technical resemblance of the two channels, further reinforces this hypothesis. If, then, the 1965 conduit descends from the central distribution reservoir of the aqueduct in a northerly direction, then this *castellum* must be located between the last mapped point of the channel (point 4) and the hill of Haghios Georgios. The candidate area (indicated with hatched line on Figure 2) where the *castellum* must be sought is *ca.* 100 m long and presently cultivated with olives. Such a location could also explain the number of pipes and drain cuttings discovered In the course of the excavations conducted both by the American School of Classical Studies and the Greek Ministry of Culture in the area covered by the sanctuary of Demeter and Kore, which seem to be unrelated to the sanctuary. According to Nancy Bookidis's brief description, ' these descend from the southeast to the northwest, taking water from somewhere higher up the hill down to the west of Hadji Mustapha' .[11]

The public fountains of the city and its baths would have been the main beneficiaries of the Stymphalian water. Pausanias (2.3.5) clearly associates the numerous fountains he saw in Corinth with the abundance of water and especially of the water that Hadrian brought from Stymphalos. Our knowledge of Corinthian fountains is inevitably lacunose, but even for those that have been studied to a larger or smaller extent, their chronological phases are far from being securely established.[12] This holds even for the emblematic fountains of Peirene and Glauke, even though the former is by far the most thoroughly studied Corinthian fountain.[13] Betsey Robinson in her authoritative re-examination of Peirene, dates the impressive marble

[9] Lolos 1997: 281-282.
[10] Landon 1994: 351-352.
[11] Bookidis 2013: 101.
[12] See the discussion of Corinthian springs and fountains in Landon 1993: 46-54, and specifically of the Roman fountains in Robinson 2013.
[13] Hill 1964: 1-115; Robinson 2011.

revetment to the late Trajanic or early Hadrianic period, followed by modifications of the *hypaithros krene* and the splendid fish paintings of the interior walls of the fountain sometime in the second half of the 2nd century AD.[14] No major addition or modification of Peirene appears to date from the later years of the Hadrianic reign or from the first decades of the reign of Antoninus Pius. Likewise, out of the nine fountains that the American Excavations have brought to light in Corinth, only one, the so-called North Nymphaeum (at the North Market), seems to post-date the construction of the Hadrianic aqueduct.[15] It is therefore more likely that the Stymphalian water was routed into some of the existing fountains, including Peirene and Glauke.

Betsey Robinson persuasively argues that Peirene with her extensive tunneling (shown with blue lines in Figure 2) and storing capacity could have well accommodated the extra mass of water which was praised by Athenaeus (2.43b) as of the 'lightest' quality.[16] Linguistic embellishment allowed, this characterization does not befit the hard water of Peirene, but certainly the water of Stymphalos and the springs of Driza which the aqueduct most likely tapped. Chemical analysis of this water and the almost total absence of calcium carbonate deposits on the floor and the sides of the channel observed at the ravine of Anapnoga (Figure 4) demonstrate the softness of the Stymphalian water, low in ions of calcium and magnesium. In the case of Glauke, its four sizeable reservoirs could have been used as some kind of secondary *castellum* meant to regulate the flow of Stymphalian water, and direct it around Temple Hill to fountains such as the North Nymphaeum.[17] A secondary *castellum* has been rightly postulated in the case of the massive, concrete tank excavated in 1933 at the northwestern corner of the South Stoa.[18] Its inner dimensions are *ca.* 2.05 m square by 1.50 m in depth and the water was channeled under pressure though two holes opened 0.50 m from the bottom with traces of the lead pipes that were once fitted into them. Although the function of the tank is rather clear, its date remains unclear and by extension its connection to the Hadrianic line is probable but not certain.[19]

Concerning the Roman baths of Corinth, our picture is also incomplete. Of the dozen of Roman and late-Roman bath complexes that have been documented so far, only three seem to date before the mid-2nd century AD, when Pausanias visited the city.[20] Yet, the traveler speaks of the existence of baths in many parts of the city, of which one was erected by Hadrian.[21] The baths of Hadrian have yet to be positively identified, but we can be fairly certain that they were fed by the Stymphalian line. A reasonable assumption is that more baths benefited from the Stymphalian water even though the evidence is lacking. The canalization feeding the fountains and the baths is the hardest to trace, since we are often dealing with many short sections in a dense network of galleries and pipes of different periods.[22] Knowing what goes with what is in most cases impossible.

In conclusion, thanks to our discovery of the aqueduct above the sanctuary of Demeter, we can extend the known route of the Hadrianic line eastwards by some 1300 m and narrow down the area where the *castellum divisorium* must be sought. In addition, if we accept that the tunnel of Anapnoga (no. 2 on Figure 2) is indeed part of the aqueduct, then we should reconsider the role of the tunnel in building this or any aqueduct of the Roman era. Instead of representing an extraordinary solution, applied when the terrain prevented the construction of a built channel, a tunnel was a solid and economic way to convey water since it required no building materials and hardly any maintenance. If so, we should expect the Hadrianic aqueduct of Corinth to have included more than four tunnels along its long course of 85 km.[23]

[14] These are Robinson's Roman Phases 4A and 4B, Robinson 2011: 206-231.
[15] Robinson 2013: 368-372.
[16] Robinson 2013: 366-367.
[17] As suggested by Robinson 2013: 367.
[18] Broneer 1935: 64.
[19] See Lolos 1997: 280 and Landon 2003: 56, with earlier references.
[20] As pointed out by Biers 2003: 317.
[21] Pausanias, 2.3.5.
[22] Peirene being the exception with her great rock-cut supply channels traced over almost a kilometer (on which see Robinson 2011: 11-17).
[23] It is interesting to note in this regard that the conduit discovered in 1965 on the lower slopes of Acrocorinth included at the southern end of the explored stretch a short section of a tunnel cut through bedrock, Landon 1994: 351.

Acknowledgments

I would like to thank the chief editor of this volume Dr Georgia Aristodemou, for inviting me to participate, as well as Professor Theodosios Tassios for kindly insisting that I submit a contribution. The assistance of Dimitris Bartzis, my companion in the field, was pivotal for tracing the aqueduct on the slopes of Acrocorinth. I am also deeply indebted to Professors Betsey Robinson and Mark Landon, perhaps the best living experts on Corinthian waterworks, for reading a draft of this article and making crucial comments and suggestions which helped me to substantially improve and enrich its content. Finally I would like to extend my thanks to Takis Papagiannopoulos of the Katagramma printing company, for providing me with a rectified aerial view of ancient Corinth used as background in Figure 2.

Bibliography

Biers, J. 2003. *Lavari est vivere*; Baths in Roman Corinth. In: C.K. Williams II and N. Bookidis (eds) *Corinth, the Centenary 1896-1996* (*Corinth* XX): 303-319. Princeton: American School of Classical Studies at Athens.

Bookidis, N. 2013. Acrocorinth: the North Slope. In: K. Kissas, W. D. Niemeier (eds) *The Corinthia and the Northeast Peloponnese: Topography and History from Prehistoric Times until the End of Antiquity* (Athenaia 4): 99-110. München: Hirmer Verlag.

Broneer, O. 1934. Excavations in Corinth, 1934. *American Journal of Archaeolgoy* 39: 53-75.

Hill, B.H. 1964. *The Springs: Peirene, Sacred Spring, Glauke* (Corinth I.vi). Princeton: American School of Classical Studies at Athens.

Landon, M. E. 2003. Beyond Peirene; toward a broader view of Corinthian water supply. In: C.K. Williams II and N. Bookidis (eds) *Corinth, the Centenary 1896-1996* (Corinth XX): 43-62. Princeton: American School of Classical Studies at Athens.

Landon, M.E. 1994. Contribution to the Study of the Water Supply of Ancient Corinth, PhD Dissertation, University of California, Berkeley.

Lolos, Y. A. 1990-1991. Το Αδριάνειο Υδραγωγείο Κορίνθου. *Ηόρος* 8-9: 311-313.

Lolos, Y. A. 1997. The Hadrianic Aqueduct of Corinth with an appendix on the Roman aqueducts in Greece. *Hesperia* 66: 271-314.

Lolos, Y. A. 2010.*'Ύδωρ βασιλεὺς Ἀδριανὸς ἐσήγαγεν ἐκ Στυμφήλου'; το Αδριάνειο υδραγωγείο της Κορίνθου και η μεταφορά του νερού στα ρωμαϊκά χρόνια* (Cultural Foundation of the Bank of Piraeus Group).

Robinson, B. A. 2011. *Histories of Peirene; A Corinthian Fountain in Three Millennia*. Princeton: American School of Classical Studies at Athens.

Robinson, B. A. 2013. Playing in the Sun: Hydraulic architecture and water displays in Imperial Corinth. *Hesperia* 82: 341-384.

Dr Yannis Lolos
Assistant Professor
Department of History, Archaeology and Anthropology
University of Thessaly, Greece
gialolos@ha.uth.gr

The Roman aqueduct of Mytilene

Yannis Kourtzellis, Maria Pappa and George Kakes

Abstract

The aqueduct of Mytilene is an enormous technical work of the Greek and Roman antiquity. Its construction intended to cover the water supply needs of the city of Mytilene, a gorgeous city, which was hymned and cited by the Roman writers for its urban design, public structures and magnificent villas. Its forepart lies at the slopes of mountain Olympus, in the centre of the island and stretching across a variable geomorphologic terrain, it ends at the east coast of the city of Mytilene (the area of the ancient Theater). The creators of the aqueduct used a wide variety of construction techniques, such as water bridges, both above and underground water channels, in order to ensure steady flow of the water. The diligence of the creators for the architectural style and the structure of the water bridges is also impressive. The water bridges are enlisted among the most beautiful bridges of the ancient world, causing even today visitors' admiration.

This study is focusing on the thorough examination of the aqueduct, dealing for the first time with issues related to the mapping of its total route, the manufacturing excellence of the water channels, the architectural and structural description of the monument, the restoration of the missed sections, the identification of the quarries from which the construction materials came from, its chronology, etc. Moreover, it is the first time that the latest data from the excavation of a vertical shaft are presented in detail.

Keywords: Mytilene, Lesbos, aqueduct, water bridges, supply networks, shafts

The city of Mytilene in the Roman era

Every study about Mytilene of the Roman period begins almost exclusively, with the highly acclaimed references of the ancient writers on its architecture and urban organization. Particularly known are Ovid's and Horace's references that Mytilene is as beautiful as Rhodes and Ephesus.[1]

The city was built amphitheatrically in an exceptional location opposite the coast of Asia Minor. A special feature of its topography was the existence of a natural strait (Euripus) with an estimated length of about 780m and a width of 30m, which divided the city into two parts (the east and west),[2] and also connected the two harbours, the northern (commercial)[3] with the southern (military) harbour.[4] The monumental formation of the city, estimated to have started at the end of the late Classical period (second half of the 4th century BC), was mainly developed in the Hellenistic period and culminated in the Roman period.[5] In the coastal zone of the northern harbour, excavations have brought to light five porticoes (stoas), among which a magnificent two-storey marble portico (stoa) of the 3rd century BC, that allowed the identification of the area as the ancient agora.[6] A significant number of baths and bath complexes that can be dated to the 2nd century AD, along with thirty public and private buildings decorated with mosaic

[1] Ovid, *Odes*, I.7,1; Horace, Epistles, I.XI,17; Kontis 1978: 197-201; Labarre 1996: 69-156.

[2] Information regarding Euripus: Diodorus Siculus, *Bibliotheca historica* 13.79.6.; Kontis 1978: 211-216 and Kourtzellis 2013: 28-31.

[3] Strabo, *Geography*, XIII, 2.2. Regarding the name 'Maloentas', see Thucydides (Thuc. 3.3.3-5).

[4] Strabo, *Geography*, XIII, 2, 2; Kourtzellis 2008: 213-222; Kourtzellis 2010: 185-200; Theodoulou and Kourtzellis 2010: 129-147; Kourtzellis 2013:11-15, 28-31, 46-49.

[5] Koldewey 1890, 6-8; Kontis 1978: 221-231; Kourtzellis 2010: 191-192.

[6] At the location Epano Skala, to the west of the Social Security Fund (IKA) services building, excavations have brought to light the sea defense wall of the city of Mytilene, along with commercial stoas, late Roman baths, etc. Indicative bibliography on Pistorius 1913: 149; *Archaiologikon Deltion* 11, 1927-8, Χρονικά: 17; *Praktika tēs en Athēnais Archaiologikēs Hetaireias* 1961: 211-212; *Archaiologikon Deltion* 41,1986, Χρονικά, B΄2: 200; 42,1987, Χρονικά, B΄ 2: 481; 53,1998, Χρονικά, B΄2: 772; 54, 1999, Χρονικά, B΄2: 749-759; Kontis 1978: 224-225; Inscription, IG XII 2,14.

floors, have come to light.[7] Many of them have a complex water system with cylindrical clay water pipes, underground tanks, etc.[8]

Although the excavation has not yet revealed public fountains of the Roman period, there is strong evidence for their existence, like the marble figure of a Silenus reclining on a perforated ovoid vase, having a spouting function.9 The figure was part of the sculptural composition of a monumental fountain building and dates back to the early Antonine period – as deducted by the wide use of the drill to form the beard.10

The city's increased needs for water supply were covered by the Roman aqueduct. The existence of an earlier aqueduct, which covered the needs of the Classical, Hellenistic and early Roman period is attested by the evidence from the plots of the Hellenic Telecommunications Company (OTE)11 and Papazoglou land property,12 the marble fountain in Theophrastus str., which dates from the 3rd century BC until the 1st century BC,13 the fountains of the wider area,14 and also the dozens of wells that were found during several excavations in the city and are preserved today15. There are also important inscriptions that refer to the aqueduct of Athanada, which channeled water to the *intra muros* sanctuary of Asclepius and the *είσω πόλιν*, already from the 4th century BC, bringing water, probably from the area of Kratigos.[16]

Mytilene aqueduct: the selection of sources

The aqueduct carried water from the water sources of Mount Olympus to the east coast, where the city of Mytilene was built.[17] The identification of the aqueduct's main water source is extremely difficult, due to the lack of stable architectural remains of the original drain tank.[18] The aqueduct began its journey from the Tsigos surface water sources (Map 1.2), which are used until today for the water supply of Agiasos village.[19] Especially interesting is also the case of the Great Lake, the *Pyrraion Lake* (Map 1.1) as referred to by the ancient authors. During antiquity, the lake would have had a large water potential,

[7] Indicative reference on *Archaiologikon Deltion* 46,1991, Χρονικά, B΄2: 366-367, pl. 142d; 53,1998, Χρονικά, B 2: 771; 54,1999, Χρονικά, B΄2: 755-756; 55, 2000, Χρονικά, B΄3: 927-928 mainly, 918. Most recently on Roman villas and public buildings with mosaic floors, see Kyriakopoulou 2013: 17-21.

[8] Regarding the reasons for which clay pipes are preferred over lead pipes for internal city water supply networks, see Kaiafa 2008: 71-72, note 71.

[9] Museum of the City of Mytilene, Inv. No. MM 229. Despite the fact that the Catalogue of the Archaeological Museum of the City of Mytilene refers to this finding as of 'unknown origin', we believe that it almost certainly originates from the city of Mytilene.

[10] Stefanidou-Tiveriou 1995: 67-75, note 23, pl. 30b, 33b; Aristodemou 2012: 124-130, cat no 216, note 899, 336 with extensive bibliography, pl. 32.8. This sculpture is similar to a group of sculptures from Athens and Thessaloniki, each thought to derive from an original fountain complex of the middle 1st century BC.

[11] Land property of the Hellenic Telecommunications Company (OTE) at the junction of Vournazon and Faonos str.,Chatzi 1971: 449-454,drawing 2 and Chatzi 1972: 584-588, figs. 6-8.

[12] The Papazoglou land property, although never published in the *Archaiologikon Deltion,* is known to us from photographic material and drawings kept in the records of the Ephorate of Antiquities of Lesbos. In 1979-1980, at the eastern half of this plot, excavations revealed part of a 9.70m long conduit with1.10m external width and 0.30m internal width, that -as far as construction is concerned- is similar to the Pergamon aqueduct conduit, Fahlbusch 1981: 143-144,145.

[13] Archontidou 1997: 900-902, drawing 6, table 333 d.

[14] The existence of a well-built isodomic, flat roofed, fountain is recorded at the location Chasan Aga, NW of the city airport (Charitonidis 1961/62: 263, drawing 3).

[15] A large number of wells were brought to light at the city of Mytilene. Indicatively, *Archaiologikon Deltion* 28,1973, Χρονικά, B΄2: 515-517, drawing 10, pl. 484a-485d (D. Chatzi); 30,1975, Χρονικά, B΄2: 314 (Tsiravakos); 46,1991, Χρονικά, B΄2: 363, drawing 3 (A. Archontidou-Argyri); 47,1992, Χρονικά, B΄2: 529 (A. Archontidou-Argyri); 50,1995, Χρονικά, B΄2: 689, pl. 207a (A. Archontidou-Argyri); 52,1997, Χρονικά, B΄2: 903 (A. Archontidou-Argyri); 55, 2000, Χρονικά, B΄3: 916 (L. Acheilara) and 919 (A. Archontidou-Argyri).
On well water, Strabo, *Geography*, IB, 568; Herodotus, *Historiae*, VI, 119. Also Hodge 1992: 53-54. On well digging, Vitruvius, *De Architectura* VIII, 6,12-13. On the classification of well water into phreatic and spring water, seeTheophrastus' *Enquiry into Plants* B, VI, 3,also including the theory that the latter do not affect human health (cf. Aristotle, *Problemata*, ΚΓ, 30; Diodorus Siculus, *Bibliotheca historica,* II, 48; Pliny, *Historia Naturalis*, XXXI, 23.).

[16] Inscriptions: IG XII,2 4; SEG 26:874. Also, Chatzi 1972: 586; Kontis 1978: 430, fr. 2294; On the evolution of large scale water supply works in the Greek mainland from Classical to the Roman period, see Kaiafa 2008: 65-72.

[17] Mount Olympus is the second higher mountain of Lesbos Island (967m).

[18] On the difficulties in identifying water sources (location, intake points, integration with the aqueduct system) during the Hellenistic and Roman period, see Kaiafa 2008: 171, note 127.

[19] This study includes a survey of almost all Tsigos springs, which are the most important water springs on the island. We have discovered that water springs are located at an altitude ranging from + 289m to + 200m. All the altitudes given in this study are measured against the sea level. Regarding the role of water quality in the choice of intake springs, see Kaiafa 2008: 47-58,171-172 (referring to the springs of Amphipolis).

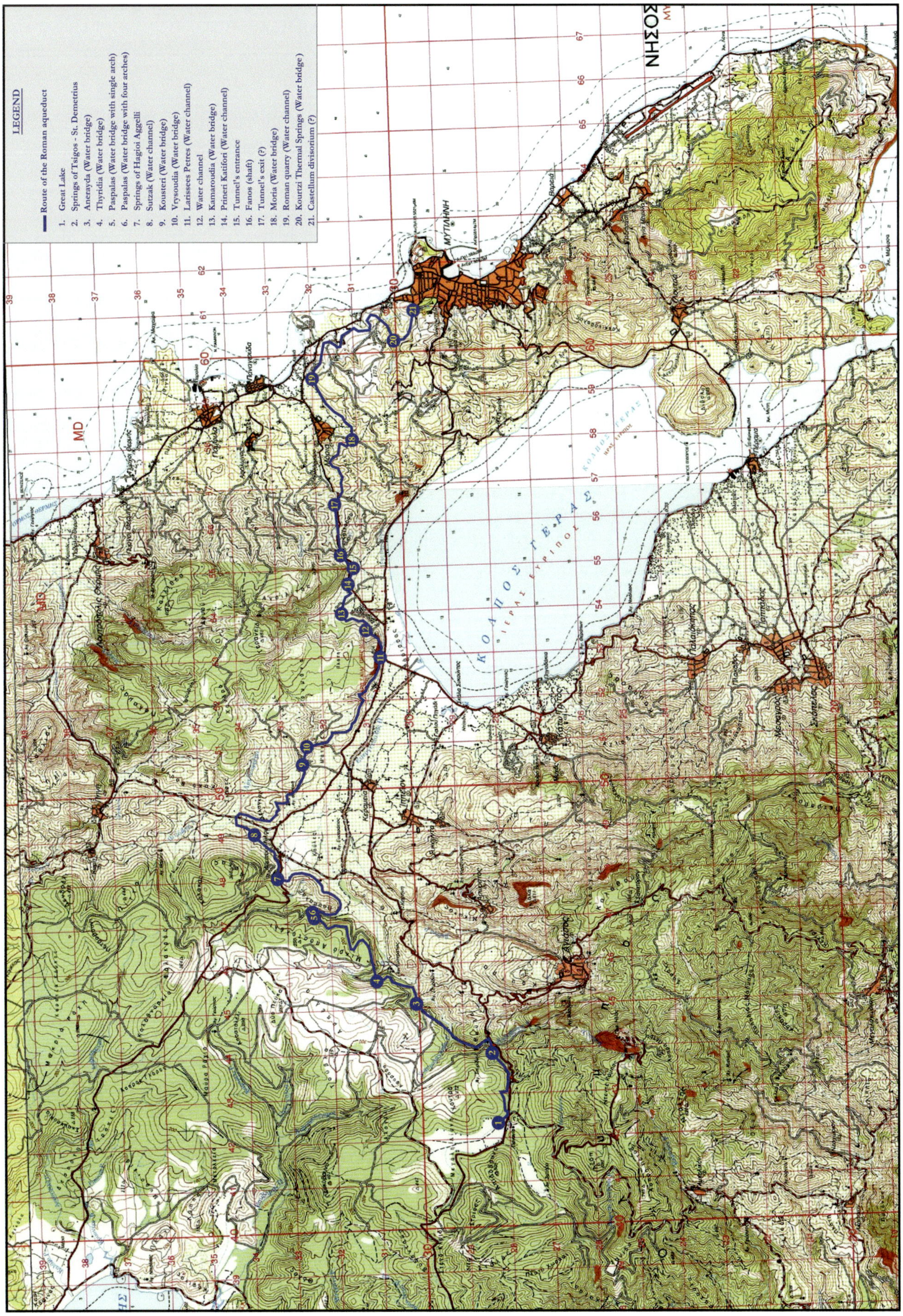

Map 1. The Mytilene Aqueduct route.

which obviously contributed to the hydrological reserves that fed the aqueduct. Initially it covered an area of 1200 acres; it was dried in 1929 to be yielded to the landless inhabitants.[20]

Mytilene aqueduct water course

The research area mapping was based on Koldewey's study,[21] as well as on the comprehensive study by I. Chatziioannou, who roughly marked upon a makeshift map the aqueduct water course.[22]

According to Vitruvius (*De Architectura*, XIII, 6) and Sextus Julius Frontinus (*De Aqueductibus Urbis Romae*), the main concern of ancient engineers was to ensure the supply of the best possible water quality, irrespective of distance,[23] as well as the maintenance of the appropriate water flow gradient using levelling techniques, such as masonry or freestone-built arched ducts (underground and/or over ground), tanks, bridges, etc.[24] For large scale construction projects during antiquity, the methods used were determined to a considerable extent by the relief of the ground.[25]

The first aqueduct bridge is located at an altitude of + 122.629m, in the forest area of Anerayda.[26] (Map 1.3) The water channel heading towards the bridge is still discernible for a couple of meters under a modern farm masonry wall. (Figure 1) A single arch bridge spans the homonymous torrent. Its span is 5.40m and its width at ground level is 2.16m. This water bridge presents similar characteristics with the other bridges, especially regarding pillar construction by the so-called *pseudo-isodomic* system.[27] Despite its remote location, the use of marble on the arch pillars reveals the intention for monumentality. Each block is decorated with a recessed perimeter strip (*peritaenia*). The arch itself is destroyed, although part of its springer is visible to the south, consisting of overlapping compact bricks which, in our opinion, may have been used to repair and replace the stone voussoirs.[28] The arch marble revetment is preserved for a length of *c*.1.50m, after which it continues in the form of cast masonry. Beyond this point, the construction extends on the slopes to support the water conduit until it went underground. The method used is a compact wall made of small stones linked with solid lime mortar (*opus incertum*). An impressive, *c*.40m long, part of the water conduit is preserved at a distance of approximately 200m eastwards. In order to achieve the appropriate gradient, it was considered necessary to build an outer cast-stone masonry wall, whose height occasionally reaches 3m. A significant part of the water channel's inner wall is stone carved. Impressive construction features of that part of the channel are the square shaped openings, located at regular intervals. They were used to drain the rain water of the mountainous land mass rising north-west of the hill, thus ensuring the construction's stability.

The next identified part of the aqueduct is located in a wooded area, at Thiridia (Map 1.4) and at an altitude of + 113,836m. The water bridge is southeast-northwest oriented and spans the underlying torrent (Figure 2).[29] Only the edges of the initial bridge are preserved. They are made of local ophiolite rubble stones linked with lime mortar. An 8.80m long (1.75m thick and 2.90m high) section of the bridge is preserved to the south-east. The channel section upstream of the bridge has been demolished to enable

[20] Kontis 1978: 340, note 607. Lake water was generally considered not healthy. Regarding good quality lake water, cf. Pliny, *Historia Naturalis*, XXXI, 21.

[21] Koldewey 1890: 65-68, pl. 29.

[22] Chatziioanou 1985: 153-162; Papakonstantinou *et al.* 2015:11, fig.1. The local Ephorate of Antiquities, despite the fact that it had -as early as 1991- declared three sections of the aqueduct as protected archaeological areas, had at its disposal the coordinates of the protected area and not those of the architectural remains. Therefore, it adjusted a map of the Hellenic Military Geographical Service (HMGS) (scale: 1:5000) to approximately 'trace' the aqueduct's course.

[23] Mays 2010: 116-117. According to Plutarch (*Causes of Natural Phenomena* I, 911) human health is influenced by water.

[24] Regarding the technical part of an aqueduct construction, the tools and instruments used, the marking method and slope maintenance procedures, Vitruvius, *De Architectura*, VIII, 5,1. Also, Hodge 1992: 126-170 and 197-208; Lewis 1999: 147-150.

[25] Kaiafa 2008: 29, note 41.

[26] 'Anerayda' in the Agiasos' local dialect stands for 'the water fairy' and is a direct reference to water nymphs, the so called Nereids, creatures of the Greek Mythology residing in locations with fresh, spring water.

[27] The pillars *facade* is built according to the *pseudo-isodomic system* while the gap between the faces was filled with smaller stones linked with solid lime mortar (*opus incertum*). Koldewey 1890: 67, pl. 29.4 and 29.6.

[28] On issues regarding water quality control throughout the water course, maintenance, cleaning, along with the relevant legislation, see Kaiafa 2008: 98-112.

[29] This torrent is the origin of Evergetoulas River, one of the most important rivers of the area; it springs from the northern part of Mount Olympus and flows into the gulf of Gera.

Figure 1. The first single arch water bridge at the wooded area of Anerayda, close to the water sources (photo by the authors)

Figure 2. The southeast edge of the water bridge area of Thiridia (photo by the authors)

Figure 3. The small single arch water bridge at the wooded area of Paspala (photo by the authors)

the passage of a forest road. The north-western section, 15.40m long, is also preserved. Of these, 5.40m are preserved only at ground level. According to measurements, the bridge width varies between 1.80 and 1.84m at its lowest level and 1.60m at its highest level. The north-west section's maximum height is 3.95m. Very few traces of the aqueduct's floor and wall springing line are preserved.[30] The central section of the bridge is completely disappeared and only few lime mortar traces are still visible along the river bed. The original length of the bridge is estimated at 54.69m. We believe, nevertheless, that a four arched bridge would have been necessary to span this torrent.

At the west side of the monument, we have detected at least two shallow lime kilns, which may have been used to turn the bridge construction material to lime.

The channel course continued towards the rural region of Paspalas (Figure 3). At *c.*100m southwest of a large, four arched bridge (to be described in the following paragraphs) and at an altitude of + 103,977m there is a single arched bridge (Map 1.5), 1.66m wide at its uppermost part. It is oriented NW-SA and spans the underlying small stream. The water conduit and the upper part of the bridge were demolished as early as Koldewey's time.[31] The arch span at the base is 5.40m and has a maximum width of 1.67m. The bridge is preserved at a height of 2.47m. The arch is built of small rubble stones linked with lime mortar. Although this is one of the smallest bridges, it preserves the same characteristic features also observed

[30] Assuming that the width of the bridge conduit was the same at all points, i.e. 0.60m, the restored conduit walls should have an approximate width of 0.50m, each.

[31] Koldewey (1890: 67, pl. 29.3) measures the arch span at 5.56m. Cited also at Chatziioannou 1985:153-162.

Figure 4. The four arch bridge that spans a river known as 'The Fountain of Achlia' at the area of Paspala (photo by the authors)

at the other bridges. The arch's external sides consist of voussoirs embedded in the cast stone masonry. Most of the stone voussoirs are now lost, with only few left at the arch base.[32]

This bridge is followed by an impressive four arched bridge that spans a river known as The Spring of Achlia and at an altitude of + 101,735m (Figure 4; Map 1.6). Only one of the four arches has collapsed, as early as Koldewey's time.[33] The water channel stands upon three, almost square shaped piers. The pier dimensions are 3.10m X 3.30m (at the base), while the distance between them, inside the ravine is 6.80m (central arch) and slightly less for the lateral arches. The piers and arches are meticulously built in marble according to the pseudo-isodomic system, with regular alternation between narrow and high series of blocks. The bridge is attached to the terrain slope via a compact structure made of small stones and strong mortar, exactly as the other remaining bridges. Rectangular and circular through holes which are visible at regular intervals in the cast-stone masonry wall, were used to fix and mount the scaffoldings. The inner sides of the piers have triangular holes, for the same reason. The bridge is preserved at a height of 24.95m and a width of 2.10m.[34] Only traces of the conduit floor are preserved.[35]

The water channel continues its course and goes underground towards the spring of Haghioi Aggelli (+92,135m, Map 1.7), where, according to certain scholars, the central conduit was also supplemented

[32] Stone voussoirs thickness: 0.27m. Visible length: 0.40m.

[33] However, native forest vegetation has considerably worsened the monument's stability.

[34] As measured at the bridge, at the level of the conduit's bottom.

[35] Koldewey 1890: 67, pl. 29.5; Chatziioannou 1985: 159. The total thickness of the conduit eastern wall is measured 0.60m. At Koldewey's time, the conduit was 0.60m high (at the eastern edge of the bridge), while today it is considerably shorter. In addition, a 0.07m thick brick fragment was detected among the debris.

Figure 5. The small single arch water bridge at the area of Kousteri (photo by the authors)

(filled up) with water from other sources.[36] Then, the channel passes over the settlement of Lambou Myloi and turns northward towards the location Sutzak (Figure 7c; Map 1.8).[37] At that point, it surfaces again on the rocky hill slope and runs for hundreds of meters following the road that leads to the Pigi settlement. The water channel is well-preserved, almost to its vaulted crowning, and per sections is preserved at a height of *c.* 2.00m; it is mainly masonry built and its external facet is particularly elaborated. The area's bedrock has been treated to provide a stable bearing for the channel's foundation.[38]

The conduit reappears at the Kousteri area (Map 1.9), where it crosses the small creek via a single arch bridge (+76,764m, Figure 5). This bridge is similar to the Anerayda bridge; its piers are riveted with marble blocks decorated with a recessed perimeter strip (*peritaenia*).[39] The bridge is founded to the valley slope via small irregularly shaped rock stones mixed with concrete. It has a northwestern orientation. The arch has a span of 2.10m and its central part has collapsed. Each of the piers bears a 0.23m high projecting moulding (impost). Only the springing trapezoidal stones of the marble arches are preserved (two running blocks along the width of the arch, 0.47m in height). The pier's south section is preserved at a length of approximately 2.70m and the north section at a length of 3.00m. Thus, the pier's total length must have been at least 7.80m. The bridge width is 1.45m and projecting moulding (impost) width 1.50m. The bridge is preserved at a height of 2.40m. A diagonally mounted clay water pipe, whose diameter is 0.25m, is

[36] The filling up of a central conduit by secondary (*vamus*) and subsidiary (*ramus*) conduits was a common practice in Roman aqueducts, Mays 2010: 117, fig. 7.2.

[37] In the Turkish language 'Sutzak' stands for thermal spring.

[38] The conduit's outer wall width is 0.57 (+ 0.03m mortar thickness). The inner wall width is 0.55 – 0.60m, while the conduit width is 0.50 – 0.51m. A conducted test dig ascertained that the preserved height of the conduit is 0.50m.

[39] The peritaenia's width is 0.05m.

Figure 6. The edge of the water bridge area of Vrysoudia (photo by the authors)

attached to the western bridge façade. It is lined with lime mortar and does not seem incorporated to the original construction.

The next visible part of the aqueduct is another bridge at the location known as Matses or Vrysoudia (Figure 6; Figure 7b; Map 1.10). The water channel is visible, carved on the rocky valley slope for several meters, both upstream and downstream of the bridge.[40] Despite the fact that this bridge is not as well preserved as the others, one can still detect its basic features; however, its restoration requires further research. Only a few lateral pier blocks and parts of the cast-stone masonry used to set the bridge on the valley slope are preserved.[41] Numerous rectangular blocks can be found in the ravine and are thought to be voussoirs or pier blocks. The bridge coupling span width is 21.50m at the bottom of the ravine and 54.50m at the water channel height. At the centre of the valley there are traces of a central pier, 2.50m wide. On each side of the latter there are traces of two more piers with a 4.05m distance between them. The bridge width at the water channel height is 2.20m.

Given the existed traces to the south of the valley, it seems possible to restore another half-pier, 1.45m wide, on each of the sides (the distance of the half-piers from one another is also 4.05m). Consequently, the Vrysoudia bridge is similar to the four arched bridge of Paspalas, although it remains unclear how the end arches are connected to the 'wings'. Based on measurements, the channel ground is located at an altitude of +20.50m (north) and +20.40m (south) from the valley level or + 71,402 m above the sea.[42]

[40] Downstream of the bridge, the carved channel has a width of 0.55m.

[41] Two typical features of the bridge are the embedded clay pipes (max. diameter: 0.28m) deploying on three rows, and the stone-built rectangular pipes (0.43 X 0.23m); in both cases, they were probably used to drain rain water and to support the scaffolding while the bridge was being built.

[42] Eleutheriades 2010: 48, figs. 34, 42 and 43.

Further, the water channel is detected, carved on the vertical calcareous mass known as Larissees Petres (Map 1.11). This *c.*1000m long part is an impressive section of the aqueduct, evidently the most difficult and demanding to construct since the channel has been carved on the hard crystalline limestone at an altitude of approximately 78m from the sea level. The channel is oriented E-W. It has a width of 0.60m, while its rectangular section height varies between 0.80 – 0.85m.[43] The conduit's total height, including the vaulted crowning, is 1.15m. The inner wall is mostly stone-carved; the outer wall is either stone-carved supplemented with the built-masonry technique, or exclusively constructed by built-masonry technique. Although the channel has been destroyed, mostly by earthquakes or rock falls, its trace is still detectable.[44]

After passing the contemporary aggregate quarry, the 830m long conduit, continues its course, almost intact, either masonry-built or carved on the rocky slope (Map 1.12) until the next bridge at Kamaroudia (Map 1.13) (+60,331m). This section of the conduit has already been mapped with a fixed topographic system which revealed the minimal inclination given to the conduit by its constructors. The bridge at Kamaroudia is oriented NW-NE, its width varies between 2.15-2.10m, while the channel width remains 0.60m. The maximum preserved height of the bridge is 5.70m. It is supported by two plain arches whose span is 2.95m. A blind, superimposed arch is incorporated in the masonry indicating restorations done at an unspecified time period. Only one of the original two arches is preserved, the façade of which has been altered due to the loss of its external stones (voussoirs).[45]

Passing the Kamaroudia area the water conduit turns eastwards. A new section of it has been recently located and brought to light during road widening works at the Prineri Katifori location (Map 1.14). This newly-revealed, 65m long, section, oriented NW-SE, presents the same construction features with the so far known conduit sections. Its gradient is interesting since its westernmost point lies at an altitude of + 60.31m, while the easternmost point lies at an altitude of +60.16m. This reveals the diminutive required inclination for a smooth water flow.[46] The conduit was unearthed at a depth of -0.50 to -1.00m from the modern ground level. Its crown consists of small and medium size rubble stones and slabs, loosely connected without binder material. At this point, it is worth noticing that the Prineri Katifori conduit has a characteristic, not yet detected or confirmed in other locations; at least three square shaped openings (0.75 X 0.70m) have been detected on its upper surface; they were roughly covered with irregular slabs and must have been used to inspect and clean the conduit. The conduit's interior is filled with an approximately 40cm thick layer of fine-grain sandy soil and small rubble stones that have fallen from the conduit's crowning.[47]

This conduit, although extensively destroyed by the modern road to its SE, continues running underground; it is detected both carved on the rocky slopes, and masonry-built along the torrent. This section must have been particularly crucial for work progress since the engineers had to cope with a large rocky mass aptly known as 'Fanos'. The selected solution was to lead the conduit underground using a masonry-built tunnel (Map 1.15).[48]

---

[43] These measurements correspond to the eastern 'end' of the conduit, just before the latter reaches the contemporary aggregate quarry.

[44] In the wider area (including the Mytilene-Kalloni main road), see Papakonstantinou *et al.* 2015: 12, fig. 3.

[45] Papakonstantinou *et al.* 2015: 12 with fig. 4. For the possible bridge reconstruction see Eleutheriades 2010: fig.73.

[46] Papakonstantinou *et al.* 2015: 12-13 with fig. 5.

[47] The openings (manholes) at the Prineri Katifori conduit are used to inspect the network, which in that area was surface built. An underground network requires masonry built shafts, such as the shafts of the Eifel aqueduct (Cologne), Gier (Lyon), Toumba and Thermi (Thessaloniki). Certain similarities to the conduit openings at Philippi are also detected, although those of the Mytilene conduit are simpler in construction, Kaiafa 2008:108, pl. XXII.

[48] Chatziioannou1985:158.The tunnel entrance was detected and cleared up after information provided by the architect Th. Eleutheriades while its exit remains unknown. Eleutheriades (2007:155) estimates the tunnel's length to be from 800m to 1km (Map 1.17) and its exit at a distance of approximately 40m from the location known as Tsesmes tis Batzakenas, to the north of the torrent. Cf. the Hadrianic aqueduct in Corinth that has three known tunnels, two of which are respectively 1.07km and 750m long. The tunnel mouth entrance dimensions are respectively 1.50 X 1.90m (Siouri tunnel) and 1.60 X 1.80m (Prathi tunnel), Lolos 1997: 289-290. Vertical shafts were used to remove excavation residues, to enable the simultaneous work of many teams, to provide air during rock removal and, upon completion of the works, to allow the inspection of the water conduit, the removal of alluvial deposits, as well as repair works following rock precipitation. Further information on Kaiafa 2008: 64-65. The term 'φρεάτια' is recorded in Xenophon (III, 1, 7) and Polybius (X, 28, 2). On the use of vertical shafts as tunneling method, Korres 2002:15; Hodge 1992: 127, fig. 78. Regarding various forms of vertical shaft at the Gier Aqueduct (Lyon), also Kaiafa 2008: pl. XXII d.

Figure 7a. The upper part of vertical shaft at the area of of Moria with the characteristic name 'Fanos' (photo by the authors)

Figure 7b. An example of entirely carved water channel in the limestone rock at the area of Vrysoudia.

Figure 7c. An example of the water channel built from freestones joined with lime mortar at the area of Sutzak.

Figure 7d. View of the upper vaulted not water-proof crowning of the water channel at the area of Kako Lagadi.

Figure 7e. An example of entirely carved water channel in the limestone rock at the area of Koutsuk Loutza.

At a distance of about 320m from the vaulted tunnel entrance, an impressive vertical shaft has been revealed and brought to light. (Figure 7a; Map 1.16) The altitude of this shaft is +95.268m. This indicates that in order to ensure a smooth inclination – in relation to the previously visible conduit section (at +60.16m) – the total height of this vertical shaft should be 37.79 m. estimating the relevant aptitude as well.[49] Currently, the shaft interior is visible at a depth of 6.62m. The remaining part is filled with earth, tree branches, etc. The masonry built walls are made of rubble stones of various sizes (lime stone, slate, etc.) and bound with a strong whitish lime mortar. The maximum overall dimensions of the shaft are 2.54 X 2.24m and internal dimensions 1.32 X 1.14m while the wall thickness varies between 0.46 and 0.78m.

It is actually one of the more vertical, masonry-built in natural soil, shafts – and probably, the deepest one. Such vertical shafts were necessary to provide natural lighting and fresh air (air shafts or ventilators) to the underground tunnel. Thus, they were known as 'fanoi', that is, light shafts or roof-lights. According to Vitruvius, the ideal distance between two shafts is one *actus*, i.e. approximately 35m – however, this distance was not always respected by the engineers.[50]

The inner SE side of the shaft is endowed with three, still discernible, superimposed, 0.09m thick, horizontal groves. The latter were, in all probability, used to fix wooden beams in order to reinforce the walls or facilitate access to the shaft bottom.[51]

From that point the conduit continues underground for approximately 2.685m. We have not yet been able to trace its course towards the area of Moria (Figure 8; Map 1.18). From there, it continues towards the city of Mytilene set upon the 170m long (at the water channel's height level), homonymous bridge whose maximum preserved height at the centre of the valley is 24.46m. This constitutes the most important section of the Roman aqueduct and potentially the greatest rural Lesbian monument (Figure 8).[52] The bridge, according to Koldewey's restoration, consists of 17 arched openings with three superimposed arcades at its central part, all supported by sixteen piers. [53] Compact, cast-stone masonry structures, exactly as in the previously discussed bridges, were used to transfer the conduit from one valley end to the other.[54] The central piers are today preserved in a rather good condition (+58,24 m above the sea). Regarding the end piers, only their lower courses are preserved.[55] The bridge of Moria is not included in the biggest and more impressive Roman water bridges.[56] It is however one of the most beautiful ones, due to its exceptional architectural design that transformed a large scale utilitarian structure into a monument of unique architectonic design. According to Ioannis Kontis, 'the structure resembled the façade of a building with three superimposed galleries'. A comparison of this bridge's architectural form with that of the other bridges will be presented in the following paragraphs. Few of its characteristic measurements are as follows: the distance between the piers varies between 6.85 and 9.38m. Their length (along the bridge axis) varies between 2.30 and 2.45m. Similarly, their width varies between 3.25 and 4.35m.[57]

[49] According to Eleutheriades (2007: 155), this shaft was sealed with a concrete slab in 1990 by the then Mayor of the Moria Commune in order to prevent the disposal of dead animals and other waste within. On vertical shafts, Kaiafa 2008: 68-69, note 54.

[50] Kaiafa 2008: 68-69, note 54. According to Pliny (*Historia Naturalis*, XXXI, 3.1), the most appropriate distance is 2 *actus*, i.e. 75m. Further on the subject, Hodge 1992: 415-416; Lewis 1999:151; Hodge 2000: 57. Three vertical inspection manholes (*putei*) have been discovered in the Corinth aqueduct, with a varying distance between them *c.* 30-40m, Lolos 1997: 289-290. Cf. also Lolos' paper in this volume.

[51] A 0.70m wide wall oriented NE-SW was discovered at a small distance to the west. It is constructed of small and medium size rubble stones (slate, limestone) and mud as binder material. Its preserved length is 5.90m. It is suggested that it was built at the same time as the shaft and consisted an additional protection wall or part of the shaft's outer wall.

[52] Kontis 1973b: 77-81; Koldewey 1890: 66-67, pl. 29.1.

[53] The piers are numbered from NW to SE. The rural road passes between piers 3 and 4.

[54] The conduit's support base is also preserved. Its width, mainly at the SE end of the valley, is 1.40m. Measurements were taken at the ground level of the conduit's stone built supporting wall.

[55] Only 12 of the original 16 piers are still preserved, along with the bases of three more. Today, one stone arch of the lower arcade is preserved and traces of a second one exist. Protruding blocks are visible at the lowest level of the inner sides of the piers; the use of them unfortunately is quite difficult to explain because they are not homogenous. They might have been used either to support the scaffoldings (and were never removed) or to accurately calculate the height of various construction elements. The second arcade consisted of 13 stone arches, only 8 of which are currently preserved. Koldewey's survey, 100 years ago, recorded 9 arches. The third arcade, which hosted the water conduit, is made of bricks. Regarding its upper part, the sides are made of small stones while the inner walls are built of rubble stones and binder. According to Koldewey (1890: 66, pl. 29.1), of the 15 or 17 original arches, only two and the foundations of a third are currently preserved; on his drawings, four preserved arches are recorded, plus the foundations of a fifth one. Also, Kayer 1995: 4.

[56] The biggest and one of the most imposing Roman bridges is the *Pont du Gard* (275m long, 48.4m high) in France, cf. Mays 2010: 118-119.

[57] The distance between the pier axes vary between 9.36 and 9.60m, while the distance between piers varies between 2.36 and 2.40m (Kayer 1995: 5).

Figure 8. The monumental multiple arch bridge of Moria.

The arcade that supports the water conduit is made of bricks and cast-stone masonry. Few traces of the conduit floor are still visible at the central part of the upper structure. The conduit continues its course eastwards, underground, for 1560m (Figure 7d; Map 1.19). It re-surfaces inside the Roman quarry at the location of Kako Lagadi. At that point, the conduit is visible for more than 100m, oriented towards E-W, preserved in good condition and with a clear architectural form.[58]

During the final section of the water's course towards the city of Mytilene, the conduit goes yet again underground in order to acquire the appropriate gradient. It surfaces again, carved on limestone, at the location Koutsuk Loutza (or, Kourzi Thermal Springs) and is visible for approximately 150m.[59] The construction of the water conduit follows the terrain's relief and it's typologically similar to the conduit at 'Larissees Petres', i.e. combined masonry/carved sections and, occasionally, exclusively carved sections. According to measurements, the conduits inner width is 0.65m, while the masonry-built outer wall width varies between 0.50 and 0.70m.[60] The very impressive stone-carved section has an inner width of 0.53m.[61] The external carved surface is 1.00m high and 0.60m wide. Its construction required carving the rock's 'inner' wall at points reaching a height of almost four meters (Figure 7e).[62]

[58] Papakonstantinou *et al.* 2015: 13, fig. 7.

[59] It is highly probable that a small bridge would have existed (and should be searched for) to enable the conduit passage over the previously mentioned torrent.

[60] The outer wall dimensions vary (cf. 0.90 – 1.40m) depending on the increased bracing requirements of the conduit. Hydraulic mortar was traced throughout the conduit course.

[61] The conduit's bottom is not rectilinear during its entire width. At the junction with the lateral walls, the conduit is carved slightly higher (width ranging between 0.05 and 0.06m and height 0.01m) probably to facilitate hydraulic mortar adhesion.

[62] The fully rock-carved section of the conduit was followed by a section of combined masonry and carved conduit that probably collapsed into the valley after an earthquake. Similarly, rocks falling from the vertical limestone rock mass destroyed part of the conduit at Larissees Petres.

The homonymous bridge (+44,017m, Map 1.20) spans the rough valley supported on a central, 2.40m long and 2.55m wide (at its axis) central pier and two half-piers at each end that support two arches. The central pier is decorated with an impost, like those at the Moria piers. According to Koldewey's restoration, the water conduit is situated at a height of 12.60m. The bridge width, at the conduit floor level, is 1.95m – of which 0.60m corresponds to the conduit width and approximately 0.66m corresponds to each lateral wall width. This construction is attached to the valley slopes via courses of cast masonry, whose preserved height at the northern side is 3.50m. Small square-shaped through-holes (putlogs) are visible on its façades and must have been used to support the scaffoldings during construction. The conduit exits the bridge and continues its course southwards, carved in stone along a couple of meters until again it gets disappeared.[63]

According to Koldewey, the terminal section of the aqueduct's course was *intra muros*, under the ancient theatre's so called Teke.[64] Further investigations are however required in order to accurately calculate the position of the water collection and distribution tank (*castellum divisorium*) (Map 1.21) that supplied the baths, tanks and fountains, as well as the wealthy private houses of the city with running water.[65] The majority of Roman villas have been located on the northeast slopes of the ancient theatre hill, at a privileged location that offered excellent view to the East (ancient city, Euripos, North harbour, etc.) and to the Asia Minor coast.[66]

Architectural description of the bridges: the Moria and Paspalas examples

all the aqueduct bridges present some common construction characteristics.[67] The architect opted for simple, single arched bridges at the locations Anerayda, Kousteri and Paspalas, double arched bridges at Kamaroudia and Koutsouk Loutsa, four arched bridges at Paspalas and possibly at Vrysouda and Thyridia and for a multiple arched bridge at Moria.[68]

At the Paspalas bridge, the architect designed and implemented a technically challenging solution consisting of uniform, high arches supported by strong marble piers built according to the pseudo-isodomic system (Figure 4). The outcome is a 24.95m high structure. Each block has a recessed perimeter strip (*peritaenia*) on its outer surface and a boss at its centre.[69] The blocks are laid without mortar and connected with joints. The piers' interior is filled with small stones and tiles connected with strong lime mortar (*opus caementicium*), according to the Roman 'emplekton' system.[70]

Along their height the piers are decorated with imposts, which improve the structure's aesthetics. The marble arches are not free-standing, as in the case of the Moria bridge, and they are built with smaller, rectangular voussoirs.[71] Each arch is mounted directly on the piers and imposts are used to distinguish it from the rest of the structure. The inter-arch space (spandrel) is filled with small rubble stones bound

---

[63] Koldewey 1890: 66, pl. 29.2.

[64] Koldewey 1890: 65, pl. 1 and 2. Cf. the examples of the central tank at Dion (Kaiafa 2008:181, with literature, pl. CXIX), in Pompeii (Hodge 1996: 261-276), as well as at the city of Apamia from the late 4th century to the late 5th century AD (Haut and Viviers 2010: 139-169). Regarding the construction of the *castellum divisorium* at the edge of the city, Hodge 1992: 279-291.

[65] Cf. Vitruvius, *De Architectura*, XIII, 6, 1-2. Also, Kaiafa 2008: 68; note 50. Recent excavations at the Euthymia Syreli's private plot enhance the view that the central tank was located close to the ancient theatre. These excavations have revealed a square-shaped building of the Hellenistic period mounted on the adequately carved natural rock. This building was demolished in the Roman period and replaced by a system of four cylindrical, clay water pipes (Pipes 1, 2, 6 and 3) consisting of interconnected clay pipes, whose diameter was 0.25m. These central pipes were turned towards the city and were evenly mounted and supported on a substrate of clay tiles and hydraulic mortar.

[66] Kyriakopoulou 2013: 18-21. The fact that their construction is dated between the 2nd and the 4th century AD may consist a helpful indication in dating the aqueduct.

[67] For general information regarding bridges, see O' Connor 1993:150-162.

[68] Koldewey 1890: pl. 29.

[69] Indicatively, the *peritaeniae* thickness was measured horizontally as 0.09 – 0.11m and vertically as 0.01 – 0.10m.

[70] This is a common feature of all Mytilene aqueduct bridges (Vrysoudia, Koutsouk Loutsa). The half-piers of the bridges of Anerayda, Kousteri, Vrysoudia, Paspala and Koutsouk Loutsa are also built using the same system, i.e. with stone blocks at the exterior and cast masonry (opus caementicium) at the interior. The lower courses of the Paspala half-piers are lined with marble for a length of 2.04 to 3.04m.

[71] The lower courses consist of four voussoirs, while the upper courses consist of five (or six) and of smaller size voussoirs. The voussoir dimensions -as deducted from a voussoir having fallen from a collapsed arch- are 0.70 (length) X 0.27 (width) X 0.53m (height).

together with strong lime mortar (opus caementicium), again according to the 'emplekton' system, faced with *opus incertum*.[72]

The bridge that stands out -compared to the other bridges- for its excellence is the bridge of Moria. As a result of an original architectonic solution chosen by its engineer, this work is unique in terms of monumentality and advanced structure configuration.[73] While other water bridges consist of superimposed arcades that produce a bulky result (i.e. the bridges of Aqua Anio Vetus, Aqua Claudia, Pont du Gard, Segovia, Tarragona, etc), the Moria bridge has been implemented as a single carrier structure, with piers connected and supporting each other thanks to three rows of superimposing arches (Figure 8).[74]

This is achieved by the fact that the central pier consists the highest point of the structure (25.46m to the conduit level),[75] while the lateral piers are shorter and gradually adapt to the terrain's relief, thus giving the impression that the bridge slowly 'dives' towards the valley slopes. The architect maintained uniform and at the same height all the structural and decorative elements (arches, imposts), while he modified the dimensions of each piers' base dimensions.[76]

The final elegant form is the result of a series of 'optical refinements', that can briefly be described as follows: The piers have a truncated wedge-like shape. Their longitudinal dimensions are fixed while their transversal dimensions are constantly degrading.[77] The central piers are decorated with three imposts in succession, consisting of an oblique moulding and an abacus.[78] The impression is that each section of the pier 'emerges' from the impost. Moreover, the first two-marble arches are of a smaller width than the piers so that the vertical elements can rise uninterrupted to their full height. The piers are built using the pseudo-isodomic system with large dry laid blocks on the outer sides without visible signs of binding or joints.

Only the block tops are joined together with Π- shaped clamps and lead injections.[79] All four sides of the pier present square grooves used to vertically join the blocks.[80] Certain blocks, especially those of the arches, have rectangular grooves, most probably for the use of 'lewis' lifting devise.[81] In addition, each block has a central boss and an outer recessed perimeter strip (*peritaenia*). It is worth mentioning that boss carving presents various forms, from blocks having a rough quarry surface, to blocks with a rather rustic aspect (rustica), or to a more elaborated side.[82] The grey-blue marble with the golden shades help so that the construction smoothly integrates to the beautiful olive-grove Lesbian landscape.

The arches of the middle arcade are of a smaller width than those of the piers. These arches consist of trapezoidal voussoirs, whose dimensions are 0.70-0.75m (height), 0.80-0.90m (width) and 0.60-0.70m (bottom side). The external side and bottom arch block surfaces follow the same configuration as that of the pier blocks (*peritaenia* and boss). The centre of each of the semi-circular arch follows the virtual horizontal line of the base impost; its radius varies between 3.42 and 3.58m. The arch width is 2.14m. Each arch consists of 1-2 arch-stones of variable dimensions (interwoven joints).

---

[72] The external façades of the spandrel are lined with 5 rows of medium size blocks, which are not decorated with *peritaeniae*. More on Roman concrete (*opus caementicium*), Lamprecht 1987: 69-101; Kaiafa 2008: 183, note 185.

[73] Hodge 1992: 129-147, mainly 130-131 and 142.

[74] Hodge 1992, with relevant bibliography.

[75] At least 1.15m should be added in order to restore the bridge to its full height.

[76] Kontis 1973b: 77-81 (regarding the different pier base dimensions according each time to the ground elevation). Further on Kayer 1995: 5.

[77] The central pier transversal base dimension is 4.20m on the ground, and 2.93m at the upper arch base, Koldewey 1890: 67.

[78] Total height: 0.40cm. The molding protrudes from the pier by 0.30cm.

[79] Clamp (joint) dimensions: 0.22 X 0.03 X 0.03m.

[80] Groove dimensions: 0.065 X 0.065m.

[81] Lewis socket dimensions: 0.125 X 0.04 X 0.08m. Kayer 1995: 6. The lewis lifting system was also used for the piers at the Paspala bridge (socket dimensions: 0.075 X 0.025 X 0.10m).

[82] Despite the fact that the bridge is intensely monumental, neither the blocks, nor the boss, or the *peritaeniae* quality are homogenous. This is an indication that despite the project's initial design for the bridge and the block forms, no further care has been given – as we believe – nor any supervision was organized to ensure that the quarry provided the best possible construction materials or that block carving was performed according to strict specifications. Moreover, our research revealed that lower course blocks have a *peritaenia* only on their three sides (the lower horizontal side and the two vertical ones).

The arches are connected to the piers as follows: the first (two or three) blocks (springers) are part of the pier blocks; the third (or fourth) arch block sinks into the pier core. Thus, the remaining arch blocks consisting of smaller stones are firmly seated. This method transfers the arch loads directly to the pier core without the arch base blocks to be shear stressed.

The blocks of the following row have a triangular section that penetrates into the pier body. They are used to prevent the first row arch blocks from shifting upwards in case of arch deformation.

The arches of the first and second arcade are semi-circular, while the upper arcade's arches are made of square bricks, whose dimensions are 0.70m (length) X 0.50m (width) X 0.06m (thickness). They are bound together with strong lime mortar and joints 0.02 – 0.03m thick. At regular intervals (approximately every fifteen brick courses) there is an interposed course of stone slabs of the same thickness (two to three placed along the arch width). The transverse diameter of the arches is 2.30m, their centre is to be found at the impost's height and their radius is approximately 3.80m.[83]

At the base of the sixth northwards pier there is a small arch, possibly used as an artificial by-pass bed during construction.[84]

A common characteristic of the cast-stone masonry is the existence of rectangular openings (putlogs) for mounting the scaffolding and, possibly, for draining the walls[85]. Moreover, the triangular sockets at the inner pier façades over the imposts identified both at Paspala and at Moria bridges are thought to have been used also to fix the scaffoldings.

Description of the main water conduit

the water conduit presents certain general characteristics that according to the transit area and its morphological and geological particularities may be locally formatted (Figure 7 c, d). In general, it consists of a water-proof rectangular section 0.60 to 0.65m wide and 0.85 to 0.90m high. Its inner walls are covered with a 0.02-0.05m thick mortar lined with a very thin, off-white, hydraulic coating.[86] The conduit's vaulted crowning, approximately 0.25m high, is built of rubble stones joined with lime without being water-tight.[87] The section that deviates largely from this general description is that of Larissees Petres, where, according to measurements, the conduit's inner width varies from 0.45 to 0.60m. A similar differentiation is found at the outer walls of the conduit; while for the major part of the route the conduit walls are 0.40m wide, at Larissees Petres the outer wall is, at certain points, 0.82m in width, thus revealing the attempts made to strengthen the structure. Correspondingly, the inner wall is not uniformly built; it consists either of masonry-built wall of normal width, or a wall only 0.12m wide, or an entirely stone carved wall with a minor adjustment for the conduit arch fitting.[88]

Although very few traces of the water conduit are preserved upon the bridges, we have discovered that, for example, in Kamaroudia, the conduit width was 0.60m and each wall's width was 0.75m, that is, identical to the bridge width at the conduit base level.

[83] Kayer 1995: 6-7.

[84] Kayer 1995: 7.

[85] Rectangular sockets, some of which are also through and through, have been identified mainly at the bridges of Anerayda, Thiridia, Paspalas and Vrysoudia.

[86] Bouras (1968: 284) described that Romans started adding ceramic powder (*opus signinum*) in mortars during the Imperial period. More on how water conduits were built, Hodge 1992: 93-105.

[87] Cf. Vitruvius, *De Architectura*, II, 4, 3; II, 5, 1; VIII, 6, 14,regarding the composition of hydraulic mortar, and Lefas 1997: 167-168 for comments upon these passages. Also, cf. the summary from Kaiafa 2008: 175, 180 (on examples of vaulted conduits from Macedonia and their dimensions). On Hadrian's Aqueduct in Corinth, Lolos 1997: 281-283, fig. 2. Also Lolos' paper in this volume.

[88] The water conduit of the Mytilene aqueduct presents similarities with those of Dion (Handrian's reign, 2nd century AD), Philippi (Antonine Dynasty, 2nd century AD), Edessa (The Tetrarchy, late 3rd century AD) and Thessaloniki (The Toumba/Thermi conduit), Kaiafa 2008: 173, drawing 3, 178-180 and Volume I, plates CXIV a-f, CXV a-f, CXVI, a and c. The Amphipolis water conduit is similar in form to that of Mytilene but it is brick crowned. Regarding its crowning, the Mytilene water conduit is comparable to the Thessaloniki (Toumba) one. However, the examination of several parts of the outer walls has revealed that although the water conduit was built as a compact structure, its rectangular and vaulted sections are clearly separated by a thin horizontal line.

Especially regarding the newly-discovered conduit section at the 'Prineri Katifori' location, as it has been measured, the wall width varies between 0.75 and 0.84m, the conduit's internal height is 0.85m, while the external height is 1.16m from the base level to the arch top. The floor is 0.05m thick. Similar measurements are gathered for other adequately preserved sections of the conduit at the locations Kako Lagadi, Sutzak and for the section upstream of the bridge, at Kamaroudia.

The conduit walls, at certain points along its course, are entirely carved in the limestone rock (Figure 7 b, e). Typical examples are the sections upstream of Kamaroudia and Vrysoudia, as well as the sections at the rocky areas of Kourtzi Thermal Springs. The water flowed directly through the water-tight rectangular section of the conduit.[89] Considerable care was also given to the preservation of the conduit's lateral walls static stability, as well as to the drainage of rain water. The latter was achieved with the help of superimposed discharging square holes (cf. the conduit downstream of Anerayda bridge, Soutzak, etc).

Finally, no carbonaceous and calcareous sediments (sinter) were detected on the conduit's inner walls, apart from some minuscule traces. On the contrary, stalagmites sediments were detected at the vaulted crowning of the conduit that crosses the quarry at the location Kako Langadi.[90]

Secondary – peripheral conduits

Recent improvement works at the Thermi Pigi – Lambou Myloi country road have brought to light, in the middle of a rocky hill slope, part of an also ancient peripheral (secondary) water supply network, which is thought to have supplied water from local sources to the Lambou Myloi area. The unearthed network section is 145m long and it consists of a clay pipe, whose diameter is 0.28m, running downhill, N to S, at an altitude varying between +78.33 to +77.70m. This conduit supplied the main network with water.[91] The unearthed section consists of a water conduit of a diversified construction (either masonry-built or clay, or a combination of both) which was locally set within a carved rock and was supplemented by a relief shaft. Based on the above, it seems that this water network lasted a long period of time, either having different operation phases, or perhaps, having two networks operating simultaneously.

At the end, at a distance of 2km N-E of the above mentioned structure, near the village of Pigi and at an altitude between +109.94 and 109.72m, researchers have recently revealed a masonry-built Roman conduit almost similar in construction to that of the aqueduct, length of 70,14m, internal width 0,35m and external width is 0,93m.[92] However, this new conduit has rectangular shape, is narrower than the central one and does not have a vaulted crowning. It is probably part of another peripheral (subsidiary) aqueduct that needs to be further studied. Note also that at the same location – upwards of the water conduit – a farm complex (*villa rustica*) of the 6th century AD was found.

Dating of the aqueduct

no direct record exists in the ancient texts regarding the aqueducts' construction, or its architect.[93] Therefore, its general dating to the Roman period is based on external criteria.[94] Various scholars

[89] Koldewey (1890: 66) measured the conduit's width at Kutzuk Loutsa to be 0.35 -0.64m and its height, 0.65 – 0.80m. However, his observation that the Larissees Petres conduit had the same width was not verified since (according to recent measurements, its width is 0.45 – 0.60m. Furthermore, at two conduit passage locations (Larissees Petres and Sutzak) researchers have identified scattered fragments of cylindrical clay pipes, 0.035 meters thick. Their origin remains for the moment unidentified.

[90] Sediment formation depends on water hardness, which forms a layer of calcium carbonate (or, lime carbonate, CaCO3). Regarding sediments and their impact on water flow, see Lolos 1997: 296 and Hodge 1992: 227-232 with examples.

[91] The unearthed findings require further study in order to clarify the conduit's operation and usage -potentially as a smaller, peripheral (subsidiary) aqueduct.

[92] The conduits orientation is N-E given its gradient; the conduit is directed towards Lambou Myloi. Compared to the preceding section of the main conduit (Sutzak, + 85,916m) and to its next section at Kousteri (+ 76,764m), there is an elevation difference of + 23.914m and + 33,066m correspondingly.

[93] Chatziioannou's (1985: 154, note 2) reference that 'The Moria arches bear the incised inscription "ΔΑΜΟΣ" indicating that the bridge was built with the city's active support', has not been confirmed so far. We believe that this misunderstanding originates from the work of St. Anagnostou (1850. *Lesbian Ode or Historical Tribute to the Island of Lesbos*, 1st Edition, Ismir, Mytilene/3rd Edition 2000, Petras Publishing House [ed. N. Christopoulos], 128). It was further perpetuated by T. Economou (1909[2], *Concise History and Topography of Lesbos*. Cairo: 56, § 4) and uncritically reproduced by subsequent authors.

[94] On hydraulic technology during the Roman period, Kaiafa 2008: 67 note 44, with extensive bibliography.

have attempted to connect it to Marcus Vipsanius Agrippa, the great engineer and Augustus' son in law, who lived in Mytilene between 22 and 14 BC and was honoured by the city.[95] Ioannis Kontis took into consideration the classicist style of the Moria arcades and the fact that the use of the so-called Roman 'emplekton' (*opus incertum*) was introduced on Lesbos at the beginning of the 1st century AD.[96] He then associated them with the first use of the *rustica* technique at Hadrian's Library[97] and secondly the occurrence of the Great Earthquake that hit the Aegean Islands in the middle 2nd century AD.[98] Based on the above, he attributed the construction of the aqueduct to the period of Hadrian's Reign – though not excluding the possibility of it being built during the late 2nd or early 3rd century AD.[99] In accordance to the above and based on epigraphical evidence, Hadrian was repeatedly honoured by the Lesbian authorities for his benefactions to the island, as Liberator, Olympian, Saviour and City Builder.[100]

The Mytilene aqueduct presents morphological similarities with two other ancient aqueducts, the Side aqueduct, dated also to the middle 2nd century AD,[101] and the aqueduct of Pisidian Antioch.[102]

The technical solution of supporting the water conduit upon high piers connected with arches, either free standing or not, has also been implemented in the aqueducts of Cherchell, Qued Ilelouine[103] and Qued Bellah (Algeria),[104] in the aqueduct fed by the Proserpina Spring in Augusta Emerita (mod. Merida, Spain), as well as in the famous Los Milagros bridge, also dated at the 2nd century AD.[105]

Lastly, the main conduit's form clearly resembles that of other aqueducts of the 2nd century AD.[106]

Based on our current knowledge, the operation period, and/or the pause, of the aqueduct are not specifically defined. Certain features, however, such as the completion of the upper arcade of the Moria

[95] On Marcus Agrippa's contribution to the water supply of Rome, Evans 1982: 401-411. Koldewey (1890: 67) reports the existence of grooves on the 8th pier from the left, which he interprets as traces for the support an embedded inscription. Both Kayer's and our own research have discovered carvings on one of the blocks of the 16th pier. These carvings may be associated with a modern era inscription; in any case they can, under no circumstances, be part of the monument's votive inscription. In addition, illegal *graffiti* of recent periods were discovered on one of the 14th pier's blocks. See, Chatziioannou 1985:156 and Dimopoulou-Pilliouni 2015: 568 note 115, on the connection of Agrippa with Mytilene.

[96] A typical feature of the classicist style is the vertical, smooth joint at the pier ends. These smooth bands were used as guides to ensure the vertical centering of the edge, a technique which appeared in defense walls since the end of the 5th century BC and in used throughout the Hellenistic period (Orlandos 1958: 221, 258, 259 with relevant examples).

[97] Bouras 1968: 285.

[98] Not all scholars agree upon the date of the 'Great Earthquake': AD 151 (Kontis 1978: 201, 229), AD 178 (Dimopoulou-Piliouni 2015: 569) and AD 148/9 (Papadopoulos 2015: 30-33). In Aelius Aristides (*Sacred Tales* 49.38-42) there is a description of a catastrophic earthquake, when Albus was proconsul of Asia (cf. SEG 4.533). Albus' rule is placed during the period of Antoninus Pius and H. Mason in his public lecture (*The aqueduct of Moria, Lesvos*, 2002; accessed from https://www.academia.edu/12001076/The_Aqueduct_at_Moria_Lesvos) suggested that the earthquake occurred in AD 147/8. He also suggests a date for the aqueduct after the earthquake, by attributing its construction to M. Pompeius Macrinus, also known as the 'New Theophanes', son of Theophanes, friend and historian of Pompey (cf. IG XII,2.235; SEG 29:740,1 and 2). Macrinus and his daughter, priestess Pompeia Agripinilla, were repeatedly honored by the Mytileneans as Benefactors (cf. IG XII,2 236 and 237; SEG 29.271). Mason proceeded on dating the aqueduct after AD 150, due to the title *hypaticos* (consular) used for the father of Cornelia Cethegilla (daughter of Pompeia Agripinilla) as recorded in IG XII, 2.237. Macrinus and his family are associated with the Lesbos aqueduct, via another source: according to an epigram (cf. IG XII.2.129 and 129[1]), a fountain, called 'Plataneis', has been dedicated to a certain Orphitos, identified by H. Mason as Gavius Orphitus, consul in AD 165 and brother-in-low of Pompeia Agripinilla. H. Mason believes that it was not a plain water structure but a monumental 2nd century AD Nymphaeum supplied by the new aqueduct. Regarding M. Pompeius Macrinus, see also Dimopoulou-Piliouni 2015: 583-590.

[99] Kontis 1973a: 77-81. Labarre (1996: 214-216) believed that the construction of the aqueduct was related to the Roman bath complexes (thermae) of Mytilene. This assumption led him to date the aqueduct at the beginning of the 2nd century AD. Indeed, according to archaeological evidence, large bath complexes (thermae) were in operation on Lesbos as early as the 2nd century AD. Regarding the connection between bath complexes and aqueducts, see Wilson 1996: 16.

[100] There is no agreement in the scholarly literature regarding whether Hadrian has visited Mytilene: some scholars are against that view (Birley 1997: 38, 164, 168, 319, note 7), while others believe that the Philhellene Emperor visited Lesbos during his first trip to Greece in the summer or autumn of AD 124 (Boatwright 2002: 31 note 72; Dimopoulou-Piliouni 2015: 567-574 note114). Irrespective of whether Hadrian has visited Lesbos (or not), the discovery of no less than 25 votive inscriptions is sufficient evidence that he was a benefactor of the city of Mytilene having financed a number of public works. On Hadrian's trips to Greece, Lolos 1997: 294 note 48 with a list of sources and bibliography.

[101] Grewe 1994:192-203; Izmirligir 1979. Their similarities are attested at the level of pier and arch configuration.

[102] The similarities between the Pisidian Antioch aqueduct and the Mytilene aqueduct are identified in the conduit's shape, the architectural form of the water bridges and the row of arches. In addition, blocks surfaces follow the same configuration with *peritaenia* and boss. In contrast, the piers' imposts in both aqueducts present slight differences. Yet, the researchers of the Pisidian Antioch aqueduct date its initial construction at the early 1st century AD (during Augustus' reign), along with a contemporary nymphaeum, associating its construction with the urban development of the city; it renovation (mostly at the arcades and piers) is placed at the 4th century AD, Burdy *et al.* 1998: 175-200. Having said the above, one could indeed support a date of the Pisidian Antioch aqueduct into the 2nd century AD, based on the dates of the other aqueducts previously discussed in this study.

[103] Hodge 1992: figs. 97, 98b.

[104] Leveau and Paillet: 1976. The aqueduct dates to the period of Augustus, but was repaired and restored during the 2nd century AD.

[105] Mays 2010: 130-132; Hodge 1992: 80-90, fig. 117; Grewe 1993.

[106] Also above, note 87.

bridge with bricks (*opus testaceum*),[107] are thought to have been part of subsequent additions or repairs that enabled the use of this technical work for many centuries.[108] A fact supporting this assumption is the different bearing method of the arches of the 8th, 9th and 11th pier, compared to those of the 10th pier. While the brick arches of the 8th, 9th and 11th piers are set upon one row of marble blocks, the arch of the 10th pier is set upon two layers of marble blocks;[109] this could be an indication that the upper arcade may have originally been stone-built and was rebuilt of bricks (perhaps after having collapsed due to a natural disaster).[110]

Repair works are also visible at the bridges of Anerayda and Kutzuk Loutsa.[111] In the case of the latter, the engineer has also employed rough stone blocks without recessed perimeter strips (*peritaeniae*) for the central pier, an indication of either reduced financial means or emergency repair works.[112] In addition, the bridges at Moria and Paspala are endowed with strong retaining, cast-stone masonry walls between some of the piers, indicative of the effort to retain the latter.[113] Lastly, the Kamaroudia bridge, whose form differs from the others, may have been extensively repaired at a later period.

It is also worth mentioning that the aqueduct's section crossing the quarry at Kako Langadi has been built from rubble stones and abundant gravel (Figure 7d). This is an indication either that the aqueduct was built after the quarry or that they existed simultaneously for some time in late antiquity. Mining fields and rough construction material are still identifiable on both sides of the masonry-built conduit. The height of the conduit's external walls (including the foundations) reaches 4m at several points, thus indicating that care was taken to reinforce the structure and to ensure that it was firmly founded in the loose gravel substrate.[114]

Given all the above, we believe that the Roman aqueduct of Mytilene must have been built at the earliest during the second half of the 2nd century AD and, certainly, after the earthquake of the mid-2nd century AD.

Acknowledgements

We would like to warmly thank the Director of the Ephorate of Antiquities of Lesbos Island, Dr Pavlos Triandafyllides for facilitating our research work. Thanks also to our work group member, the Geographer, Dr Apostolos Papakonstantinou and our colleague, Mrs. Thalia Kyriakopoulou. We are also grateful to the Architect Themis Eleutheriades, for information regarding the 'Fanos' and the tunnel; to Panagiotis Chlimbos, for his invaluable contribution and Panagiotis Koutskoudis, for his assistance in locating the springs. Lastly, we sincerely thank Dr Georgia Aristodemou, the Chief Editor of this volume, for contributing to the improvement of the English text and, Professor Emeritus Theodosios Tassios for his meticulous proofreading.

[107] On the construction of brick arches, see Bouras 1968: 380-382.

[108] Koldewey 1890: 66; Kontis 1973a: 120; Kontis 1973b: 79.

[109] In addition, all blocks used to support the brick arches are poorly carved and their dimensions differ. These characteristics do not relate to the architectural form of the construction. It is, obviously, rather difficult to place this type of repair in a specific moment. We believe that brick arches of similar form are also found in Basilica B, at Philippi. Basilica B dates at *c.* AD 550 (Koukouli-Chrysanthaki and Bakirzis 1997). They are also met in the arcade of the Selçuk aqueduct (Ephesus), which is believed to have been also built at the 6th century AD by Emperor Justinian (Foss 1997: 92 note 96 with relative bibliography).

[110] Equally interesting are the studies regarding the city of Argos and, generally, the Peloponnese, especially the attempt to date brick structures based on the thickness of the mortar used to bind the bricks. For more information on this subject, see Lolos 1997: 287 note 32. Unfortunately, no such research has been undertaken on Mytilene.

[111] Koldewey 1890: 67.

[112] The completion of the narrow marble block of the 5th course of the west pier at Anerayda bridge with a small square marble wedge is, nevertheless, not part of subsequent interventions, but a solution to cope with material failure during construction. Similar marble wedges are also to be found at the Moria bridge.

[113] A characteristic example is the retaining wall between piers 5 and 6 at Moria bridge, whose dimensions are: 5.20m (length), 2.15m (width), 1,50m (height). Retaining walls are visible also the Paspala bridge, where it is clear that the latter are a subsequent addition since they are attached on to the fully carved blocks of the inner pier façades.

[114] Further information regarding the conduit's section within the quarry, Millar and Williams 1996: 220-223, fig. 5 – although the supported view, i.e. that the conduit was not part of the main aqueduct but was built especially to serve the quarry, is mistaken.

Bibliography

Abbreviations

IG XII, 2 *Inscriptiones Graecae, XII. Inscriptiones insularum maris Aegaei praeter Delum, 2. Inscriptiones Lesbi, Nesi, Tenedi*, ed. William R. Paton. Berlin 1899.

SEG *Supplementum Epigraphicum Graecum*. Vols. 1-11, ed. Jacob E. Hondius, Leiden 1923-1954. Vols. 12-25, ed. Arthur G. Woodhead. Leiden 1955-1971. Vols. 26-41, eds. Henry W. Pleket and Ronald S. Stroud. Amsterdam 1979-1994. Vols. 42-44, eds. Henry W. Pleket, Ronald S. Stroud and Johan H.M. Strubbe. Amsterdam 1995-1997. Vols. 45-49, eds. Henry W. Pleket, Ronald S. Stroud, Angelos Chaniotis and Johan H.M. Strubbe. Amsterdam 1998-2002. Vols. 50- , eds. Angelos Chaniotis, Ronald S. Stroud and Johan H.M. Strubbe. Amsterdam 2003- .

Archontidou, A. 1997. Οδός Θεοφράστου (οικ. Μ. Κουκέλλη), *Archaiologikon Deltion* 52, Χρονικά, Β΄3: 900-902, drawing 6, pl. 333 d.

Aristodemou, G. 2012. *Ο γλυπτός διάκοσμος Νυμφαίων και Κρηνών στο Ανατολικό Τμήμα της Αυτοκρατορίας*. Thessaloniki: Kornelia Sfakianakis Editions.

Birley, A. R. 1997. *Hadrian, the Restless Emperor*. London and New York: Routledge.

Boatwright, M. T. 2002, *Hadrian and the Cities of the Roman Empire*. Princeton University Press.

Bouras, Ch. 1968. *Μαθήματα Ιστορίας τής Αρχιτεκτονικής, I*. Athens: Symmetria Editions (1999, 2nd ed.)

Burdy, J., Taşlialan, M., Waelkens, S., Mitchell, S. 1998. The aqueduct, nymphaeum and bath house. In: S. Mitchell, M. Waelkens (eds) *Pisidian Antioch: The Site and its Monuments*: 175-200. London: Gerard Duckworth & Co.Ltd.

Chanson, H., 2000. Hydraulics of Roman Aqueducts: Steep Chutes, Cascades, and Dropshafts. *American Journal of Archaeology* 104: 47-72.

Charitonidis, S. 1961/62. Μυτιλήνη. *Archaiologikon Deltion* 17, Χρονικά, Β΄1: 263, drawing 3.

Charitonidis, S. 1968. *Αι επιγραφαί της Λέσβου. Συμπλήρωμα*. Βιβλιοθήκη της εν Αθήναις Αρχαιολογικής Εταιρείας αριθ. 60. Athens.

Chatzi, D. 1971. Ανασκαφικαί Έρευναι. *Archaiologikon Deltion* 26, Χρονικά, Β΄2: 449-454, drawing 2.

Chatzi, D. 1972. Οικόπεδο ΟΤΕ επί των οδών Βουρνάζων και Φάωνος. *Archaiologikon Deltion* 27, Χρονικά, Β΄2: 584-588, figs. 6-8.

Chatziioannou, I. 1985. Το ρωμαϊκό υδραγωγείο Μυτιλήνης. *Λεσβιακά, Δελτίον της Εταιρείας Λεσβιακών Μελετών Θ́* : 153-162.

Dimopoulou-Piliouni, A. 2015. *Λεσβίων Πολιτείαι, Πολίτευμα, Θεσμοί και Δίκαιο των Πόλεων της Λέσβου (αρχαϊκοί, κλασικοί, ελληνιστικοί, ρωμαϊκοί χρόνοι)*. Athens: Eurasia Editions.

Eleutheriades, Th. 2007. Η σήραγγα του ρωμαϊκού υδραγωγείου στη θέση «Φανός» Μόριας. In: Αριστείδη Ιγν. Κουτζαμάνη (επιμ.) *Λεσβιακό Ημερολόγιο 2007*: 143-199. Mytilene.

Eleutheriades, Th. 2010. *Το προδρομικό υδραγωγείο της αρχαίας Μυτιλήνης και η Εφαρμοσμένη Πυθαγόρεια Αριθμοφιλοσοφία*. Mytilene.

Evans, H. 1982. Agrippa's water plan. *American Journal of Archaeology* 86: 401-411.

Fahlbusch, H. 1981. Wasserversorgung Griechischer Städte, Dargestellt am Beispiel Pergamon. In: *Vorträge der Tagung Wasser im antiken Hellas in Athen, 4-5. Juni 1981*: 135-173. Mitteilungen. Leichtweiss-Institut Für Wasserbau Der Technischen Universität Braunschweig, Heft 71.

Foss, C. 1979. *Ephesus after Antiquity: a late antique, Byzantine, and Turkish City*. Cambridge: Cambridge University Press.

Grewe, K. 1993. Augusta Emerita/Mérida. Eine Stadt römischer Technikgeschichte. *Antike Welt* 24: 79-95.

Grewe, K. 1994. Die römische Wasserleitung nach Side (Türkei). *Antike Welt* 25: 192-203.

Haut, B. and Viviers, D. 2010. Analysis of the Water System of the Ancient Roman City of Apamea. In: L.W. Mays (ed.) *Ancient water technologies*: 139-169. Dordrecht, Heidelberg, London/New York: Springer.

Hodge, Tr. 1992. *Roman aqueducts and water supply*, London: Duckworth.

Hodge, Tr. 1996. In Vitruvium Pompeianum: Urban Distribution Reappraised. *American Journal of Archaeology* 100: 261-276.

Hodge. Tr. 2000. Water supply-Collection of water, Wells, Qanats, Aqueducts, Engineering works Purity of water. In: Ö. Wikander (ed.) *Handbook of ancient water Technology*: 21-99. Leiden: Brill.

Izmirligir, Ü. 1979. *Die Wasserversorgungsanlagen von Side*. Mitteilungen. Leichtweiss-Institut Für Wasserbau Der Technischen Universität Braunschweig, Heft 64.

Kaiafa, M. 2008. *Συστήματα ύδρευσης και αποχέτευσης κατά την ελληνιστική και ρωμαϊκή περίοδο στη Μακεδονία*. Unpublished Ph.D dissertation, Aristotle University of Thessaloniki.

Kayer, K. 1995. *Αποτύπωση-Στερέωση και Αποκατάσταση ρωμαϊκού υδραγωγείου Μόριας Λέσβου*. Unpublished Study. Directorate for the Restoration of Ancient Monuments. Athens: Ministry of Culture.

Koldewey, R. 1980. *Die antiken Baureste der Insel Lesbos*. Berlin.

Kontis, I. 1973a. *Λεσβιακό πολύπτυχο. Από την ιστορία, την τέχνη και τη λογοτεχνία*. Athens: Esperos Editions.

Kontis, I. 1973b. Το ρωμαϊκό υδραγωγείο Μυτιλήνης. *Λεσβιακά, Δελτίον της Εταιρείας Λεσβιακών Μελετών ΣΤ΄*: 77-81.

Kontis, I. 1978. *Η Λέσβος και η μικρασιατική της περιοχή*. Athens: Doxiades' Athens Center of Ekistics' Archives.

Korres, M. 2002.Το Αδριάνειο υδραγωγείο. *Εφημερίδα Καθημερινή, Επτά Ημέρες, Η ύδρευση των Αθηνών*, 24 Μαρτίου 2002.

Koukouli – Chrysanthaki, Ch. and Bakirzis, Ch. 1997. *Φίλιπποι*. Athens: Ministry of Culture. Archaeological Receipts Fund.

Kouria, A. 2000. Αντικατοπτρισμοί του χρόνου στα κείμενα και τις εικόνες των Ευρωπαίων περιηγητών του ελληνικού χώρου. *Αρχαιολογία και Τέχνες*, 77: 31-39.

Kourtzellis, I. 2008. Τοπογραφία παράκτιας ζώνης πόλεως Μυτιλήνης. Από την αρχαιότητα έως σήμερα. In: *Πρακτικά του 4ου Πανελλήνιου Συνεδρίου Διαχείριση και Βελτίωση Παράκτιων Ζωνών*: 213-222. Athens: Έκδοση ΕΜΠ, Εργαστήριο Λιμενικών Έργων.

Kourtzellis, I. 2010. Νομίσματα από τις αρχαίες λιμενικές εγκαταστάσεις του βόρειου λιμένα της Μυτιλήνης. In: Π. Τσέλεκας (επιμ.), *Οβολός 9, Πρακτικά συνεδρίου 5ης επιστημονικής συνάντησης, Μυτιλήνη*, 16-19 Σεπτεμβρίου 2006, Vol. I (ancient period): 185-200, Athens: 'The Friends of the Numismatic Museum' Association.

Kourtzellis, I. 2013. The Northern Harbour, The Commercial Stoa of the Hellenistic period, The Euripus-Ermou Street, The Southern Harbour. In: M. Μυρογιάννη (ed.) Mytilene. Unique Approaches: 11-15, 28-31, 46-49. Mytilene: Publication of the Chamber of Lesbos.

Kyriakopoulou, Th. 2013. Mytilene of the Roman era. In: M. Μυρογιάννη (ed.) *Mytilene. Unique Approaches*: 17-21. Mytilene: Publication of the Chamber of Lesbos.

Labarre, G. 1996. *Les cites de Lesbos aux époques hellénistique et imperiale, Collection de l'Institut d'Archéologie et d'Histoire de l'Antiquité.* Volume 1, Paris, Université Lumierè Lyon 2 : Diffusion de Boccard.

Lamprecht, H. O. 1987. *Opus Caementitium*, Bautechnik der Römer. Düsseldorf : Beton-Verlag.

Lefas, P. 1997. *ΒΙΤΡΟΥΒΙΟΣ, «Περί Αρχιτεκτονικής»*, Παύλος Λέφας (Translation-Commentary), Antonio Corso (Introduction), vol. A and B. Athens: Plethron Editions.

Leveau, Ph. and Paillet, J. L. 1976. *L'alimentation en eau de Caesarea de Mauretanie et l'aqueduc de Cherchell.* Paris: Harmattan.

Lewis, M. 1999. Vitruvius and the Greek aqueducts. *Paper of the British School at Rome LXVII*: 145-172.

Lolos, Y. 1997. The Hadrianic Aqueduct of Corinth (With an Appendix on the Roman Aqueducts in Greece). *Hesperia: The Journal of the American School of Classical Studies at Athens* 66: 271-314.

Mays, L. W. 2010. A brief history of Roman water technology. In: L.W. Mays (ed.) *Ancient water technologies*: 115-138. Dordrecht/ Heidelberg/London/New York: Springer.

Millar, R. and Williams, H. 1993. The Roman quarry at Moria, Mytilene. *Échos du monde Classique, Classical Views* 37, n.s. 12, No. 2: 211-224.

O' Connor, C. 1993. *Roman Bridges*. Cambridge: Cambridge University Press.

Ορλάνδος, A. K. 1958, *Τα υλικά δομής των αρχαίων Ελλήνων κατά τους συγγραφείς, τας επιγραφάς και τα μνημεία*, Μέρος Ά , Τεύχος 2, Athens: Estia Editions.

Παπαδόπουλος, Γ. 2015. *Λέσβος, Χίος, Ψαρά. Οι σεισμοί και τα τσουνάμι από την αρχαιότητα έως σήμερα. Athens: Oselotos Editions.*

Papakonstantinou, A., Kourtzellis, I., Pappa, M. and Kakes, G. 2015. Εφαρμοσμένη Γεωπληροφική και Γεωοπτικοποίηση στην Αρχαιολογία. Μελέτη της πορείας του ρωμαϊκού υδραγωγείου Μυτιλήνης. In: C. Papadopoulos, E. Paliou, A. Chrysanthi, E. Kotoula and A. Sarris (eds) *Archaeological research in the Digital Age. Proceedings of the 1st Conference on Computer Applications and Quantitative Methods in Archaeology*

Conference - Greek Chapter (CAA-GR), (Rethymno, Crete, 6-8 March 2014): 10-17. Rethymno: Institute for Mediterranean Studies- Foundation of Research and Technology (IMS-FORTH).
Pistorius, H. 1913. *Beiträge zur Geschichte von Lesbos im vierten Jahrhundert v.Chr.* Bonn: Druck von A. Kämpfe.
Stefanidou-Tiveriou, Th. 1995. Σιληνός καθεύδων. *Archaeologike Ephemeris.* 134: 67- 75, pl. 30-34.
Theodoulou, Th. and Kourtzellis, I. 2010, Λιμάνια και ναυτικοί δρόμοι στην αρχαία Λέσβο (Harbours and nautical routes in ancient Lesbos). In: Π. Σκορδάς (επιμ.) *Λεσβιακό Ημερολόγιο 2011 Γράμματα Τέχνες Πολιτισμός*: 129- 147. Mytilene: Aiolida Editions.
Wilson R. 1996. Tot aquarum tam multis necessaries molibur... Recent studies on aqueducts and water supply. The *Journal of Roman Archaeology* 9: 5-29.

Yannis Kourtzellis
Archaeologist, MSc, PhD
Hellenic Ministry of
Culture and Sports,
Archaeological Ephorate of Lesbos
ctma05005@ct.aegean.gr

Maria Pappa
Archaeologist,
Hellenic Ministry of
Culture and Sports,
Archaeological Ephorate of Lesbos
pappamaria615@gmail.com

George Kakes
Civil Engineer, MSc
Hellenic Ministry of
Culture and Sports,
Archaeological Ephorate of Lesbos
gkakes@hotmail.com

Roman aqueduct of Samos

Telauges N. Dimitriou

Abstract

The Roman Aqueduct of Samos is one of the most important technical projects made on the island during its Roman period. It was constructed to complement the Tunnel of Eupalinos, which could no longer meet the inhabitants' increased water needs. Approximately 15,500m long, the aqueduct begins at the sources of Imvrassos River and reaches the ancient city of Samos (today's Pythagorion). A relatively substandard construction and the natural wear it suffered over time led to its gradual abandonment and disuse.

The present paper is a first attempt at tracing the relics and charting the course of this extensive project. Emphasis was placed on the technical aspects, as evidenced by the extensive drawings and the detailed topographic mappings. The whole venture was made possible after the devastating fire of 2000 on the island revealed some surviving sections which indicated the course of the aqueduct.
In addition to mapping, the main objective of this work is to serve as a useful basis for further research on the monument.

Keywords: Samos, aqueduct, Eupalinian Tunnel, Imvrassos River

History

One of the most important technical projects from the Roman period of Samos is the Roman aqueduct, relics of which survive along its entire course.

A lot of important data on the creation of the Roman Aqueduct remain unknown to this day. Its total length is estimated at 15,500m, but the absence of historical sources means that we do not know the exact time of its construction and operation.

Hieromonk Synesius in his treatise on the Eupalinian Tunnel dismisses the idea that the tunnel had fell in disuse during the island's Roman occupation, based on the fact that its conduits survive intact. He believes that the aqueduct was either constructed much later, when the Eupalinian Tunnel had been destroyed, or that it was meant to complement the Tunnel to meet the city's increased water requirements.

The latter view is promoted also by Epaminondas Stamatiadis in his 1884 work on the Tunnel of Eupalinos; he says that the water from the spring of Agiades may have become insufficient for the needs of the city, and that the locals used the Roman aqueduct to carry water from the springs of the river Imvrassos.

Of course, it could be the case that over time the Eupalinian Tunnel had developed problems which rendered its maintenance increasingly difficult or impossible. At the same time we must consider also the city's increased water requirements in Roman times – requirements which the Eupalinian Tunnel could no longer meet and which the aqueduct was called upon to meet.

The Eupalinian Tunnel is believed to have been built in the mid-6th century BC and abandoned around AD 400-500. Given that the problems in water supply must have started earlier, construction of the Roman aqueduct must have begun before AD 400; therefore, it must be a project of 1600 years or more.

Sources of water

The engineers of Roman times, well-versed in hydraulic projects, opted for an open conduit of rectangular section, running above ground and following the terrain from the source(s) of water. The available springs at a short distance from the city were Nerotrouvia on the one hand and the springs of the river Imvrassos

Figure 1. The spring of Zastano

on the other. Detailed analysis and in-situ research has garnered sufficient evidence on these springs that the builders may have used for the city's water supply in Roman times.

Nerotrouvia lies NE of the village of Myloi, at an elevation of approx. 60m. Although rich in water, this spring cannot have been used to feed the Roman aqueduct: on the one hand, it was only active in spring and summer and thus could not meet the city's needs throughout the year; on the other, the spring lies lower than the Roman conduit, as seen by the remains of the aqueduct that run above it, and continue above the village of Myloi to the torrent Imvrassos, west of Nerotrouvia.

Zastano lies north of Myloi, at a horizontal straight distance of approx. 1800m. The elevation is 122m from sea level, and it comes out of the riverbed, with a mean supply of 52m^3 of water per hour for 15 years, based on measurements by the Agricultural Service of Samos (Figure 1).

Karitsa lies 650m to the NW of Zastano, at an elevation of 180m from sea level. The water springs from the west bank towards the village of Pagondas, at a height of approx. 7m from the riverbed and a distance of 20m from its axis. The mean supply for 15 years was 52m^3 per hour.

Stournar lies to the west of Karitsa, at a horizontal straight distance of approx. 1000m from it and at an elevation of 268m. The water springs at a short distance from the river bed, and the mean supply for 15 years was 136m^3 per hour.

Fleva lies at a horizontal straight distance of 750m east from Stournar and at an elevation of 260m. The water comes out inside the riverbed of Imvrassos at 54m^3 per hour.

Kryo Nero lies at a horizontal straight distance of approx. 600m to the NW of Fleva at an elevation of 260m, it provides 142m^3 of water per hour within the riverbed.

Of these springs, it appears most likely that the aqueduct was fed primarily by Zastano and, if the needs in water were even higher, by Karitsa and Stournar. In this case it would be necessary to build an arched aqueduct bridge at the appropriate location, because these two springs lie on the west bank of Imvrassos whereas the main spring of Zastano is on the east. Two likely locations for this were identified in the region of Kamares.

New research

The bulk of the evidence on the Roman aqueduct came from our research in situ. The aim was to locate the aqueduct's remains along its entire length and perform a tacheometric survey of its course. It was also deemed expedient to calculate additional data of the aqueduct's function, such as its flow capacity.

In sections where no traces were found, the tacheometric survey continued by estimating the likely course of the aqueduct on the basis of the locations of previous remains and the measured inclination of the conduit. Data from the ground near the conduit were collected at every stop, the positions of any surviving sections were noted and then these sections were measured, charted and photographed. Also mapped were the areas with major technical works and where the aqueduct connected to the sources of water, based on the gradients and the elevations of the springs.

A specially made stadia rod enabled us to mark the kilometric positions along the aqueduct and facilitate the finding of sections and other information from the survey. After collection of the data the horizontal and vertical alignment of the project was mapped, including as much information as possible about the names of the various locations.

In-situ research was a difficult undertaking, with most of the data being collected after many hours of hiking, often over rough terrain.

The survey began in the area of Imvrassos river as the most likely starting point for the aqueduct. A levelling measurement taken in the area, approximately at the location of the reservoir, led to the conclusion that the aqueduct was fed by the spring of *Zastano*, the one that lies closest to the village of Myloi. Of course, this does not mean that additional supply came from other springs in the area, such as *Karitsa* or *Stournar*, but no remains to confirm this were found. The survey thus started from Zastano, and record was made of the relics found in the broader area.

Near the spring and going downstream along the east bank of the torrent, there is a mass of almost perpendicular rock that rises to a height of 300-400m. Access to that was extremely hard and dangerous, considering that the few narrow passages are inundated by the torrent's waters. Nevertheless, it was found that the aqueduct was routed along the east side of the torrent.

The task of mapping the project and measuring cross-sections where relics were found was also very difficult. Charting the aqueduct's course from the spring to the reservoir (Km 0+610) was one of the most arduous undertakings in this survey, carried out over a terrain with transverse slopes in the order of 250 – 350%.

In any case, it was found that no uniform gradient was maintained for the conduit (it started at 64‰ and went as low as 3‰). On the horizontal plane the terrain was generally followed, otherwise it would require huge additional works at a very high cost.

Some points near sections Δ16-Δ17 suggest that the project must have been rebuilt at a later stage. This is based on the evidence of a different —and inferior— method of construction, as there are points where the original conduit, which was hewn in the rock, has been replaced with a roughly-built wall which is a mere 25cm thick, while on the horizontal plane the course has been changed for the worse. Moreover, the conduit's cross-section in this area has variable dimensions, with a width between 35 and 60cm and the height of the side walls at around 40cm. At many points inside the conduit there are traces of pointing and plastering.

Finally, several lime kilns were found along the entire course, and if they are proven to be of the same age as the aqueduct they may be assumed to have been built and used for the needs of the project.

The course

Description

The description of the project follows the horizontal and longitudinal section. (**Appendix:** Drawings 1- 6).

The Roman aqueduct starts at the spring of Zastano (Δ1 Km 0+000), at an elevation of 122m, and continues its course along the west bank of Imvrassos river. No traces of the aqueduct were found between the spring, site Δ0, and site Δ1; remnants appear again in the lower part of Xerolithia (Δ1 Km 0+102) and after that in site Δ3–Δ4 (Km 0+127.5/ 0+135.5), where some remains of the aqueduct's wall were located. A 40cm-thick section of the conduit was found in site Δ5 (Km 0+158.5). The next traces along the course are located on rocks in site Δ6 (Km 0+158.5). Between points Δ7 (Km 0+202) and Δ8 (Km 0+217) there are several stones with mortar (from the aqueduct wall), and more remains continue along the side of the hill up to point Σ3. After site Δ9 there is a host of remains which determine the course of the conduit. Between site Δ10 (Km 0+356) and point Σ7, the many relics that survive make it easier to track the Aqueduct's course. Remains of aqueduct wall, several hewn rocks and traces of mortar were found at points Δ10 (Km 0+356), Δ11 (Km 0+398.6), Δ12 (Km 0+467.1), Δ13 (Km 0+472), Δ14 (Km 0+478), Δ15 (Km 0+485.4) and Δ16 (Km 0+541). Moreover, the remains of a lime kiln which may be from the period of the aqueduct's construction were found near Δ11 (Km 0+398.6), while sections of the bottom of the conduit were found at points Δ15 (Km 0+485.4) and Δ16 (Km 0+541). In the area around Σ6 (near Δ16 (Km 0+541)) a trough was cut into the rock but not used (Figure 2), as another one was built on a different elevation and horizontal plane. Also, the section of the conduit varies in this area depending on the terrain's morphology.

From Σ1 onwards, above the reservoir (approximately up to point 47), the entire aqueduct has been destroyed by construction work in the first section. In section 48 up to point Δ17 (Km 0+829), where a small bridge whose pedestals are natural rocks was located (Figure 3), no traces of the aqueduct were found.

Figure 2. Conduit on a rock (Δ16. Km0+541)

A section of protective stonewall was discovered at point Δ19 (Km 1+010.7), after which there is a constant presence of aqueduct remains. Traces from the bottom of the conduit were found at Δ20 (Km 1+029.8), Δ21(Km 1+045.9), Δ22(Km 1+057.9) and Δ23(Km 1+071.2). After point Δ19 (Km 1+010.7), as well as after point Δ23(Km 1+071.2), where the conduit is continuous, it runs high above the rural road whose construction destroyed a sizeable part of the aqueduct. A host of relics survive between points Δ24 (Km 1+189) and Δ37 (Km 1+493.4), albeit in very bad condition as sections of the project have been subjected to shifts. A conduit runs high above site Δ32 (Km 1+357.1), where a trough is cut into the rock. Between site Δ37 (Km 1+493.4) and shortly beyond point 175, there are no traces of the aqueduct; they reappear in the area of Myloi (Km 2+450), where the aqueduct bridge was found (Figure 4).

Figure 3. Small bridge with natural megaliths as pedestals (Δ17. Km0+829)

From site Δ38 (Km 2+479.6) onwards there are sporadic remains such as hewn slate, a section of conduit (point 200, Km 2+623.3), stones with mortar (point 201, Km 2+662.8) and a 1.50m-long section of wall (point 200, Km 2+674.3). Another section of wall is found in site Δ39 (Km 2+808), while site Δ41 (Km 3+033.3) shows some arches, although their presence there is somewhat difficult to explain (Figure 5).

There follows another sizeable section with no finds of the aqueduct up to site Δ42 (Km 3+581.6) and points 232-233, where we find remains of a bridge with a 2.20m span. Remains of the Aqueduct's wall and a bridge were found at points 239 – 244. Between site Δ44 (Km 3+901.4) and point 307 there is a continuous section of aqueduct wall, whose remains continue intermittently up to the first aqueduct bridge in the area of Nerotrouvia (Km 4+311.1) (Figure 6).

Remains survive from all three aqueduct bridges in this area. After Nerotrouvia bridge all traces of the aqueduct are lost again until point 435-436, where we find heaps of small stones and collapsed segments of wall. Various remains continue up to the aqueduct bridge over the torrent Mavratzas. A major part of it is completely destroyed, but several traces survive, too, as sections from its stonework are scattered in the vicinity. After that bridge all traces of the aqueduct are lost again until point 514, where various remains of wall survive all the way through point 635. Sections from the bottom of the conduit are found along this section —specifically at points 532. 543/Δ53 (Km6+150), 546/Δ54 (Km6+157), 555/Δ55 (Km6+170), 567/Δ56 (Km6+191), 571/Δ57 (Km6+217.3), 601/Δ58 (Km6+270.7) – as well as a lime kiln (point 539). In the Papaevangelinos area (Km 6+250) we find traces of a bridge that replaced an older, problematic one. At

Figure 4. Aqueduct bridge remains in the area of Myloi (Km 2+450)

Figure 5. Relics of the bridge (Km 3+033.3)

Figure 6. The first bridge of Nerotrouvia (Km 4+311.1)

point 692 the new conduit meets the old one. All traces are lost again for a while (points 708-710), after which some remnants of aqueduct wall are found again. Remains of an aqueduct bridge are found in the riverbed at Paleo Garage (Km 7+000). Parts of wall can be found along the entire section between points 755 and 886. After that, most traces of the aqueduct are lost for most of its course up to point 1013: some remains were found near the stream of Aghios Panteleimon (points 891-898), but none in most of the settlement of Chora. A small aqueduct section was found outside Chora (point 931), and another one after Aghios Athanasios (points 984-990); additionally, a piece of built-n ceramic pipe was found in this locality. Aqueduct sections appear around the rural road to Agiades, where we find also the remains of the original aqueduct bridge over the stream of Ai-Gianniou. This bridge was replaced with another one at a distance of 100m upstream. Two arches from this bridge – the largest one of the aqueduct, at about 90m long and 21.7m high – survive as its most important relics (Figure 7).

From that point onward, the entire slope of the hill shows several sections of the aqueduct up to the Moraitis area (Km 12+900), where some arches were found in quite good condition. The finds continue along the length of the hill, while at point 1317-1318 the aqueduct crosses the wall and goes on towards the city (Figure 8).

Plenty of ceramic fragments used in the masonry and the mortars were found in this area, as well as some small arches (point 1336). At point 1349 the aqueduct branches out to the Thermae and continues its course towards the city. The branch to the Agora is found at point 1398, shortly before the ancient reservoir (Δ97 Km 14+778 – Δ98 Km 14+784). From there on and as we approach the city, the aqueduct's stonework is appreciably improved. The finds of the aqueduct stop in the vicinity of the High School of Pythagorion, where there are relics of an aqueduct bridge. After that, the town of Pythagorion begins.

Materials and construction

Despite its impressive length, the project required no extraordinary special operations, and the materials used in its construction were rather simple. The operations include digging for the foundations, masonry and plastering.

Figure 7. Remains of the aqueduct bridge over Ai-Gianniou to Rema (Km 14+300)

Excavations

The technicians of the time must have used the ordinary tools that have been in use from ancient times to this day, such as hoe and shovel, with additional equipment employed for digging in rocky ground: pickaxes, hammers, chisels, crowbars, sledgehammers, wedges, etc.

As research has shown, not enough care and attention was paid to the digging for foundations to ensure the stability and longevity of the project, which resulted in the need for frequent repairs or even in abandoning some sections and shifting the project to another location.

Masonry

It involved three main materials: *stone, sand* and *lime*.

a) The stones used in the project came mainly from the vicinity of each section, since there was no hewing was necessary and the work would thus be completed much sooner. Nevertheless, hewn stones can be seen at several points, especially near the city (Pythagorion): it seems that more emphasis was given to appearance as the project neared its destination. In such cases there are even examples of carefully pointed stonework. The processing of stones before building was minimal and only limited to specific technical elements such as corners, arches, etc.; in no way can we talk about hewing or any other serious work. In a few cases of destroyed sections which were reconstructed, the stones were replaced with bricks measuring approximately 50 x 50 x 4 cm. On the whole, the stonework was shoddy and of inferior quality. Mortar played a key role. In several instances we can observe the use of "stone pitching", i.e. large quantities of mortar mixed with stones, and elsewhere there are voids for which there is no technical justification, or small segments of large cross-sections without mortar (drywall). The inferior stonework of the project is one of the explanations for the few surviving segments.

b) Sand was a major factor in the construction of the aqueduct, given its extensive use in the mortar. Since there were no organised sand quarries, of course, they made do with river sand and, secondarily, with sea sand.

c) Lime was a key material in the construction of the aqueduct, used together with sand in both mortars and plaster. The prevailing type of limestone in the region is *galazopetra,* which is suitable for efficient lime production. Lime was either purchased or made by the building crews. Production and slaking into a paste took place in lime kilns. In view of the considerable quantities of lime

Figure 8. The aqueduct meets the city walls (point 1317-1318)

required for the project, it is postulated that several lime kilns were set up at suitable locations. This is corroborated by the lime kilns found in the proximity of the aqueduct's course.

As several finds confirm, the conduit was internally plastered on both sides and the bottom. Plastering helped ensure a smoother water flow and reduce leakages.

The plaster mix comprised approximately one part lime to two parts of sand (per volume).

Finally, in some parts it was found that some of the sand was replaced with powdered ceramic tile to form a mortar known since antiquity (the subsequent *kourasani*). In these cases the plaster was pressed onto the surface with a spatula.

Types of conduit

The engineers of the time opted for a free-flow aqueduct with an open-top conduit made of stone, since the relatively smooth geography permitted the construction of a conduit above ground which could follow the terrain at a certain slope. The necessary materials (mainly stone) were readily available in situ, ensuring a quick and above all economical construction. No major section of underground routing of the aqueduct was found. The sole exceptions may have been the section that went through the settlement of Chora and one more location (Km 8+750) where it could have been used to avoid crossing an inaccessible area or building a considerably longer conduit to bypass it.

In terms of the choice of conduit, an open type was opted for; it would make construction easier and more economical and would also facilitate cleaning and maintenance. The solution of using ceramic pipes could be envisaged to ensure cleanliness of the water; however, no traces of such pipes were located at any point.

Crossings of aqueduct

It is natural for a project of this aqueduct's scale, as it runs from the river Imvrassos to the ancient city of Samos, to come against existing roads, riverbeds and perhaps inhabited areas as well.

Roads and riverbeds

The course of the aqueduct follows approximately one contour line. Obviously, this line could not be parallel to the ground, so the aqueduct would have different heights along its course. Moreover, it had to be difficult to access and so it was built quite higher above ground, this height varying depending on the terrain. Where the project crossed roads or other conduits, provision would be made to facilitate its crossing by pedestrians, animals and carts. When the aqueduct stood higher than approx. 2.50m above the ground, there would probably be arches built into the wall aqueduct to provide an underpass. For heights of about 1.50m or less, it would be easier to build a mound on either side of the aqueduct and take the road across on an 'overpass'.

Another case is the aqueduct bridge over the stream of Ai-Gianniou. The project had to be carried over to the opposite bank at that location because it was the narrowest point, but mainly because upstream of that spot the mountain's gradients were too steep.

The part of the project over the naissances of the arches is more recent and construction is neater, while bricks have been used in the arches (uniquely among the surviving relics of the aqueduct); the bricks measure 50x50x5cm and are joined with wedge-shaped lime mortar of considerable thickness (4cm, on average). It is obvious that the project was partially destroyed while in use, by flooded torrents or, more likely, by a major earthquake (perhaps that of AD 358, which devastated the region), and the section above the naissances had to be rebuilt.

Inhabited areas

Although no remains have been found to attest to this, it is almost certain that the Aqueduct must have gone through the settlement of Chora. This would necessitate the construction of an underground conduit, because otherwise it would hinder traffic, split the settlement in two and risk contamination of the water.

An underground conduit would travel on a horizontal plane along roads with a constant gradient towards its destination, its depth under the road surface compensating for any small changes in the terrain. Inspection wells would be built at certain intervals to enable cleaning and repairs.

The underground route would end towards the east side, where the pipe would reach the torrent of 'Kalathi', cross it over an aqueduct bridge and continue on its course to the city with a normal conduit above ground.

Ridges

Research has shown that in the area of Haghios Panteleimon, at Km 8+750, the aqueduct deviated from its course because of a small hill. The options at this point were two: the conduit could run around the hillock of Avgo, or follow an underground path under the ridge. No traces have been found to confirm either option. However, a study of the pros and cons of each solution suggests that an underground pipe would be more likely for clearing this obstacle.

The option of a peripheral course would render the project costlier and more difficult, given the morphology of the ground, while its length would be 290m. Conversely, an underground conduit would be just 50m long.

Branches of the aqueduct

A project of this scale and importance as this aqueduct would not have stopped at simply meeting the city's water needs but would be expected to supply additional functions as well. Indeed, research has located two branches of the aqueduct, of which one fed the baths and the other took water to the area of the Ancient Agora.

The Thermae were housed in a grandiose complex designed to meet all of their patrons' needs. The baths required huge quantities of water to operate; hence the main conduit had to branch out to feed the facility. An effort was thus made to locate the starting point of this branch. Remains were found as some point where the aqueduct starts and up to the Samos-Pythagorion-Pyrgos road that runs past the Thermae. The fact that this point marks the shortest distance between the aqueduct and the baths indicates that this was indeed the path of the conduit. Systematic research along the entire length of the likely route uncovered small sections of aqueduct wall among the stones of cairns. Indeed, some of these were found in the lower section of the conduit near the Thermae, corroborating the existence of a branch that joined the baths with the main conduit.

The course and the longitudinal section were mapped. The steep gradients, observed particularly in the mountainous north section, must have caused difficulties which may have been resolved with the use of ceramic pipes, as suggested by the relics found in the area. Ceramic pipes would mean a smoother flow, without water spilling out at the slightest obstacle; there would be less wear in the pipe, and maintenance would also be easier by replacing the pipes.

The difference in elevation between the main conduit and the Thermae is considerable (33.5m), but the lower part of the conduit runs on low-lying ground, at the same level as the floor of the bath facilities. The various water tubs and pools must have stood at least some way higher than the floor, and the same may have been true of the conduit's lowest point. Therefore, the conduit must have run over small arches to meet the Thermae building at a level appreciably higher than that of the floor. However, no evidence from the building's interior plumbing survives to confirm any of this.

The water that reached the building may have been fed to an elevated tank which supplied the network of bathing facilities. The quantity of water would be adjusted with a flow regulator at the branching-out point. Any excess water would overflow into the main drain and thence to the sea, which is quite near.

The second branch of the aqueduct was found at Point 1404-1405, where there is a tank whose two sections are roofed with semi-cylindrical domes built with mortar in a similar way as the aqueduct. A conduit runs parallel and close to the tank's southern side, but it is smaller in dimensions than the rest of the conduit. Initially it was thought to have supplied a specific are with water from the tank, but when the aqueduct was mapped and the 'contour lines' were drawn, it was found to coincide with the Aqueduct's main pipe. In-situ research revealed (a few metres away from the tank) a N/S oriented conduit that led water to the Forum in the centre of the city.

Several impressive remains of the aqueduct have been located along this conduit. In the lower part there are sections of masonry from a row of arches that carried the conduit. This may have been necessary in order to elevate the water for easier distribution to the various points of consumption (fountains, nymphaea, etc.). Techniques in the Forum branch are likely to have been the same as in that to the Thermae, since conditions were much the same.

Estimation of water supply

An attempt at estimating the conduit's supply is necessary despite the inevitably approximate nature of the calculation.

- Effective rectangular cross-section 0.50x 0.50m (E= 0.25 m^2)
- Prevailing minimum longitudinal slope J=3‰
- An unfavourable assumption of wear-and-tear in the plaster and roughness of the stonework leads to a value of n= 0.025 for the roughness coefficient in Kutter's formula.

- Wet cross-section by wet perimeter R= 0.167
- Kutter's supply coefficient is calculated at C=26.14
- Supply $Q=C\cdot E\cdot\sqrt{RJ}=0.14m^3/sec$, i.e. approx. 500 m^3 per hour or 12,000 m^3 per day.
- We also calculated that water evaporation along the open-top conduit would be less than $200m^3$ per day.
- Finally, if we assume seepage losses due to sealing defects at 20%, we are left with an effective supply of over $9000m^3$ per day.

This supply appears to be greater than the combined supplies of the three likely springs that fed the aqueduct (6,000 m3) if we take today's hydrological measurements, i.e. Zastano $50m^3/h$, Karitsa $50m^3/h$ and Stournar $140m^3/h$ (total 240 m^3/h).

However, we ignore the actual supply of these springs during the Roman period, and we cannot be sure that all of them were connected to the Roman aqueduct. As a rather rough estimate we could take as a likely supply 50% x 240 =120 m^3/h = $2800m^3$ per day or, after losses, something in the region of 2,000 m^3 of water per day.

It would be equally expedient to make even a rough estimate of the water needs of the inhabitants of the ancient city of Samos:

- Assumed population 40,000 inhabitants (lower than the earlier, excessive estimate of Herodotus).
- Assumed daily consumption per person (in a civilised environment in antiquity): 50 lit per person per day.
- Likely daily consumption 40,000x0.050= $2.000m^3$ per day.

Thus at first sight our rough estimates of water supply and consumption in the city do not clash.

Nevertheless, the one thing that can be said with certainty is that the 0.50x0.50 m3 dimensions of the aqueduct's cross-section (which lead to a nominal maximum supply of $9{,}000m^3$ per day) were not arrived at through a calculation of water needs but were a minimum constructional requirement for two builders working face-to-face.

Acknowledgments

We had expressed our warm thanks to all those who had contributed their valuable help in researching and publishing our book *Το Ρωμαϊκό Υδραγωγείο της Σάμου* in 2003. Specifically for this paper, many thanks are due to Ms Vicky Lekati and Ms Matroni Ktistou for their unstinting cooperation.

Bibliography

Hieromonk Synesius, 1899. *Πραγματεία περί του Ευπαλινείου Ορύγματος*, Published by Spyridon Aeginitis, Samos.

Stamatiadis, Ep. 1884. *Περί το Όρυγμα του Ευπαλίνου εν Σάμω*. Samos: Hegemonic Press.

Dimitriou, T., 2003. *Το Ρωμαϊκό Υδραγωγείο της Σάμου*. 'N. Dimitriou' Samos Cultural Centre.

Telauges N. Dimitriou
Chairman of the 'Nicolaos Dimitriou' Cultural Centre
Civil Engineer
pnedim@otenet.gr

Appendix: drawings

Horizontal plans of the Roman Aqueduct

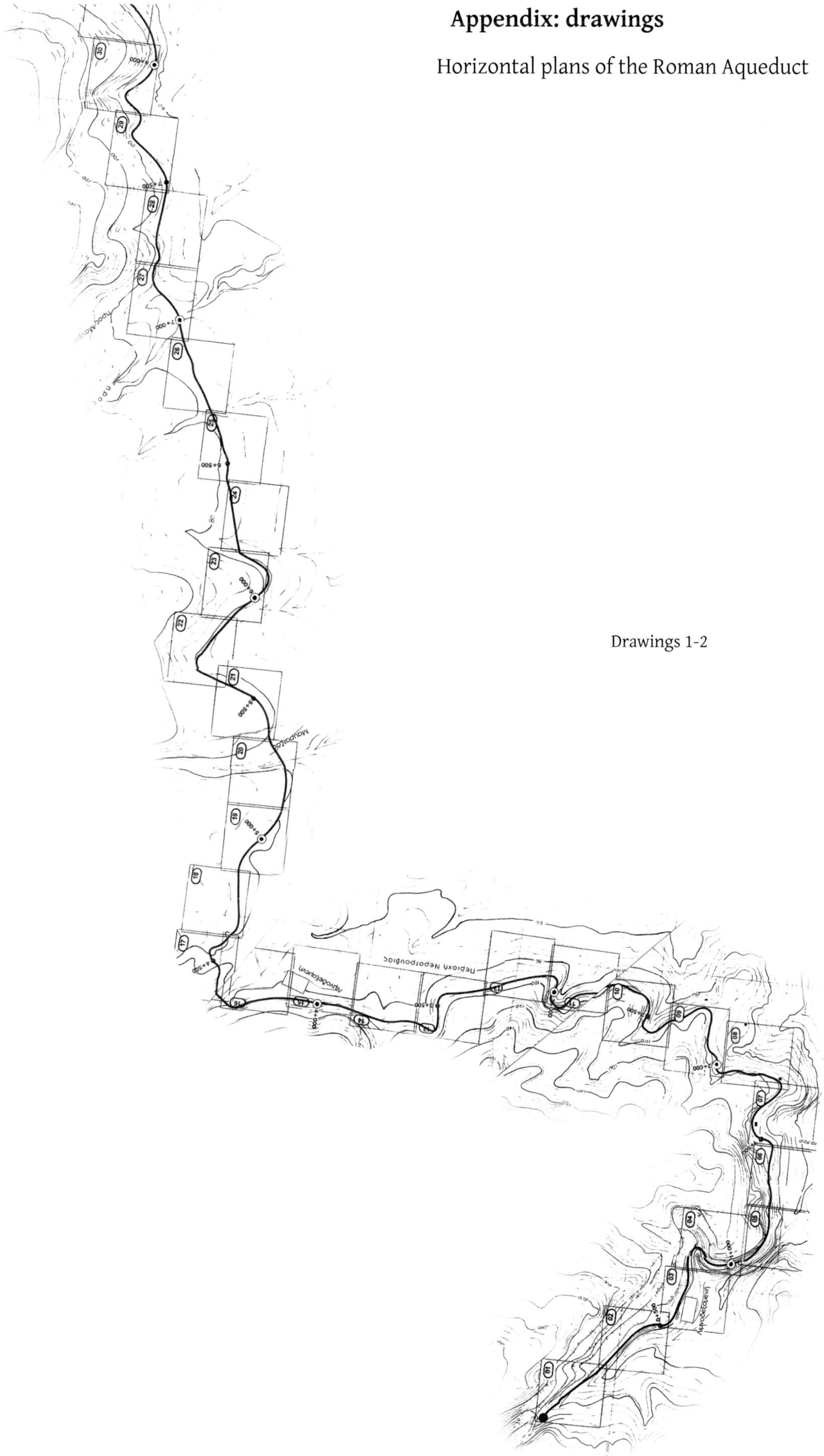

Drawings 1-2

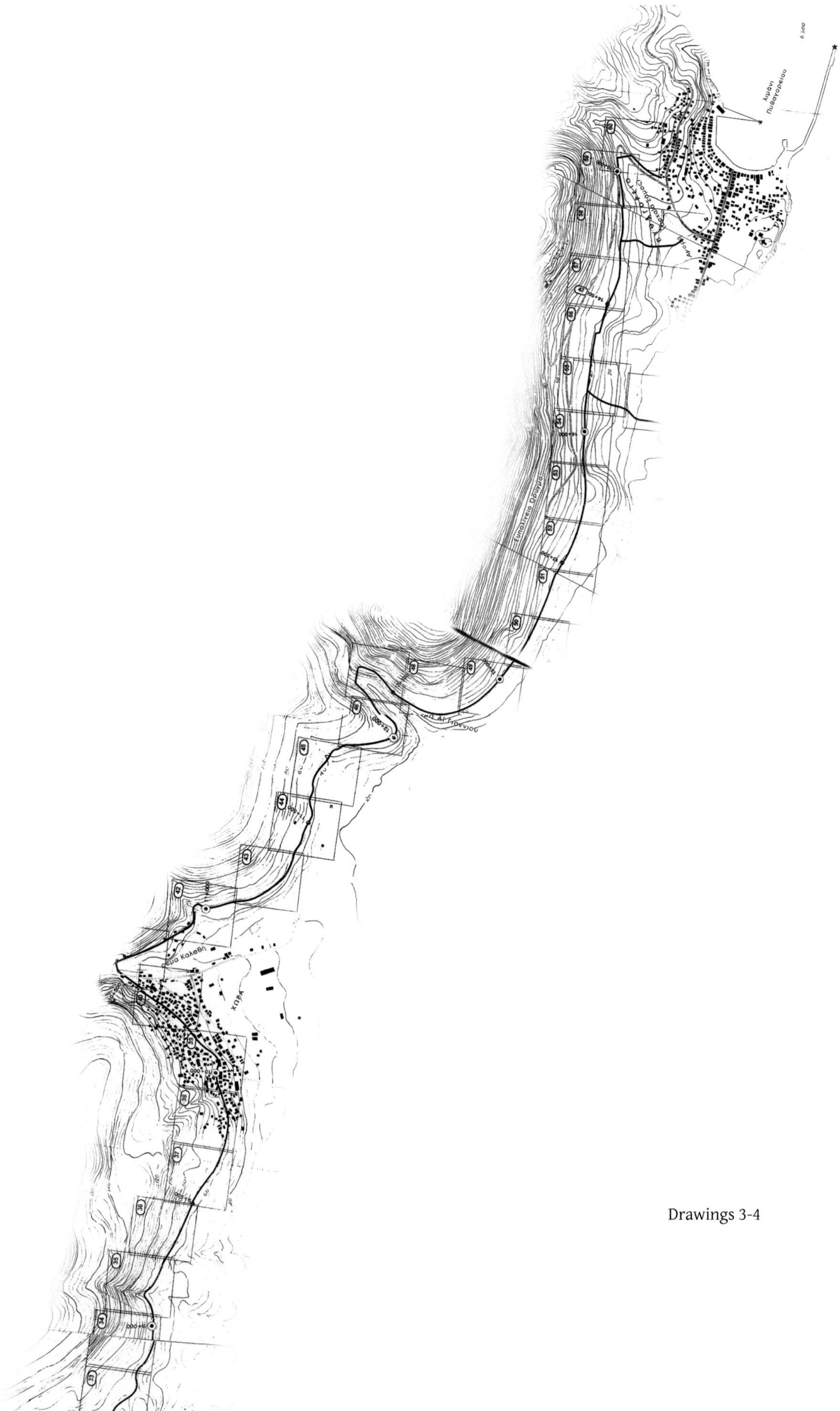

Drawings 3-4

Longitudinal sections of the Roman Aqueduct

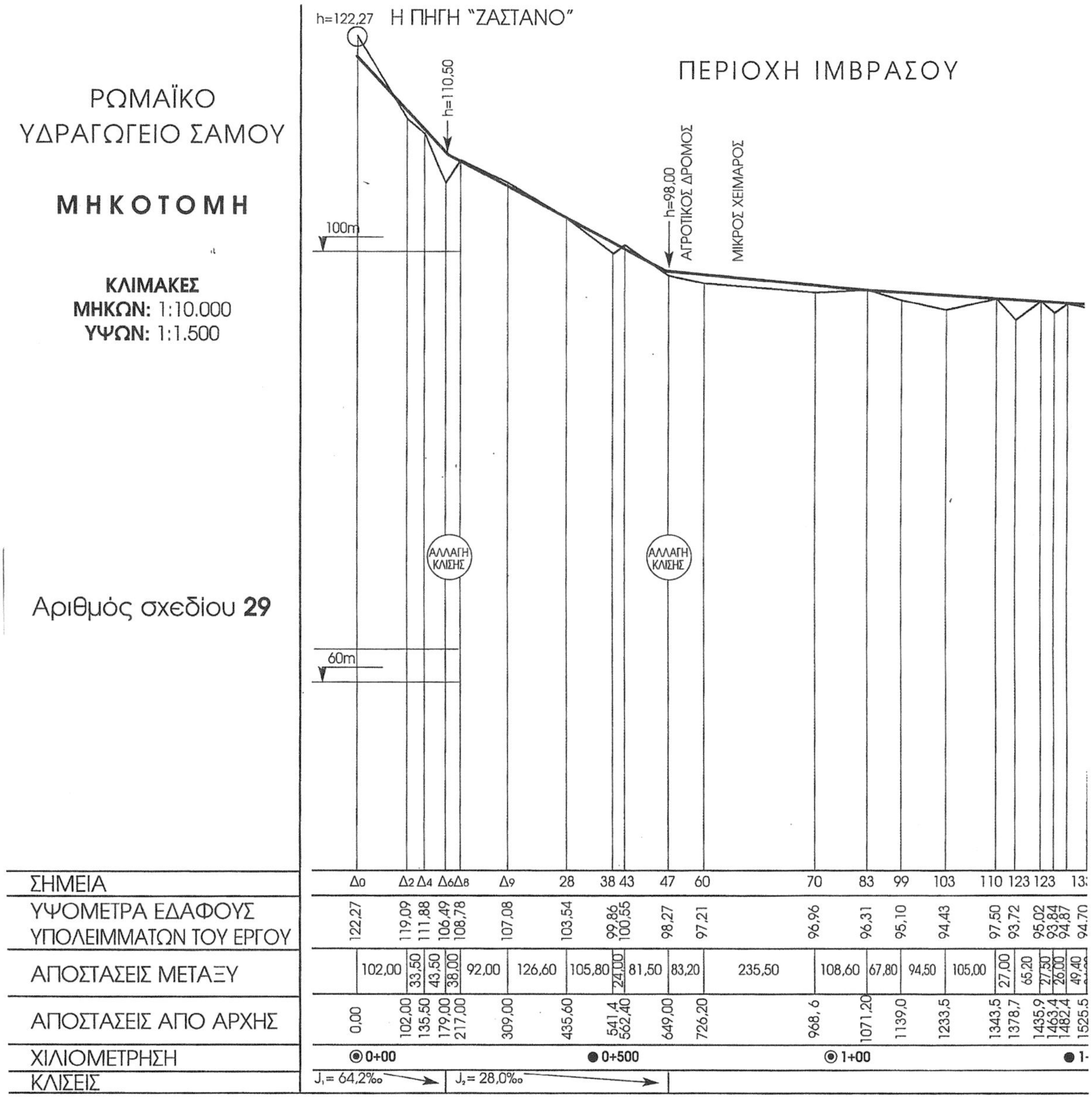

Drawing 5

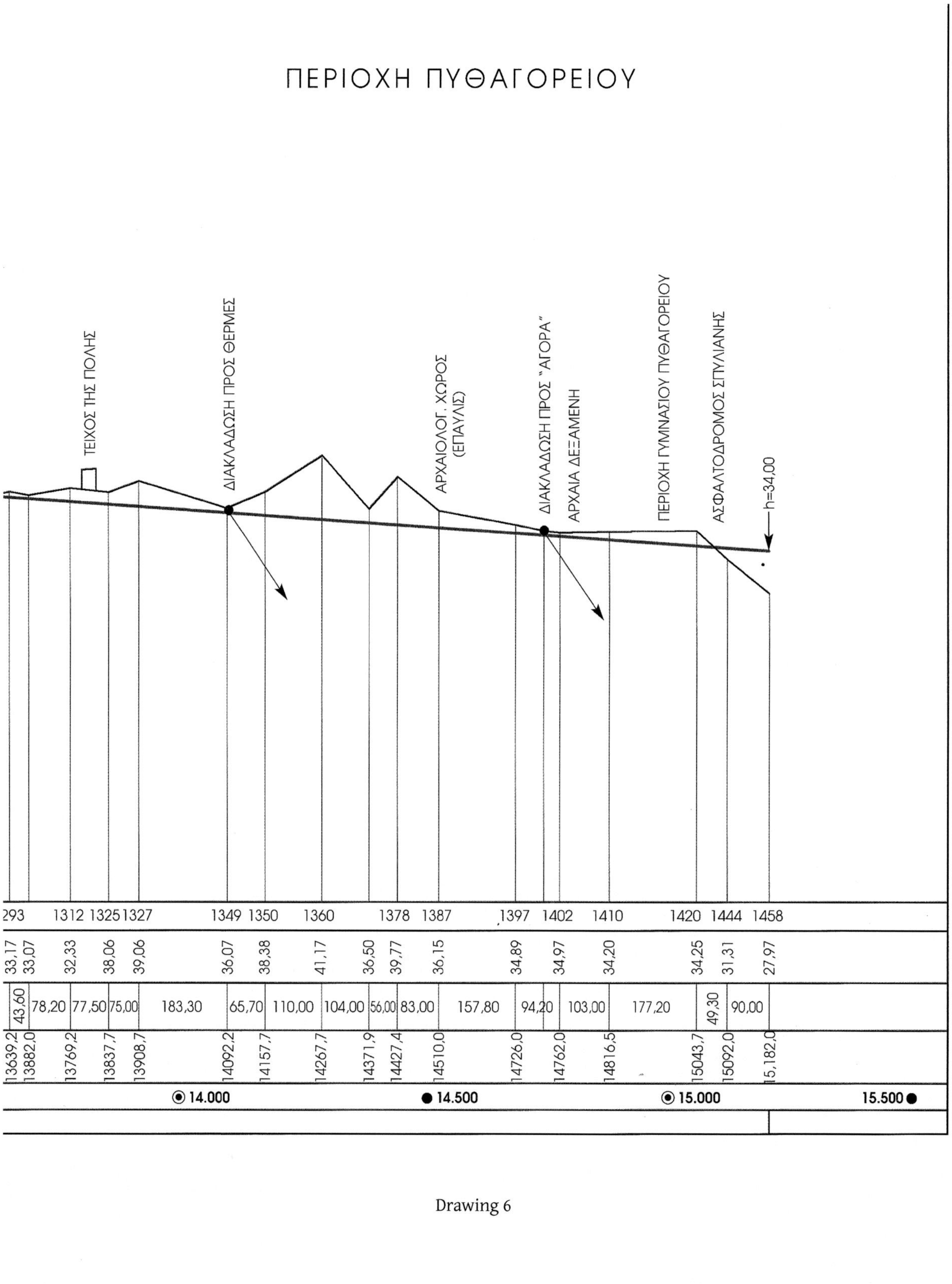
ΠΕΡΙΟΧΗ ΠΥΘΑΓΟΡΕΙΟΥ
ΤΕΙΧΟΣ ΤΗΣ ΠΟΛΗΣ
ΔΙΑΚΛΑΔΩΣΗ ΠΡΟΣ ΘΕΡΜΕΣ
ΑΡΧΑΙΟΛΟΓ. ΧΩΡΟΣ (ΕΠΑΥΛΙΣ)
ΔΙΑΚΛΑΔΩΣΗ ΠΡΟΣ "ΑΓΟΡΑ"
ΑΡΧΑΙΑ ΔΕΞΑΜΕΝΗ
ΠΕΡΙΟΧΗ ΓΥΜΝΑΣΙΟΥ ΠΥΘΑΓΟΡΕΙΟΥ
ΑΣΦΑΛΤΟΔΡΟΜΟΣ ΣΠΥΛΙΑΝΗΣ
h=34,00
293 1312 1325 1327 1349 1350 1360 1378 1387 1397 1402 1410 1420 1444 1458
33,17 33,07 32,33 38,06 39,06 36,07 38,38 41,17 36,50 39,77 36,15 34,89 34,97 34,20 34,25 31,31 27,97
43,60 78,20 77,50 75,00 183,30 65,70 110,00 104,00 56,00 83,00 157,80 94,20 103,00 177,20 49,30 90,00
13639,2 13882,0 13769,2 13837,7 13908,7 14092,2 14157,7 14267,7 14371,9 14427,4 14510,0 14726,0 14762,0 14816,5 15043,7 15092,0 15,182,0
14.000
14.500
15.000
15.500

Drawing 6

A Roman aqueduct through the Cretan highlands – securing the water supply for elevated Lyttos

Amanda Kelly

Abstract

In this paper I wish to examine the difficulties encountered in securing the water supply for the Roman city of Lyttos in east central Crete. The city, set on an elevated spur in the western foothills of the Lasithi range, represents one of the relatively few examples of a flourishing upland Roman city on the island. Lyttos was both an inland centre and one of the most prosperous cities of Roman Crete. Its lofty position, simultaneously overshadowing the Pedhiadha plain and controlling the main pass into the Lasithi plateau, secured its control over a wide agricultural area. At this inland, and relatively inaccessible site, economics (as manifested in viticulture), as opposed to geographical accessibility *per se*, connected the city with the broader Roman world. Despite the relative inconvenience of the city's topography, the city remained on its perch in order to control the pass into the lucrative Lasithi plain. The city's strategic placement undoubtedly presented a challenge for its Roman planners, yet the city survived (and continued to flourish into the Byzantine period), by virtue of its hydraulic surveyors taking full advantage of the city's mountainous surrounds in designing its aqueduct.

Keywords: Roman Crete, inverted siphons, aqueducts, Lyttos, Lasithi

Introduction

Lyttos is widely held as one of the most prosperous Roman cities in Crete despite its relative physical inaccessibility, high in the mountain ranges of east central Crete (Figure 1). Broader onomastic studies suggest that Lyttos featured alongside other major Cretan cities (such as the capital at Gortyna, the colony at Knossos and the vibrant western hubs of Chania and Aptera) as a centre best known to the outer world and, therefore, also potentially one of the largest on the island.[1] Epigraphic evidence demonstrates that Lyttos was just as receptive to Roman influence as the poleis of Gortyna, Hierapytna and Knossos.[2]

Due specifically to its elevated position (rather than in spite of it) the city gained great prominence, as this siting clinched its dominance over a wide agricultural area, simultaneously overlooking the plain and controlling the pass into Lasithi.[3] By the mid-1st century AD its fertile hinterland produced much of the wine prominent in the trade network of the wider Empire, with the city's insignia stamped on the amphora being shipped to Pompeii and Herculaneum in Campania.[4] In addition to its early access to the north coast, originally via the port at Chersonisos,[5] the city possibly gained access to the south coast via Keratokambos where a pottery production facility was in operation,[6] producing the type of amphora found in Pompeii in the 1st century AD.[7] Crete maintained its leading position in the wine trade for the first three centuries of Empire with Cretan amphorae representing more than a third of the Aegean and Eastern imports so far discovered in Ostia in the 2nd century AD.[8]

[1] Baldwin Bowsky 1995: 53; Pumain 1997: 99.
[2] Baldwin Bowsky 1995: 54; Baldwin Bowsky 2006.
[3] While its far-flung amphorae attest that Lyttos prospered in viticultural terms, the city's territory may also have been significant for herbiculture under Augustus. See, Rouanet-Liesenfelt 1992: 184, 187-188; Chaniotis 1999: 209-210; Baldwin Bowsky 1999: 338, note 47; Baldwin Bowsky 2006: 410, note 18; Vogeikoff-Brogan 2012: 91.
[4] Chaniotis 1988: 75; De Caro 1992-1993: 307-312; Baldwin Bowsky 1995: 50 and 57; Marangou-Lerat 1995: 131-134.
[5] Viviers 1994: 252-254, for the city's early association.
[6] Baldwin Bowsky 1999: 339; Coutsinas 2013: 261, map 20.
[7] Empereur *et al.* 1991: 510-511; Baldwin Bowsky 2006: 410.
[8] Marangou 1999: 271, 278.

Figure 1. Aerial photograph overlaid with Oikonomakis' trajectory (after Oikonomakis 1984: 68)

Lyttos: a ghost town or developed cityscape

Van Effenterre and Gondicas refer to pre-Roman Lyttos as a 'ville fantôme',[9] yet the city's dominant role in the broader east Cretan region during the Archaic to Classical period is attested by the ceramic record in the principal shrine in the area: Kato Simi.[10] Furthermore, excavations conducted by the Greek Archaeological Service have revealed an active civic centre in the Hellenistic period, albeit earlier activity may have been intermittent, as the city's ceramic record for the 6th, 5th, and 4th centuries BC would suggest.[11] Any Hellenistic activity on the acropolis of Lyttos, however, affirms a continuum of place in terms of civic development, presenting a spatial persistence which is fundamental to the appraisal of the Roman aqueduct presented here. That a sense of geographical connectivity was entrenched in the psyche of Lyttos' citizenry is expressed in the Late Hellenistic tradition of annually celebrating the city's re-foundation in the late 3rd century BC (after its destruction in 221 BC).[12]

It was specifically during the Roman imperial period that the city entered its floruit, a florescence which we can now document architecturally. Archaeological investigations conducted during 1981-1986 indicate an urban design of some magnitude, where the core of the Roman civic centre is marked by the city's bouleuterion lying north of the Byzantine church of Haghios Georgios. The complex included a large room (13.90m x 11.40m), paved with grey limestone slabs, incorporating a platform along its west side.[13] Notable among the finds were three honorary imperial inscriptions dedicated to Hadrian and Sabina.[14] Both epigraphic evidence and architectural style confirm that the structure was in use in the early 2nd century

[9] Van Effenterre and Gondicas 2000: 129-139.
[10] Erickson 2002: 82-86.
[11] Catling 1976: 30; Catling 1984: 64-65; French 1991: 69; Also, Erickson 2002: 48, note 20.
[12] Chaniotis 2009: 33.
[13] Rethemiotakis 1984: 49-65; Chaniotis and Rethemiotakis 1992: pl. 3.
[14] Chaniotis and Rethemiotakis 1992: 27, notes 1, 2 and 4.

AD (e.g. the Corinthian capitals from this structure are comparable to those found in the Villa Dionysos at Knossos).[15]

Another substantial complex, with an imposing 75m long western wall, was uncovered 120m northeast of the bouleuterion. A paved road (*c.* 5m in width) ran parallel to this structure from which at least 9 apsed features could be accessed through narrow entrances. It is also probable that both churches of Haghios Georgios and Timios Stavros located on the summit of the acropolis, were founded over earlier basilicae and possibly Roman foundations.[16]

The city's theatre is celebrated as the largest on the island, yet it is only known through a plan drawn by Onorio Belli, who was sent to Crete as a staff doctor in 1583 and remained there until 1599. The plan Belli sends to his uncle, Valerio Barbarano, in 1586 presents a grandiose monument,[17] with an external diameter of 435 Venetian feet.[18]

Faith in Belli's reports varies among subsequent visitors to the site: Spratt and Taramelli find no trace of the large theatre which casts some doubt on Belli's claims.[19] Taramelli proceeds to include a drawing of the acropolis, by Mariani, in which a possible location for the theatre is marked on the northern slopes of the hill.[20]

In the last century, Spanakis also regrets that the theatre's location had not been secured in the field, but still refers to Belli's theatre as the largest in Crete.[21] Moreover, he contests the positioning of the theatre championed by Taramelli and Mariani, favouring a natural niche on the southern slopes where the open-air facility would be sheltered from harsh northern winter winds at this altitude.[22] Sanders too, while acknowledging Vitruvian influence in Belli's report (de Architectura 5.5), accepts the validity of his general description of the theatre.[23]

The city grounds also served as a source of statuary as early as the 16th century. Francesco Barozzi, in his *Descrittione dell'Isola di Creta* (1577–1578), remarked on the numerous antiquities, countless marble, sarcophagi and ancient buildings on the site.[24] He recalled that in 1567 a sarcophagus of high craftsmanship was found in the region of Lyttos and transported to the hamlet of Diavaide (located between Kastelli and the modern village of Lyttos) to decorate the gardens of nobleman Marcus Cornaro Borgognon.[25] Papadaki highlights that although the account is vague, it still constitutes the earliest testimony to

[15] Chaniotis and Rethemiotakis 1992: 28. The finds from the bouleuterion provide a chronology for the prosperity of the Roman city – and hence its aqueduct construction – during the 2nd century AD. A 2nd-century AD date for the construction of the Lyttos aqueduct is, however, in keeping with the monument's incorporation of comparative construction techniques and styles such as stone bonding courses, a stone-piped inverted siphon, brick facing and brick buttressing (as seen in the *castellum*). When the aqueduct ceased to function is uncertain, but Gigourtakis (2013: 56, note 12) believes that the citadel fortification precluded the continued use of this public amenity. Tsigonaki (2007: 274) suggests a 7th century AD date for the fortification of the acropolis at Eleutherna (Sector II), and compares its construction mode with those of the walls of Gortyna and Lyttos; in light of Gigourtakis's observation and Tsigonaki's comparanda we can propose a *terminus ante quem* in the 7th century AD for the functional life of the Lyttos aqueduct. An earlier date could be argued with reference to the abandonment of the bouleuterion at Lyttos which, while probably destroyed in the earthquake of AD 365, may have been deserted earlier, as indicated by fragmentary evidence underlying the rubble layers pertaining to the seismic event.

[16] Sanders 1982: 104, 147; Gigourtakis (2013: 52, pl. 1) publishes a photo of masonry under Haghios Georgios which he believes could be Roman in date.

[17] A detailed study of Belli's plans would perhaps reveal more of his familiarity with the imperial monuments of Rome and the 16th-century architectural schools of Venice, than reflect the actual ancient architectural structures of Crete (Heinrichs 2013: 209; Kelly 2011: 85, note 9). Belli's reference to niches holding three ranges of copper vases, for acoustic purposes, clearly echoes Vitruvian recommendations (de Architectura 5.5).

[18] In a review of Belli's work on the ancient theatres in Crete, published in The Builder (1901: 499), it was noted that '*The original drawings were made to fit pieces of paper of a uniform size; in our [sic] translation they have been all reduced to one uniform scale of the 'foot' of Vicenza, i.e., the old Venetian foot which was in the sixteenth century somewhat larger than the modern English measure (3 Venetian feet = 1 meter)*'. This consideration would be consistent with an outer diameter of 145m (personal observation), although Sanders (1982: 61) reports measurements of 167m for the outer diameter, and 55m for the inner.

[19] Spratt 1865: 97; Taramelli 1899: 390. Belli reports that the seats of the theatre were carved into the bedrock (a soft tufa), which may contribute to the structure's faint footprint archaeologically (cited in *The Builder* 7/12/1901: 499; Di Napoli 2010: 817-818).

[20] Mariani 1895: 390, fig. 39.

[21] Spanakis 1968: 158, pl. ΚΓ.

[22] Spanakis 1968: 159, note 52 and pl. AG.

[23] Sanders 1982: 61.

[24] Barozzi (ed. Kaklamanis) 2004: 309.

[25] Barozzi (ed. Kaklamanis) 2004: 226-227, 311 and notes 110-112.

Figure 2. The aqueduct's elevated substructure (photographer facing north)

the removal of antiquities from the ancient city.[26] Belli also informs us that much of the fine statuary from the theatre area had been brought to the home of his patron, the Venetian provveditore generale, Alvise di Antonio Grimani.[27] The abundance of surface finds still remaining over two and a half centuries later, however, was sufficient to give Spratt the impression of a rich and densely-populated city.[28]

Finally, the most impressive example of monumental architecture on the site is the elevated tract of the city's aqueduct approaching the acropolis from the southeast (Figure 2). In light of this imposing civic facility, the report relating to a monumental theatre, the rich remains from the Roman bouleuterion, and the accounts attesting an abundance of statuary scattered throughout the site, we might assume that the complete urban layout included all the trappings of a Roman city.[29] Yet, the absence of any bathhouse at Lyttos is particularly surprising and can only be viewed as an anomaly in the archaeological record. That a bathhouse must have existed at Lyttos is incontestable, not least because daily bathing was a definitive Roman practice, but because the aqueduct descending to this otherwise elevated city could easily have conveyed water for such an establishment and was probably designed for this purpose.

Lyttos – a challenging topography

The ancient city of Lyttos was founded on a precarious ridge in the western foothills of the Lasithi range, *c.* 1km to the northeast of the modern eponymous village, previously known as Xidas (Figure 1). The settlement survived as a relatively rare example of an upland Roman city in Crete by taking full advantage of its mountainous surrounds to support its water supply system.

[26] Papadaki *et al.* 2010: 6.

[27] Cited in *The Builder* 7/12/1901: 499; see also Heinrichs 2013: 205, note 6.

[28] Spratt 1865: 97.

[29] Kelly 2013: 133.

Belli remarked that the hill ridge on which Lyttos was founded presented the worst topography he had ever encountered for an urban setting, with a remarkably constricted area of level ground.[30] Similarly, Spratt reported that the remains of the site

> 'exist upon the summit of a narrow but tortuous ridge, overlooking the plain upon its eastern margin, just over the village of Xidhia or Xidhi ... Thus situated upon the summit of a western shoulder of the Lasethe, the city of Lyctus or Lyttus occupied occupied a very commanding position; yet the hill has no striking form of outline or feature. The site is, however, a remarkable one; for it is at the point of junction or branching-off of numerous narrow ridges that descend from it to the south, west and north, the sides of all of which are very abrupt' (1865, I: 94-96).

Spratt's overall account of the setting conveys the prominent strategic location of the city, but also alludes to practical problems faced by the inhabitants of the steep slopes of the hill.[31] Four centuries earlier, the Italian priest Cristoforo Buondelmonti noted in the preface of his work *Descriptio insule Crete* (*c.* 1417) that the site was located at a considerable elevation whereby the route to it became almost impassable.[32] Papadaki adroitly attributes the lack of ancient travellers to visit Lyttos to the difficulty in getting there;[33] certainly, Belli found it easier to ship material back to Venice from low-lying coastal sites like Hierapytna.[34]

The position of this Roman city directly above its Hellenistic predecessor, must have been of great importance in symbolic, economic and commercial terms, as, despite the relative inconvenience of Lyttos' topography, the city stayed on its perch in order to control the pass into the Lasithi plain, the harvest from which fuelled the city's lucrative economy. The survival of Lyttos' location from the Hellenistic to the Byzantine period stands as testimony to its durability of place and the ingenuity of Roman engineering which sustained its position.[35] The populace's refusal to relinquish its ancestral place must have presented a challenge to its engineers who were charged with securing the city's water supply.

Aqueduct trajectories on Crete – an overview

In general, aqueducts on Crete availed of the river contours which provided natural trajectories from the uplands of the interior towards the Roman settlements at low-lying – and often costally located – river mouths.[36] Public bathing establishments in coastal locations are represented by the sites of Chania, Kastelli Kissamou, Stavromenos Chamalevri, Aptera,[37] Chersonisos, Lebena, Knossos (not directly on the coastline, but still located on the southern fringe of the low-lying Herakleion-Mallia coastal plain), Sitia, Kouphonisi, Hierapytna, Plakias, Loutro, Souia, and Lissos (Figure 3). In this regard the engineers harnessed height from the natural contours of the topography sweeping towards the sea, which largely dispensed with the need for elaborate man-made substructures to attain the height required to sustain a constant hydraulic flow.[38]

The most common type of Cretan aqueduct descends the valley contours abundant in the natural topography of Crete. Following this premise, the distributions of aqueduct-fed bathhouses on Crete display a slight predilection for the coast (with most Cretan rivers emptying directly into the sea) over the relatively more inland and elevated pre-Roman settings; however, to say that the Romans ignored

[30] Cited in *The Builder* 7/12/1901: 499; Also cited by Spanakis 1968: 160.
[31] Spratt 1865: 94-96.
[32] Buondelmonti (transl. Aposkitis) 2002: 23.
[33] Papadaki *et al.* 2010: 7-8.
[34] Cited in *The Builder* 7/12/1901: 499; Also Beschi 1972–1973.
[35] As previously noted, on p. 148, the Late Hellenistic period, the citizens annually celebrated the city's re-foundation in the late 3rd century BC (after its destruction in 221 BC), Chaniotis 2009: 33.
[36] Kelly 2006a: 303.
[37] Aptera is a problematic addition to the list as, while it is technically coastal, it is also considerably elevated, with baths founded at altitudes of over *c.* 200m. Moreover, Drerup originally proposed that the massive L-shaped cistern at Aptera was supplied exclusively by rainwater (1951: 93), a view that has subsequently been reiterated by Ninou-Kindeli and Christodoulakos (2000: 34), with a wholly convincing model for rainwater harvesting put forward by Grizis (2014).
[38] Kelly 2006a: 303.

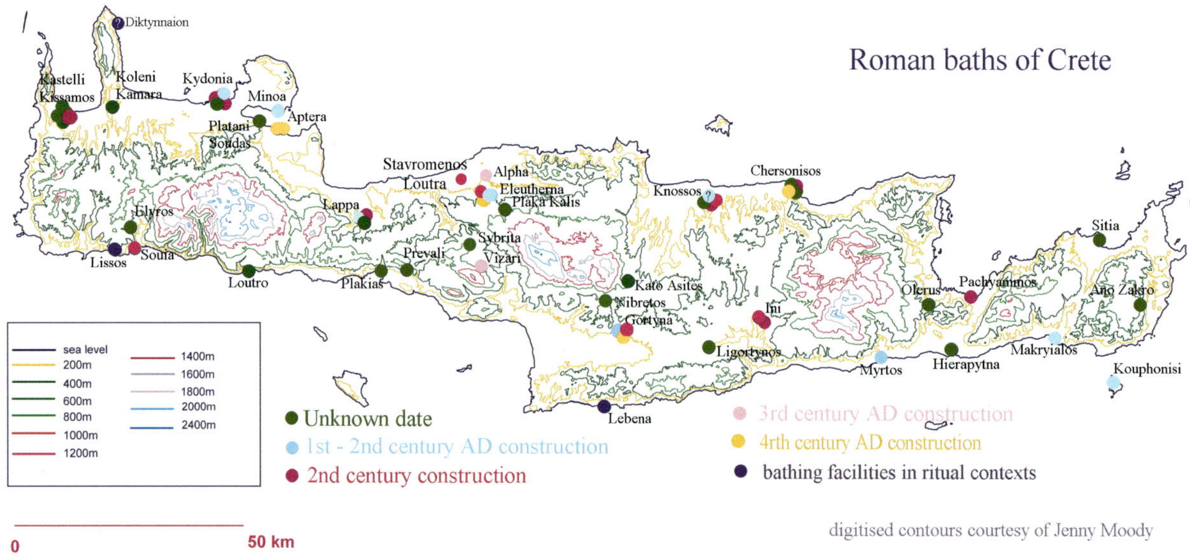

Figure 3. Bathhouse distribution in Crete

the interior of the island, as has been previously implied, is unfounded. There was a significant number of flourishing inland Roman cities across the island (*cf.* Iliad II 645-649 for the earliest foundations), including Lyttos, Ini, Gortyna, Lappa[39], Elyros, Eleutherna, Sybritos, and Axos with other sites of less renown at Plora, Vizari, Kato Asites, Ligortynos, and Oleros.[40]

If location alone is considered, as opposed to quantity, there is not such a huge disparity between the number of sites with public bathhouses in the interior of the island and those located coastally; however, the coastal sites demonstrably flourish and often incorporate a multiplicity of bathhouses. If baths, rather than cities, are counted, then the statistics lean favourably towards coastal sittings – as attested by the preponderance of establishments in Chersonisos where 6 Roman bathhouses have been recorded.[41] It can be concluded that while the island demonstrates a high degree of Romanisation percolating into the interior, there is still a slight majority of wealthy and developed Romanized centres along the coast and its complementary lowland plains, to which Cretan rivers cut courses from the upland springs of the interior.

If the correlation between aqueduct construction and urban development has repercussions for site placement (and, consequently, bathhouse distribution), it also follows logically that bath complexes would be placed at relatively low altitudes within the urban landscape where they could be easily accessed by an aqueduct system. Significant urban development in the lower city seems to be triggered once an aqueduct supply is secured. Site development can often incorporate subtle alterations in aspect, rather than locale, whereby the lower areas of the city developed and expanded. This is probably best demonstrated at Gortyna where the city expanded *c.* 1.6km to the south and east through the alluvial plain of the Geropotamos River.[42] Its sudden spread through the inner reaches of the plain seems prompted by the installation of the Roman aqueducts which descended along the contours of the Mitropolianos Valley.[43]

[39] Lappa as a spring site, has no watering problems and is still famous for its springs today, where even in the heat of August the lower village of Piyi is lush with verdant vegetation.

[40] Kelly 2013: 159.

[41] Galanaki *et al.* 2006: 269; Papadaki *et al.* 2010: 35-40.

[42] Kelly 2006a: 305.

[43] Kelly 2006a: 305. Gortyna is a particularly well appointed inland city situated at the mouth the Mitropolianos river valley, along the contours of which the aqueduct descends to the city. The river does not empty directly in the sea, but in the inner reaches of the Messara's alluvial floodplain formed by the Geropotamos River.

But even if the dominant role bathing played in daily Roman society had an undeniable impact on the attraction of the lower cities over those in the hills, there is no limit to the altitude at which a bath complex can be founded, once the landscape allows for a more elevated water source. The recent discovery of a public Roman bath high on the acropolis at Eleutherna (Sector II), founded at an altitude of over 300m, whose function required running water is a case in point;[44] here the rock-cut aqueduct tunnel emerges from the bedrock at an altitude *c.* 20m higher than the baths, at a point *c.* 30m to the south along the ridge.[45] This phenomenon is also encountered in the Anagni area near Rome where the location of many Roman villas and cisterns on the crests of ridges implies water transport by conduits from a higher point further along the ridge.[46]

The highest mountain ranges in central Crete catered for the successful development of Roman cities over certain altitudes, far from the coastal plains. MacGillivray describes the island's dramatic orography with appropriate prose:[47]

> 'Its four mountain massifs with hard, blue-grey limestone at the core explode from the sea and climb suddenly to 2,200 meters in the White Mountains at the far west, then go even higher, to 2,456 meters at Psiloritis, the pinnacle of the Idaean range, into which the White Mountains blend above Rethymnon. Psiloritis, the ancient Mount Ida is the highest peak on the island and the focal point of legends both ancient and modern: it was the obvious setting for Kingsley's Theseus to pursue the Minotaur. There's a break in the middle of Crete linking the rolling hills of the northern center, where Herakleion (Candia) sits, with the Mesara plain on the southern coast. Here, Ida's foothills mingle with the beginning of the Diktaian range, which there rises to the east in a ring around the Lasithi plain, a concealed plateau 900m above the sea'.

But, even with such a suitable landscape for river-valley aqueduct construction, when a fixed location for a particular city becomes invaluable in terms of trade and production, as can be argued for both Lyttos and Eleutherna,[48] aqueduct construction along valley contours may simply not be feasible and another solution needs to be found.

The Lyttos aqueduct – an aerial view

Lyttos, although a substantially elevated inland site (the highest Roman city in Crete), is still overshadowed by the Lasithi range, and that is a crucial consideration for the city's continued success. The city's bouleuterion, representing the core of the Roman civic centre, was founded at the impressive altitude of over 600m (nearly twice the height of the terrain on the acropolis ridge at Eleutherna, Sector II, in west central Crete). Lyttos' exceptional altitude was still only relevant for its hydraulic engineers in relation to the city's surrounding topography which would ultimately determine the feasibility of an aqueduct supply. Moreover, the height at which an aqueduct enters a city prescribes the altitudes of its bathhouses which need to be relatively lower-lying within the cityscape.

The distance between the city and the aqueduct's water source is *c.* 10km,[49] as the crow flies, yet Nikos Oikonomakis, who studied the aqueduct in detail, recorded a length of 22km for the aqueduct's winding trajectory.[50] The extensive length of the aqueduct arises from the nature of the intervening terrain between source and city, frequently indented with gorges and ravines; the aqueduct of Lyttos had to

[44] Tsigonaki 2015; Karanastasi 2015: 419-421, pls 1-3; Tsigonaki and Yangaki 2015: 430.

[45] This does not negate the fact that there is substantial development in the lower city, as is clear from the remains of villas and bathhouses in Sector I, Themelis 2002; however, a complementary development was still sustainable in the upper city due to the availability of water higher up along the summit of the acropolis.

[46] Thomas and Wilson 1994: 149, 180, 185.

[47] MacGillivray 2001: 117.

[48] Kelly 2006a: 306-307; Kelly 2006b: 248.

[49] A comparable length was recorded in the early 15th century by Buondelmonti who remarked on an aqueduct tract running down from the high mountains to Lyttos (which he referred to as Myrrina), Buondelmonti [transl. Aposkitis] 2002: 49; cited in Van Spitael 1981:171.

[50] Oikonomakis 1984: 67, note 9. Such extended trajectories are seen elsewhere, cf. for instance, the aqueduct of Nîmes, which, although 39km long, its source is located only 17km from the city.

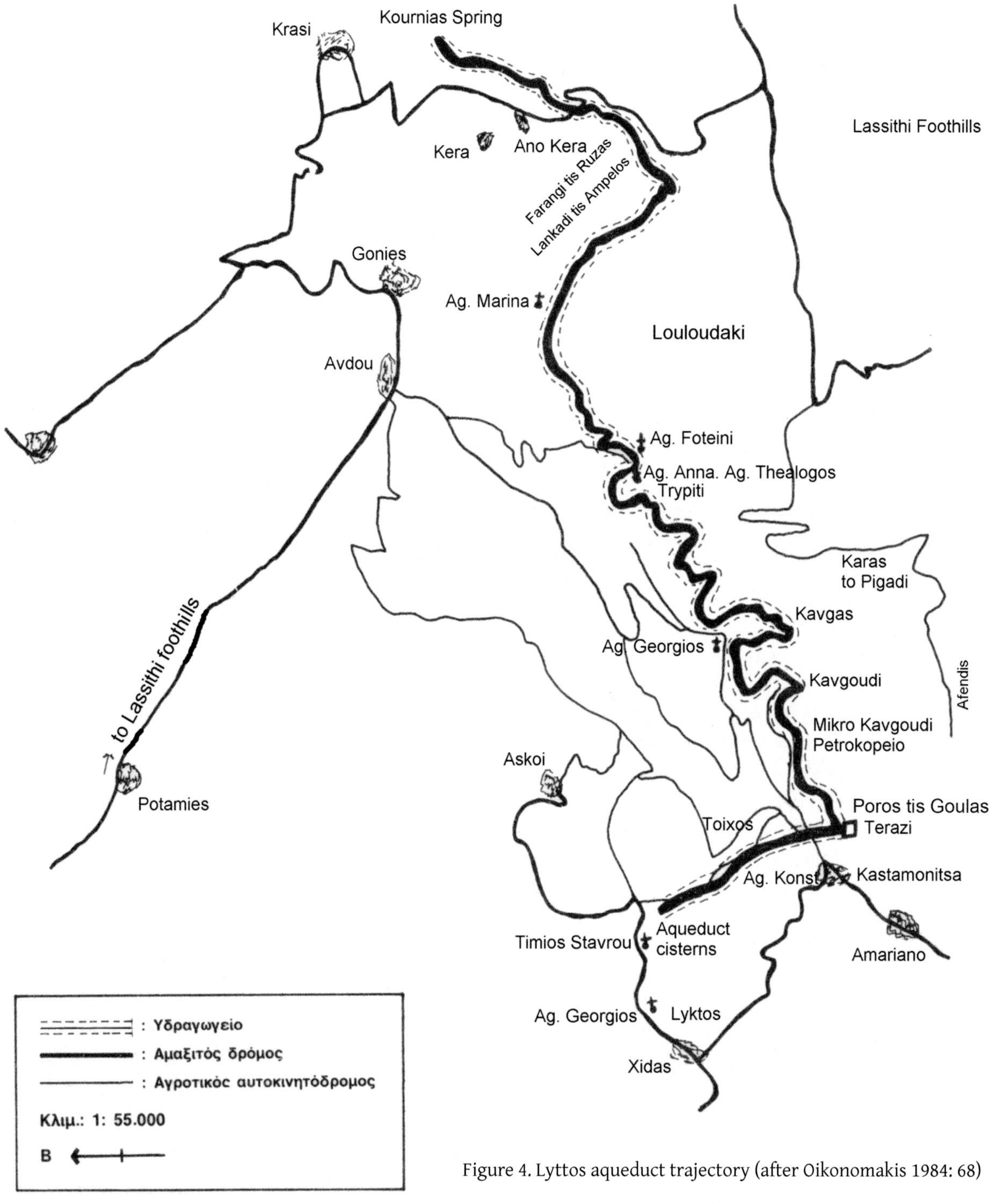

Figure 4. Lyttos aqueduct trajectory (after Oikonomakis 1984: 68)

completely avoid the deep and expansive valleys of Langadas and Astirakia lying on the direct line between the springs of Kournias and the city (Figure 4).

The aqueduct of Lyttos follows the contours of the steep west-facing slopes of the mountain range of Lasithi, which provided the necessary elevation for the operation of such a gravity-flow system (Figure 1). The karstified highlands through which the Lyttos aqueduct travelled created significant repercussions for its design and layout.[51] Its winding trajectory, following the mountainside contours, clearly increased the length of its route but reduced the need for substantial substructure walls.

[51] Fernandes *et al.* 2012: 356.

Its mountainous tract is referred to as Αμαξοδιάβαση by the villagers of Kastamonitsa, while the following section which departs from the massif is, appropriately, known as ο Τοίχος.[52] The aqueduct only relied on substantial free-standing substructures on its departure from the massif which it left *c.* 2.5km before its destination. On diverting from its mountainous contour, it travels along a lower-rising ridge in a north-westerly direction towards the acropolis of ancient Lyttos (Figure 1).[53]

In brief, the aqueduct traverses the hinterlands of the villages of Krasi, Keras, Gonion, Avdou, Kastamonitsa and Xida (Figure 4). Oikonomakis highlights the numerous settlements the aqueduct passes through along its route:[54] starting from the spring at Kournias, where it passes just above the villages of Kera and Ano Kera, into the Gorge of Roza and the ravine of Ambelos, skirting the chapel of Aghia Marina, Stolous, Aghia Fotini, Lugias, Kalastra, Tou Kuthiou to Plai, Kavga, Kavgoudi and Mikro Kavgoudi leading to the tip of the ravine of Poros tis Goulas where it departs from the mountains. Subsequently, it passes through locales under the following toponyms: Poros tou Toikhou, Koutelo, Kalyva, Platyvola and Troullia before it terminates at the ancient city of Lyttos.

Prior to Oikonomakis' definitive work, which finally proved that the aqueduct terminated at Lyttos, the aqueduct was thought also to supply Chersonisos. The two separate aqueducts (supplying Lyttos and Chersonisos) were misconceived, not only as one and the same aqueduct, but also as the longest aqueduct in Crete. The misconstruction that the aqueduct of Chersonisos formed a continuous extension to that of Lyttos provided an architectural expression of the historic connection between the two cities whereby an extensive aqueduct trajectory could visibly unite the two broad geographic civic territories. Based on this fallacy (that they represented one sweeping aqueduct linking the two cities), both aqueducts were subject to various misinterpretations.

The aqueducts of Lyttos and Chersonisos were not mentioned by the ancient sources and eluded many travellers over the centuries, as their physical remains were mostly confined to the inaccessible slopes of forested riverbeds and steep mountainous terrain. Early reports of the aqueduct of Lyttos were either brief or restricted to its elevated siphon bridge (technically a venter bridge as discussed below) approaching the city (Figure 2). Similarly, mention of the Chersonisos aqueduct focused on its bridges, at Xerokamares 1 and 2, rendering them isolated features deprived of architectural contextualisation or comparative analysis (Figure 5).

The misconception is traceable in the very earliest descriptions of the aqueducts. Buondelmonti, in the early 15th century, was the first to present the aqueducts as an integrated system. He incorporated the two aqueducts under a general discussion of the monuments of Chersonisos.[55] His support of a unified system can be deduced from the fact that, when reporting on the water supply for Chersonisos, he makes no comparison with the aqueduct of Lyttos, while noting a source high in the mountains 'ad altis montibus' at a great distance from the city of Chersonisos.[56]

Belli was more forthright in claiming that after Lyttos the aqueduct continued to Chersonisos. He observed that from Chersonisos to Lyttos

> 'extended an excellent and commodious road and that the same distance was traversed by a stupendous aqueduct, which commencing at a spring in the Lasithi mountains, four or five miles above Lyttus, conducted the water first to that city and afterwards to Chersonisos'[57]

Belli's impression of connectivity was compounded by the fact that, not only did the main ancient thoroughfare of the region run parallel to the aqueduct of Chersonisos, but, on departing from it, the

[52] Oikonomakis 1984: 73 and 91, note 114.
[53] Oikonomakis 1984: 69.
[54] Oikonomakis 1984: 69.
[55] Cited in Oikonomakis 1984: 85.
[56] Buondelmonti [transl. Aposkitis] 2002: 49; Galanaki *et al.* 2006; Van Spitael 1981: 160.
[57] Cited in *The Builder* 7/12/1901: 499; Also, cited in Falkener 1854: 16.

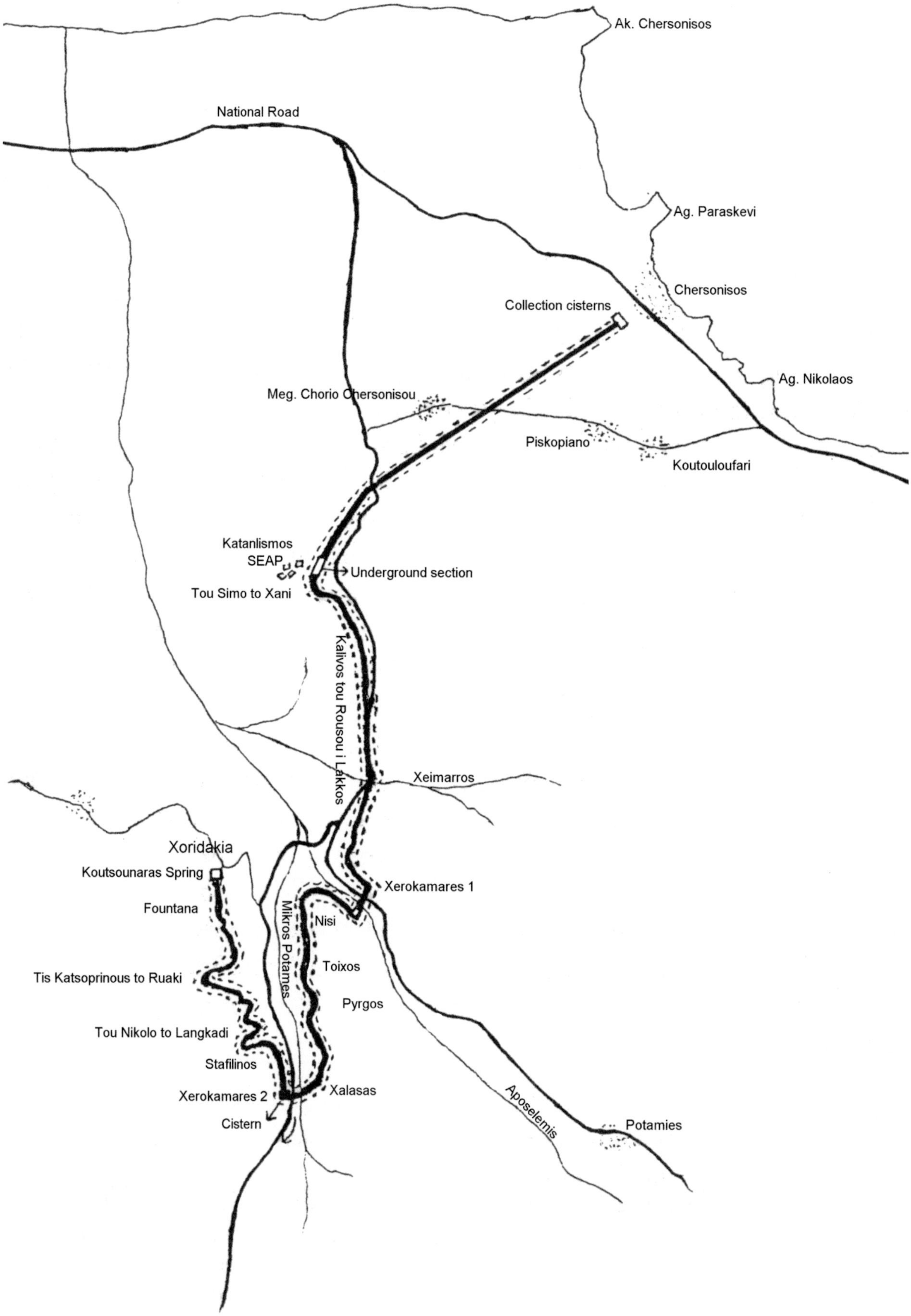

Figure 5. Chersonisos aqueduct trajectory (after Oikonomakis 1986: 19)

road also continued towards Lyttos. Consequently, the aqueduct itself became synonymous with the main communication route. Mariani encountered the remains of the Xerokamares 2 bridge along the Chersonisos aqueduct, when he travelled this road from Lyttos towards Chersonisos. Taramelli, also followed the length of this road (then known as the King's road) after the Kastelli of Chersonisos, making the same journey.[58]

Belli's description of the aqueduct emphasises its extended winding course:

> 'The aqueduct, which is about 14 feet in thickness, of solid masonry, and between 15 and 16 feet in height, winds round so many hills, and crosses so many valleys, that its length must be upwards of thirty miles [48.28km], nearly the whole of which is perfect, and seems likely to remain so as long as the world shall last. But what causes most astonishment is the prodigious height of the arches which cross the ravines, some of which are a hundred feet in height. This is the most remarkable aqueduct in Candia, though there are several others of less grandeur and extent'.[59]

Belli's over-estimated length of over 30 miles (48.28km) for the aqueduct must result from his belief that the aqueduct continued the whole way from Chersonisos to Lasithi. In reality, the direct distance between Chersonisos and the source of its aqueduct is only 6km while, according to Oikonomakis, the *actual* length of the aqueduct is *c.* 14km.[60] As already noted, the direct distance between Lyttos and the source at Kournias is 10km, yet the route of this aqueduct measures 22km.[61] But even allowing for a circuitous trajectory, the combined lengths of both aqueducts still only measure 36km, as opposed to Belli's 48.28km.[62] Belli, although over-enthusiastic in his descriptions and misled in his estimate, accurately conveys the undeniable grandeur of the aqueduct of Chersonisos.

The distorted merging of these aqueducts into one unified system is further perpetuated by local modern folklore. Oikonomakis relates a story, familiar to the villagers of Avdou and Gonies, of a foreman who lost his trowel at Kera (at the start of the Lyttos aqueduct) during the aqueduct's construction.[63] By the time he had realised his loss, the missing trowel had already reached Chersonisos, having been relayed along the aqueduct tract from person to person. The safe return of the trowel, passed hand-to-hand from Kera to Chersonisos, presumably by members of the construction teams, enshrines the popular belief that the monuments were physically integrated.

That this misconception also persisted in modern archaeological reports is a more serious matter. In 1982, prior to the publication of Oikonomakis' articles, Watrous claimed that a vast aqueduct fed from a sink-hole at Kastamonitsa first supplied Lyttos and then descended towards Chersonisos.[64] Even Sanders provides conflicting accounts, as, while he clearly states that Chersonisos was supplied by a long aqueduct descending from the Lasithi mountains and that the same system supplied Lyttos, he subsequently cites a source for the Chersonisos aqueduct at Haghios Pandeleimon, following Spanakis.[65] More recently, Raab also supported the possibility of one continuous aqueduct.[66] The persistence of the error into modern scholarship seems to be reliant on the structural similarities between the two aqueducts, their geographical settings and the intertwined histories of the two cities. These factors contribute to the illusion of a unified sweeping water supply system, as extolled in the writings of early travellers and perpetuated by the popular folklore of the region.[67]

---

[58] Mariani 1895, VI: 238; Taramelli 1899: 372.
[59] Cited in Falkener 1854: 16.
[60] Oikonomakis 1986: 52.
[61] Oikonomakis 1984: 67.
[62] Following Oikonomakis' (1984: 67; 1986: 52) estimates of 22km for Lyttos and 14km for Chersonisos.
[63] Oikonomakis 1986: 81. That the mountainous tract of the aqueduct in the hinterland of the village of Kastamonitsa is known as Αμαξοδιάβαση (Oikonomakis 1984: 73 and 91, note 114), suggests that the structure was used as a thoroughfare after its original function ceased; today, the aqueduct forms a terraced path along the side of the mountain providing a direct passage northwards towards Gonies and Avdou.
[64] Watrous 1982: 24. It should be noted that Watrous' research focus was far from Roman and that he had completed, and published, his work on Lasithi by 1982, prior to Oikonomakis' definitive publication on the aqueduct of Lyttos in 1984.
[65] Sanders 1982: 19; Spanakis 1982: 146.
[66] Raab 2001: 27.
[67] Oikonomakis 1986: 76.

In spite of such discrepancies, there were some early authors who viewed the aqueducts as separate and functionally independent. Unfortunately, the few authors who supported the autonomy of the Chersonisos aqueduct restricted their references to specific locations with little attempt at tracing the trajectories in the field in order to establish their independence. It is not certain from Spratt's accounts whether he connected the trajectories as his comments are confined to the aqueduct bridge in the Potamies Valley; however, the fact that the aqueducts are depicted as detached entities in his accompanying British Admiralty Map intimates that he considered them to be distinct.[68]

Mariani also seems to distinguish between the aqueduct of Lyttos and that of Chersonisos since, in his description of the Xerokamares 2 bridge along the Chersonisos aqueduct, he claims that the aqueduct brought water to Chersonisos from a source at Leontari in the Aposelemi Valley.[69] Taramelli follows suit, also recording the bridge, in reporting a source at Leontari and providing a topographic plan in which the aqueduct starts just north of Lyttos.[70]

Spanakis also treats the aqueducts as two separate entities but suggests that the Chersonisos aqueduct tapped the springs of Haghios Pandeleimon and also the spring of Spyridon.[71] Presumably Spanakis is referring to the Byzantine Church of Haghios Pandeleimon, which dates back to the 10th century AD,[72] located between the hamlets of Tzigkouna and Mpitzariano (to the east of Kato Karouzana), near to which there is also a spring site at the chapel of Haghios Spyridonas in the lowlands northwest of ancient Lyttos (personal observation). While Spanakis correctly distinguishes between the two aqueducts, he locates the source incorrectly; as Oikonomakis' later work shows, the source lies in the district of Kalo Chorio some 6km to the north of Haghios Pandeleimon. The course of the Chersonisos aqueduct doubles-back on itself from the free-flow bridge of Xerokamares 2 (Figure 5).

Lolos is also explicit about the distinction, but, follows Spanakis in stating that the aqueduct of Chersonisos[73]

> ... 'is now definitely disassociated from the aqueduct supplying Lyttos. The source was probably tapped at Aghios Panteleimon (Piyi)'

Mandalaki, who excavated the massive cisterns at Chersonisos, revises her earliest suggestions of a source at Haghios Pandeleimon and correctly relocates the source in the wider area of Kalo Chorio, in line with Oikonomakis' earlier findings.[74] On the map of Crete in the 2000 edition of Barrington's atlas, the aqueducts of Chersonisos and Lyttos are depicted as separate and the Chersonisos tract accurately doubles-back to the northwest.[75]

But the absolute credit for establishing the aqueducts as two distinct monuments must be awarded to Oikonomakis who compiled the first thorough studies of these aqueducts in 1984 and 1986. While their architectural articulation (and combined functionality) was always improbable in theory and impossible in practice, it was Oikonomakis' study which ultimately affirmed that the two aqueducts are distinct entities and should be regarded separately. He traced the two aqueducts in the field, an act which finally proved that they were hydraulically and functionally independent. He followed the course of the Chersonisos aqueduct with precision from the bridge at Xerokamares 2 towards its source at the Koutsounara spring on the fringe of the Khoridakia area, which he marks on the map to the east-southeast of Kalo Xorio (Figure

---

68 Spratt 1865, I: 103; the map was originally published in 1852.

69 Mariani 1895, VI: 238.

70 Taramelli 1899: 373-376, figs 35-36 (bridge). Tarmelli (1899: 374 and 371-372, fig. 34) also marks a toponym, which seems to read 'Karuano', on his map close to the source of the aqueduct trajectory depicted, while on Spratt's map 'Karsiano' is clearly marked in a similar position (Spratt – British Admiralty Map 1852). Both cartographic labels may be versions of the toponymn *Karouzana* which marks a small hamlet due north of Haghios Pandeleimon, the position of which corresponds with the cartographic references (personal observations).

71 Spanakis 1968, 161, n. 59; 1969, I: 384; 1981: 17, n. 11.

72 And possibly earlier according to Sanders 1982: 104 and 147.

73 Lolos 1997: 31.

74 Mandalaki 1996: 634; Mandalaki 1999: 258; Mandalaki 2001. Also cited in Galanaki *et al.* 2006: 266, fig. 1. *Cf.* Oikonomakis 1986: 55, map.

75 Talbert 2000: map 60 – the Cretan material was compiled by John Bennet.

Figure 6. The rock-cut tract (Kokkino Deti tou Kavga)

5).[76] The detailed analysis he assiduously recorded regarding the route of the aqueduct from Xerokamares 2 up to the source confirmed that the system was independent from that of Lyttos. The presence of springs in the area of Xoridakia and Epano Vrisi dispensed with any need to probe the countryside for water sources further south in the direction of Lyttos (as suggested by Spanakis, albeit mooted in Taramelli's earlier map). It seems that the U-turn in the Chersonisos trajectory was rather unexpected by the later aqueduct hunters.

The rock-cut water channel

The channel of the Lyttos aqueduct was constructed in various ways along its course; it could be rock-cut, revetted with masonry or supported on a free-standing substructure wall. Along its mountainous tract, in places where the limestone massif became too steep to construct any type of substructure, a terrace was cut into the bedrock (Figure 6). A channel was then cut along the inner edge of this horizontal plane, where it could be protected from the elements and erosion.[77] Taramelli reported on a section of the rock-cut channel in the Gorge of Roza – in the highest reaches of the aqueduct trajectory. His accompanying section drawings incorporate a small indentation on the inner side of the ledge which may indicate some form of covering over the channel to prevent water absorption or any debris from blocking the channel (Figure 7A).[78]

Taramelli's sketches offer a good general impression of the channel. In one section drawing, the channel width is relatively regular throughout its depth (0.40m wide at the top and 0.35m at the base) (Figure 7B), while in another, the channel tapers gently towards its base (Figure 7A).[79] The vertical height of the inner edge of the ledge depends on the sheerness of the slope, and can reach up to 2m, while the force and

[76] Oikonomakis 1986: 55, map.
[77] Oikonomakis 1984: 71.
[78] Taramelli 1899: 406-407, fig. 44.
[79] Taramelli 1899: 406-407, fig. 44.

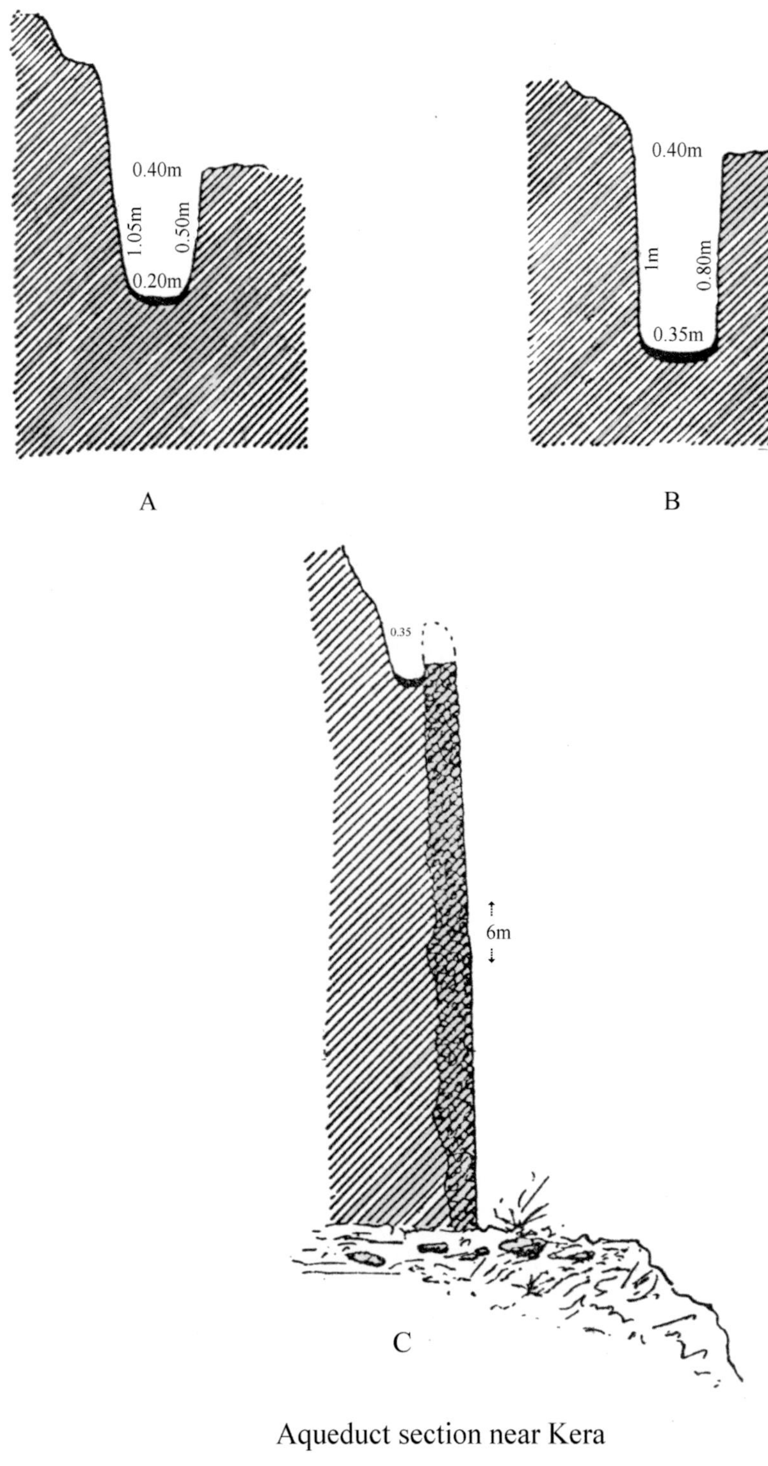

Figure 7. Rock-cut sections (after Taramelli 1899: 406-407, figs 44 and 45)

direction of the water flow determines the shape and dimensions of the channel.[80] Taramelli presents a range of vertical depths for the outer wall of the channel, between 0.50-0.80m, reflecting the actual depth of the water channel, while his inner measurements, of *c.* 1m, relate to protective considerations and the steepness of the mountainside.[81]

Another section in the area of Kavgoudi demonstrates the same mode of construction, even though here the channel is deeper in places. In one section, the ledge width is *c.* 1.20m, while the channel depth is *c.* 0.30m. In another, the width at the top of the channel measures 0.33m but tapers to 0.18m at its base, while the channel depth drops to 0.60m (where the ledge is narrower than 1m wide).[82] Oikonomakis noted that there seems to be some patterning in the dimensions represented by a correlation between ledge width and channel depth: when the ledge width is less than 1.20m the channel tends to be deeper. The differences in the section measurements are directly correlated with specific challenges encountered at various points along the route.

The revetted water channel

In some areas along its mountainous tract the channel had to be supported by a constructed wall or revetment.[83] The necessary level for the channel was projected along a notional line and a masonry substructure (a stone revetment of roughly cut limestone slabs) was built up from the mountain slope

[80] Oikonomakis 1984: 71.
[81] Taramelli 1899: 406-407, fig. 44; for Oikonomakis' dimensions see 1984: 71-72.
[82] Oikonomakis 1984: 71-72, fig. 2.
[83] Oikonomakis 1984: 69-70.

to meet this projected horizon. The drop from this line-level to the bedrock varied according to the unevenness of the terrain, resulting in differences (often ranging from 2m to 3m) in the height of the substructure wall.

The height could also 'drop' dramatically at specific points, e.g. in the Gorge of Kavga where the bedrock receded to over 6m.[84] Taramelli also recorded a similar drop just above the entrance to the Gorge of Roza and to the south of the Monastery of Krasi (presumably the Kera/Kardiotissa Monastery, just east of Kato Kera).[85] Here Taramelli records a long stretch of the aqueduct supported by a 6m high wall (Figure 7C).

The free-standing substructure wall

In order to reach the city the aqueduct was obliged to depart from its mountainous contour and follow a lower rising ridge connecting the acropolis of Lyttos with the steep face of the Lasithi Massif. The length of the aqueduct from Terazi to Lyttos is *c.* 2.5km and this tract is locally known by the appropriate toponym 'O Toikhos'. On its departure from the mountainside the aqueduct travels in a northwesterly direction on a substantially-elevated free-standing substructure wall for just over *c.* 0.5km, at which point (25°22'54.07"E 35°11'55.53"N), it angles approximately 40 degrees further northwestwards and continues for another *c.* 0.5km in a straight line (Figure 2). From here, it follows a more circuitous route towards the city, availing of the natural contours in the terrain to maintain a steady slope. On the final leg of its journey, it can only be traced as a fragmentary low substructure wall (often only appearing at ground level or as an elevated dirt track), with a thickness of 2.60m (personal observation).

It is the elevated section of the aqueduct which imprinted so strongly on the psyche of ancient travellers. As early as the 16th century Barozzi makes reference to the elevated substructure when he reports on an architectural elevation spanning a wide valley, conveying water from one mountain to another.[86] This section of the aqueduct's trajectory was also recorded by Spratt in the 19th century:

> 'this connecting ridge or neck divides the watershed or ravines that lead to the north and south coasts of Crete, by the Aposoleme Valley, the Pediada Plain, and the Valley of Ene; and there are the remains of an aqueduct upon this col or neck, that led from some distant part of the mountains of the Lasethe to Lyttus in its ancient and flourishing days'[87]

Oikonomakis measures the greatest thickness of the wall, 4.70m, at Poros tou Toikhou, just over the village of Kastamonitsa near the church of Haghios Constantinos (Figure 2).[88] Oikonomakis also recalls that in the late 1970s drilling exposed the base of the wall here revealing 7m thick foundations. This pass is also the lowest point in ground-level, on descending from the mountainside, at a height of *c.* 550m. Here the wall still stands to a height of 6.5m with a thickness of 4.40m where it is intercepted by the modern road (a breach which may indicate the original existance of a connecting gate) (personal observations). Along the tract of Poros tou Toikhou 2, running from the church to the north, the aqueduct's substructure wall survives to the substantial height of 8.10m (Figure 8). The wall thickness, where the wall is perforated (at Tripa tou Toikhou), perhaps reflecting the presence of an original doorway, measures 2.35m.

The elevations at Poros tou Toikhou 2 are constructed with roughly-cut blocks of local limestone embedded into a mortared-rubble core, where the flat side of the cut stone serves as the facing. This facing was divided into three tiers of half-metre bands defined by horizons of larger neatly-shaped limestone blocks (average dim. 0.80m x 0.30m). That these bonding courses are not superficial, but continue through the core of the wall, is confirmed by the section of the wall exposed by the road-cut at Poros tou Toikhou 1, where the seams visibly penetrate the core matrix (personal observations). That they continue through the wall at Poros tou Toikhou 2 is suggested by their positioning at the same height on both faces of the wall. Along

[84] Oikonomakis 1984: 70.
[85] Taramelli 1899: 407, fig. 45.
[86] Barozzi [ed. Kaklamanis] 2004: 309.
[87] Spratt 1865, I: 99.
[88] Oikonomakis 1984: 74.

Figure 8. The aqueduct's elevated substructure (photographer facing south)

the elevation at Poros tou Toikhou 2, the outer facing of stone has collapsed in the lower portion of this elevation but the core itself remains intact (Figure 8). If the absence of stone facing indicates a form of collapse, rather than active quarrying, the fact that it is confined to below one of these horizons affirms their stabilising function.[89] The wall also steps in slightly, by *c.* 0.14m, along these horizons which may reflect different constructional stages. These shallow recesses may also indicate the use of free-standing scaffolding, similar to the type demonstrated by Adam, as no put-log holes are visible.[90]

The inverted siphon

The term 'inverted siphon' describes the system employed by Roman engineers which enabled water to be pushed up the far side of a valley along an aqueduct route.[91] The typical inverted siphon forms a V- or U-shape. In essence, water which travels with free-flow up to a point, enters a collection tank or tower at the top of a valley's edge to allow it to settle. The water leaves the tank in a reinforced or robust piped system which conducts the water downhill, from here pressure forces the water up the far side of the valley. The water finally enters another tank or tower on the far side of the valley, from where it continues with free-flow. The hydraulic gradient is defined as 'a notional line drawn between the intake and the outlet'.[92] The construction of a venter bridge to support the piped supply served to ease pressure on the walls of the stone pipes and facilitated an ascent to a level only slightly lower than at the start of the system.

[89] Similarly, stone bonding courses feature in the bridge of Xerokamares 2 along the Chersonisos aqueduct (although this bridge is otherwise faced with brick). This common feature points to the contemporaneity of the aqueducts and also intimates good relations between the two cities who may have shared a trained workforce.
[90] Adam 1994: 84, fig. 182.
[91] Hodge 1983: 174-175.
[92] Lewis 1999: 163.

Such a siphon is detectable along the Lyttos aqueduct at the point at which the aqueduct departs from the mountainside on approaching the city. In order to follow the lower rising ridge, which effectively connects the mountain range to the Lyttos hill, the aqueduct incorporated an abrupt directional change. Oikonomakis records a water tower at this juncture, designed to compensate for the directional change and to regulate the necessary pressure to traverse the connecting ridge.[93] He describes the remains of a $2m^2$ water tower at the aptly-named locale of Terazi near the entrance to the gorge of Poros tis Goulas (Figure 4).[94] Its southern wall survived to a height of 0.75m and abutted the last section of the mountainous trajectory of the aqueduct, while traces of *opus signinum* were visible on its floor.

As noted, the greatest width of the aqueduct substructure wall occurs at Poros tou Toikhou 1 near the church of Haghios Constantinos and this also represents the point that would have had to support the greatest elevation of the wall (as, at *c.* 550m, it is the lowest point in the terrain along the aqueduct's course).[95] The ground level here is *c.* 50m below the level of the water tower at Terazi;[96] Oikonomakis deduces that *if* the aqueduct travelled with free-flow at Poros tou Toikhou 1, a considerably elevated substructure, complete with communicating doorways and arches, would have been necessary; while a connecting gate may have existed at Poros tou Toikhou 1 -where the modern road cuts its trajectory in the middle of the pass between the hill of Kalyvas and the locale Kamini (and perhaps even a connecting doorway at Poros tou Toikhou 2) – the wall is otherwise solid along its length.

It seems highly unlikely that the Lyttos aqueduct continued from Terazi to the city with free-flow across this low-lying area; in this case, the greatest height of the wall would have occurred at this point and the substructure wall would have to reach *c.* 50m; although statistically conceivable and attainable, Roman bridges rarely exceeded a height of 30.5m.[97] Some notable exceptions exist: the Anio Novus had a height of 47.52m, while the highest aqueduct bridge, the Pont du Gard, rises to a height of between 47.4-48.7m above stream level.[98] Following the relatively low height of most Roman bridges and the abrupt drop in terrain on leaving the Lasithi massif, it can be argued that an inverted siphon facilitated the crossing of the low-rising ridge towards Lyttos.

Following this rationale, it can also be argued that an inverted siphon also facilitated the crossing of the Aposelemi Valley along the Chersonisos aqueduct (at Xerokamares 1) on the basis of the excessive height needed if the water travelled with free-flow. Spratt who believed that the water passed through the valley with free-flow calculated a necessary bridge height of over 70m (Figure 5).[99] The use of an inverted siphon would have constituted the only hydraulic solution for crossing the Aposelemi Valley at this point, given that few free-flow Roman bridges exceed 30.5m, with the highest well below 50m.[100]

The main advantage of the use of an inverted siphon over a free-flow bridge is its considerable depth. As noted, exceptionally few bridges achieve a height of over 45m, whereas inverted siphon depths are, on average, much greater. The best-preserved siphons in France are the Soucieu and Beaunant along the Gier aqueduct. The Soucieu siphon is 1.20km long, with a 200m-long venter bridge, and reaches a maximum depth of 92.8m with a hydraulic fall of 9m, whereas the Beaunant is the deepest of all Roman siphons, descending to a depth of 123m, *i.e.* 12 atmospheres.[101] The existence of inverted siphons along the Lyttos and Chersonisos aqueducts can be added to a growing list of Cretan examples (two have been noted along the Gortyna aqueduct, one along the Knossos supply and another along the aqueduct of Elyros).[102] Their relative frequency in Crete would suggest that they were not more expensive to construct than bridges,

[93] Oikonomakis 1984: 73.
[94] Terazi is an Arabic word that roughly translates as water plumb-line referring to balance and scale.
[95] There is a definite correlation between substructure thickness and original height in aqueduct studies.
[96] Which Oikonomakis (1984: 78) records at *c.* +600m.
[97] O'Connor 1993: 154.
[98] Fabres *et al.* 1991: 67.
[99] Spratt 1865, I: 103-104.
[100] O'Connor 1993: 154.
[101] Hodge 1983: 221.
[102] Kelly 2004; for Elyros see Christodoulakos 2011: 171-176.

a factor which has often been offered to explain the seeming scarcity of inverted siphons throughout empire.

The proposal that the use of an inverted siphon was used to traverse the lower-rising ridge to Lyttos is supported by the discovery of a stone pipe, in an area littered with fallen debris from the aqueduct itself, near the aqueduct's substructure wall at Poros tou Toikhou 2.[103] Stone water-pipes constitute elements of inverted siphons where the pipe is comprised of a series of large stone blocks with cylindrical perforations. The stone pipe represents a unique find on Crete and, while it may seem surprising that only one pipe of the Lyttos system would survive, in the case of the Aspendos siphon, where 3,400 stone pipes were used, only 250 have survived, and even then, only as they had been built into the elevations of the Seljuk bridge; this amounts to a less than 8% survival rate.[104] The stone pipe from Lyttos is a perforated cube with a slightly tapering cylindrical conduit measuring *c.* 0.22m in diameter and 0.50m in length. The face of the pipe in which the perforation is widest, measures approximately 0.63m x 0.52m, while the dimensions for the opposite surface, i.e. with the small end of the perforation, measure 0.57m x 0.57m.[105] It might be deduced from these dimensions that the amount of water transported was relatively small and while there must be some correlation between pipe size and capacity on a general level, judging water capacity from stone pipe dimensions is not a particularly adequate assessment as it does not allow for the rate of flow.[106] The size is roughly comparable to that of the pipe drums from Oenoanda in Lycia, which measure 0.53m x 0.53m with a perforation diameter of 0.17m.[107] The stone pipes of the Patara siphon, also in Lycia, are slightly bigger, measuring 0.90m x 0.90m with a perforation diameter of 0.33m.[108] The pipes of the Aspendos siphon are similar in scale to those from Patara, measuring 0.90m x 0.90m, with a perforation diameter of 0.28m.[109]

The discovery of a stone pipe at Poros tou Toikhou 2, in association with the Lyttos aqueduct, the application of the toponym Terazi and the considerable descent and span presented by the pass all point to the use of an inverted siphon on the aqueduct's departure from the mountainside. The inverted siphon must have started where the aqueduct was obliged to depart from the mountainous contour in order to traverse the lower ridge. Its starting point is anchored by the ruined structure at the aptly-labelled toponym of Terazi (meaning balance or plumb level) where Oikonomakis identified the system's header tank at a height of *c.* +600m. [110]

On the far side of the pass, the ground-level only reaches a height of 600m at the following coordinates (25°22'34.90" 35°12'4.98"N) in the region of Plativola. This level marks the end of the siphon system, indicating a length of over 1km for the pressurised system (personal observation). It is no coincidence that this point also marks the end of the straight (albeit angled 40 degrees mid-way along its length) tract, as after this 1km stretch, the aqueduct continues on a circuitous route availing of the natural topography – clearly indicating that it is being conducted using free-flow (personal observation).

The aqueduct slope

The source of the aqueduct is thought to lie at an altitude of *c.* 700m at Kournias, *c.* 0.5km south of the village of Krasi, near the valley of the same name, on the west flank of the foothills of the Nissimos chain.[111] Over the course of its length the aqueduct drops c. 100m. The aqueduct ran overground and functioned with free-flow for its entire length (with the exception of its pressurised system of *c.* 1km from Terazi until Plativola).

[103] Oikonomakis 1984: 75 and pl. 6.
[104] Kessener and Piras 1997: 171.
[105] Oikonomakis 1984: 75 and pl. 6.
[106] Taylor 2000: 34-35.
[107] Coulton 1987: pl. viii.
[108] Işik 2000: 84-85.
[109] Kessener and Piras 1997: 167 and 187, fig. 28.
[110] Oikonomakis 1984: 78; Oikonomakis 1986: 60; see also Taramelli 1899: 401.
[111] Taramelli 1899: 406; Sanders 1982: 146.

Oikonomakis provides the following heights of the channel at the subsequent locations:[112] the surviving elevation of aqueduct trajectory, at a distance of 450m from the spring, is c. 705m. At Haghia Fotini (above the rock on which the cave chapel of Haghia Anna and Haghios Ioannis Theologos was founded), the height is c. 645m. At Petrokopeio (Mikro Kavgoudi area), the channel height is *c.* 605m. At Terazi the estimated height is over 600m. Here a drop is facilitated by a water pressure tower υδροζυγίο, which allows for a sudden drop of 2m. The final height of the terminal cistern at *Filakes* is just under 600m.

Oikonomakis identified the ruined structure, known as *Fylakes* or *Fylakes tis Aretousas*, as the aqueduct's *castellum*.[113] This structure is constructed with brick-faced mortared rubble with buttresses similar to those seen at Kastelliana and the L-shaped cistern at Aptera. Sanders, who located the cistern just off summit, describes a structure measuring 12m x 7m, with 1m thick walls, and four buttresses positioned along its downward facing wall.[114] The water collected in this cistern, allowing it to settle, before dispersal for use in the city. The difference in height between the estimated source and the *castellum* is *c.* 100m. Oikonomakis places the altitude of the castellum cisterns at just under 600m, set at only a slightly lower altitude than that recorded at the start of the inverted siphon (at Terazi, 2.5km to the southwest).[115] The exact altitude of the *castellum* is important, as it marks the horizon over which no bathhouse could easily be built in the city. We can say, for now, that any bathhouse fed by the aqueduct of Lyttos would have to be founded lower than the 600m contour within the cityscape, i.e. not on the summit of the city's acropolis.

In terms of an overall slope for the aqueduct length, Oikonomakis recorded a length of 22km for the aqueduct's winding trajectory. As the aqueduct functioned as a siphon from Terazi, the final c. 2km route to Lyttos should be omitted from gradient calculations. The height for the remaining 20km mountainous stretch dropped approximately 100m giving an approximate slope of 1:200 (Vitruvius' minimum requirement, *De Arch. 8.6.1*), although this shallowness of slope is not evenly maintained.

The comparative capacities of the aqueducts of Chersonisos and Lyttos

The fact that the Lyttos aqueduct was built solely to supply the city of Lyttos with water and served this city exclusively for all its functional life is supported by the quantity of water transported by the aqueduct which was simply not sufficient to support two cities.[116] This conclusion is based on the relatively small dimensions of the aqueduct channel and the diameter of the stone pipe associated with the inverted siphon (Figures 7A-B-C). The size of the channel is relative to the amount of water transported and, consequently, its size was a product of its purpose.

Oikonomakis records a section of the Chersonisos channel, preserved *c.* 200m from the bridge of Xerokamares 2, and demonstrates that its conducting capacity was over twice that of the channel at Lyttos.[117] Moreover, the rectangular cistern complex, which constitutes the *castellum* of the Chersonisos aqueduct, is the largest cistern complex on the island, measuring 55m x 18.50m x 5.5m, yielding a capacity of 5,596.25m^3, thereby affirming the immense potential of this aqueduct.[118] Mandalaki maintained that parallels for the massive dimensions of the cisterns at Chersonisos could only be found outside Crete, in Italy and North Africa.[119]

While it is difficult to compare the cities in terms of scale (largely due to the lack of evidence for the urban layout of Roman Lyttos), the fact that Chersonisos boasts at least 6 bathhouses (Figure 3),[120] a

[112] Oikonomakis 1984: 78.
[113] Oikonomakis 1984: 77, pls 9 and 14. Spanakis (1968: 159, n. 52 and pl. AG) had suggested that the brick-built remains, known as *Fylakes* (which he locates on the western edge of the south-facing hollow for the theatre), could belong to the theatre.
[114] Sanders 1982: 147.
[115] Oikonomakis 1984: 77-78.
[116] Oikonomakis 1984: 84.
[117] Oikonomakis 1984: 87, note 94.
[118] Galanaki *et al.* 2006: 267-268; Mandalaki 2001.
[119] Mandalaki 1999: 263; 2001; Galanaki *et al.* 2006; The scale of these cisterns also expresses a concern for water, as they allow for a buildup of supply for the arid summer months. The use of running water for bathing was often limited to times when the baths were open, at other times the supply could be interrupted with the use of a stop-cock, see Wilson 2001, 93; Kelly 2013: 148.
[120] Galanaki *et al.* 2006: 269; Papadaki *et al.* 2010: 35-40.

possible amphitheatre and a substantial harbour (features not shared by Lyttos) suggests that this city was home to the substantially larger population of the two.[121]

Nonetheless, the aqueduct of Lyttos is still impressive in its own right, and even with probably only half the capacity of that of Chersonisos, it is still indicative of a large populace – a population density which might be further supported by any evidence for the extensive theatre which Belli claimed to be the largest in Crete.[122] Ducrey and van Effenterre (in their discussion of the relative prosperity of Ini) regard Lyttos and Chersonisos as the most prosperous Roman sites in east central Crete, a claim which is probably heavily reliant on the presence of the cities' impressive aqueducts.[123] And while it is perhaps noteworthy that Taramelli considered the aqueduct of Lyttos on a par with that of Gortyna, by far the most impressive detail to consider when regarding Lyttos' gravity-flow water-supply is the city's exceptional altitude.[124] That this Roman city prospered – despite its unforgiving altitude, dramatic topography and physical remoteness – is a tribute to its resourceful and enterprising citizenry. This immutable Roman metropolis in the Cretan highlands was undoubtedly a city worthy of its lofty name.[125]

Bibliography

Adam, J. P. 1994. *Roman Building: Materials and Techniques*. London: Batsford.

Baldwin Bowsky, M.W. 1995. Roman Crete: No Provincial Backwater. In: Πεπραγμένα *7ου Διεθνούς Κρητολογικού Συνεδρίου*, vol. I.1, Rethymnon, 25-31 August 1991: 41-67. Rethymnon: Municipality of Rethymnon.

Baldwin Bowsky, M.W. 1999. The Business of Being Roman: The Prosopographical Evidence. In A. Chaniotis (ed.) *From Minoan Farmers to Roman Traders: Sidelights on the Economy of Ancient Crete*: 305-347. Stuttgart: Steiner Vlg.

Baldwin Bowsky, M.W. 2006. Roman Patronage, Greek Identity: How Lyttos finally won the war against Knossos. In: Πεπραγμένα 9ου Διεθνούς Κρητολογικού Συνεδρίου, Elounda 1-6 October 2001, vol. Γ 2: 407-419. Herakleion: The Society of Cretan Historical Studies.

Barozzi 2004 – Kaklamanis, S. 2004. *Francesco Barozzi, Descrittione dell'Isola di Creta*. Περιγραφή της Κρήτης 1577/8. Herakleion.

Beschi, L. 1972–1973. Antichità cretesi a Venezia. *L'Annuario della Scuola Archeologica di Atene e delle Missioni in Oriente* 50–51: 479–502.

Buondelmonti 2002 [transl. Aposkitis]: Aposkitis, M. (translation) 2002. *Buondelmonti C.* Περιγραφή της Νήσου Κρήτης. Ένας Γύρος της Κρήτης στα *1415*. Herakleion: Mikros Nautilos.

Catling, H. W. 1976. Archaeology in Greece, 1975-76. *Archaeological Reports* 22: 3-33.

Catling, H. W. 1984. Archaeology in Greece, 1983–84. *Archaeological Reports* 30: 3-70.

Chaniotis, A. 1988. Vinum Creticum excellens: Zum Weinhandel Kretas. *Münstersche Beiträge zur antiken Handelsgeschichte* 7.1: 62-89.

Chaniotis, A. 1999. Milking the Mountains. In: A. Chaniotis (ed.) *From Minoan Farmers to Roman traders: sidelights on the economy of ancient Crete*: 181-220. Stuttgart: Steiner Verlag.

Chaniotis, A. 2009. European Identity: Learning from the Past? In: A. Chaniotis, A. Kuhn and C. Kuhn (eds) *Applied Classics: Constructs, Comparisons, Controversies*: 27-56. Heidelberger Althistorische Beitrage und Epigraphische Studien. Stuttgart: Steiner Vlg.

Chaniotis, A. and Rethemiotakis, G. 1992 Neue Inschriften aus dem kaiserzeitlichen Lyttos, Kreta. Originalveröffentlichung. *Tyche* 7: 27-38.

Christodoulakos, Y. 2011. Ρωμαϊκά υδραγωγεία της Δυτικής Κρήτης. In: *10th Cretological Congress* A4: 147-169. Πεπραγμένα του Ι' Διεθνούς Κρητολογικού Συνεδρίου (Χανιά 1-8 Οκτωβρίου 2006).

121 Noted by Belli; cited in Falkener 1854: 16.

122 Cited in *The Builder* 7/12/1901: 499.

123 Ducrey and van Effenterre 1973: 290.

124 Taramelli 1899: 407.

125 Kaczyńska 2001: 6-7.

Coulton, J.J. 1987. Roman Aqueducts in Asia Minor. In: S. Macready and F.H. Thompson (eds) *Roman Architecture in the Greek World.* Occasional Papers (new series) 10: 72-84. London: The Society of Antiquaries of London.

Coutsinas, N. 2013. *Défenses crétoises. Fortifications urbaines et défense du territoire en Crète aux époques classique et hellénistique.* Publications de la Sorbonne.

De Caro, S. 1992-1993. Vino di Cnosso dei Campani: Un Nuovo Documento Epigrafico per la Storia del Vino Cretese in Età Romana. *L'Annuario della Scuola Archeologica di Atene e delle Missioni in Oriente* 70-71: 307-312.

Di Napoli, V. 2010. Τα Θεατρικά Οικοδομήματα της Ρωμαϊκής Κρήτης: Προσέγγιση για μια νέα Ερμηνεία. In: Μ. Ανδριανάκης – Ι. Τζαχίλη (επιμ.) *Αρχαιολογικό 'Εργο Κρήτης 1, Πρακτικά της 1ης Συνάντησης, 28-30 Νοεμβρίου 2008, Ρέθυμνο*: 814-826. Rethymnon: University of Crete Press.

Drerup, H. 1951. Zweizelliges Heiligtum in Aptara. In: F. Matz (ed.) *Forschungen auf Kreta 1942*: 89-105. Berlin.

Ducrey, P. and van Effenterre, H. 1973. Un règlement d'époque Romaine sur les bains d'Arcadès. *Khretika Chronika* 25: 281-290.

Empereur, J.-Y., Kritzas, Ch. and Marangou, A. 1991. À la recherche des ateliers d'amphores en Crete centrale. *Bulletin de correspondance hellénique* 115: 481-523.

Erickson, B. 2002. Aphrati and Kato Syme: Pottery, Continuity, and Cult in Late Archaic and Classical Crete. *Hesperia* 71: 41-90.

Fabres, G., Fiches, J.-L. and Paillet, J.-L. 1991. Interdisciplinary Research on the Aqueduct of Nîmes and the Pont du Gard. *Journal of Roman Archaeology* 4: 63-88.

Falkener, E. 1854. *A Description of some Important Theatres and other Remains in Crete, from a MS History of Candia by Onorio Belli in 1586. Museum of Classical Antiquities.* Supplementary Volume 2. London: Trübner.

Fernandes, R., Geeven, G., Soetens, S. and Klontza-Jaklova, V. 2012. Mapping the probability of settlement location for the Malia-Lasithi region (Crete, Greece) during the Minoan Protopalatial period. In: S. J. Kluiving and E. B. Guttmann-Bond (eds) *Landscape Archaeology between Art and Science From a Multi-to an Interdisciplinary Approach* (Landscape and Heritage Series): 353-368. Amsterdam: Amsterdam University Press.

French, E.B. 1991. Archaeology in Greece 1990-91. *Archaeological Reports* 37: 3-78.

Galanaki K., Grigoropoulos D., Kastanakis A., Mandalaki S., Papadaki C., Triantafyllidi I. 2006. The Management of Water Resources in Chersonissos, Crete, Greece, during the Roman Period. In: A.N. Angelakis and D. Koutsoyiannis D. (eds) *1st IWA International Symposium on Water and Wastewater Technologies in Ancient Civilizations:* 265-270. Herakleion.

Gigourtakis, N. M. 2013. Η οχυρωμένη βυζαντινή ακρόπολη της Λύκτου. *Κρητική Εστία* 14: 49-64.

Grizis, K.S. 2014. Αναπαράσταση Λειτουργίας Αρχαίων Ρωμαϊκών Δεξαμενών Απτέρων Χανίων (Representation of ancient Roman cisterns' operation. The case of Aptera Chania). Diploma Dissertation, National Technical University of Athens.

Heinrichs, J. D. 2013. The Topography of Antiquity in Descriptions of Venetian Crete. In: N. Avcioglu and E. Jones (eds) *Art, Architecture and Identity in Venice and its Territories. Essays in Honour of Deborah Howard:* 1450-1750. Burlington: Ashgate Publishing.

Hodge, A.T. 1983. Siphons in Roman Aqueducts. *Papers of the British School at Rome* 51:174-221.

Işik, F. 2000. *Patara: The History and Ruins of the Capital City of Lycian League.* Universal Holdings.

Kaczyńska, E. 2001. Remarks on the Mycenaęan Name of Lyktos. *Do-so-mo* 3: 5-9.

Karanastasi, P. 2015. Νεα Στοιχεια για τη Ρωμαϊκη 'Ελευθερνα. Π. Καραναστάση, Α. Τζιγκουνάκη, Χ. Τσιγωνάκη (επιμ.) *Αρχαιολογικό 'Εργο Κρήτης 3, Γ' Παγκρήτια Συνάντηση για το Αρχαιολογικό 'Εργο στη Κρήτη, Ρέθυμνο 5-8 Δεκεμβρίου 2013*: 417-428. Ρέθυμνο.

Kelly, A. 2004. The Roman Aqueducts and Bathhouses of Crete. Unpublished PhD dissertation. Trinity College Dublin.

Kelly, A. 2006a. The Impact of Aqueduct Construction on the Demographic Patterns in Crete. In: Wiplinger, G. (ed.) *Cura Aquarum in Ephesus. The 12th International Conference on the History of Water Management and Hydraulic Engineering in the Mediterranean Region. Ephesus. October 2-10, 2004.* Vol. II: 303-310. Leuven: Peeters.

Kelly, A. 2006b. The Roman Baths of Mylopotamos: a distribution study. In: Ε. Γαβριλάκη, Ι. Τζιφόπουλος (επιμ.) *Ο Μυλοπόταμος από την Αρχαιότητα ως Σήμερα. Περιβάλλον, Αρχαιολογία, Ιστορία, Λαογραφία, Κοινωνιολογία. Πρακτικά Διεθνούς Συνεδρίου (Πάνορμος, 24-30 Οκτωβρίου 2003)*//Proceedings of

International Conference (Panormos, 24-30 October 2003): 239-252. Ιστορική και Λαογραφική Εταιρεία Ρεθύμνου. Rethymnon: Association on Historical & Folklore Studies in Rethymnon.

Kelly, A. 2011. Occide, verbera, ure! *'Kill him, Flog him, Burn him Alive!'* (Seneca *Epistles* 7); The popularity, extent and duration of Roman *Spectacula* on Crete. In: *Πεπραγμένα 10ου Διεθνούς Κρητολογικού Συνεδρίου, 1-8 October 2006*. Vol. 5 A: 77-94. Chania: The Philological Association 'Chrysostomos'.

Kelly, A. 2013. Roman Bathhouses on Crete as Indicators of Cultural Transition; the dynamics of Roman influence. In: A. Gardner, E. Herring and K. Lomas (eds) *Creating Ethnicities and Identities in the Roman World*: 129-164. Bulletin of the Institute of Classical Studies, Supplement 120. London: Blackwell.

Kessener, P. and Piras, S. 1997. The Pressure Line of the Aspendos Aqueduct. *Adalya* II: 159-187.

Lewis, M. J. T. 1999. Vitruvius and the Greek Aqueducts. *Papers of the British School at Rome* 67: 145–172.

Lolos, Y. A. 1997. The Hadrianic Aqueduct of Corinth. *Hesperia* 66: 271-314.

MacGillivray, J. A. 2001. *Minotaur. Sir Arthur Evans and the Archaeology of the Minoan Myth*. London: Jonathan Cape.

Mandalaki, S. 1996. Χερσόνησος. *Archaeologikon Deltion* 51, B΄2: 633-635.

Mandalaki, S. 1999. Χερσόνησος. *Kritiki Estia* 7: 244-263.

Mandalaki, S. 2001. Ρωμαική δεξαμενή Χερσονήσου. In: Πεπραγμένα 9ου Διεθνούς Κρητολογικού Συνεδρίου, Elounda 1-6 October 2001, vol. A 5: 347-363. Herakleion: The Society of Cretan Historical Studies.

Marangou, A. 1999. Wine in the Cretan Economy. In: A. Chaniotis (ed.) *From Minoan farmers to Roman traders: sidelights on the economy of ancient Crete*: 269-278. Stuttgart: Steiner Verlag.

Marangou-Lerat, A. 1995. *Le vin et les amphores de Crète: de l'Époque Classique à l'Époque Impériale* (Études cretoises 30). Paris: Ed. De Boccard

Mariani, L. 1895. Antichità Cretesi. *Monumenti Antichi* VI: 153-347.

Ninou-Kindeli, V. and Christodoulakos, Y. 2000. Ρωμαϊκή Απτέρα: Μια πρώτη προσέγγιση. In: M. Livadiotti and I. Simiakaki (eds) *Creta Romana e protobizantina: atti del congresso internazionale*. Iraklion, 23-30 settembre, 2000: 313-334. Padova.

O'Connor, C. 1993. *Roman Bridges*. Cambridge: Cambridge University Press.

Oikonomakis, N. E. 1984. Το Υδραγωγείο της Λύκτου. *Lyktos* 1: 66-99.

Oikonomakis, N. E. 1986. Το Υδραγωγείο της Χερσόνισου. *Lyktos* 2: 50-98.

Papadaki, Chr. 2012. Η Λύκτος και τα μνημεία της στα κείμενα των περιηγητών [Ancient Lyktos and its monuments in the travelers texts], http://www.archaiologia.gr/ Accessed August 2015.

Papadaki, Chr., Triantafyllidi, I. and Grigoropoulos, D. 2010. Λουτρικό συγκρότημα στον Λιμένα Χερσονήσου: σωστική ανασκαφή. *Αρχαιολογία και Τέχνες* 115: 35-40.

Pumain, D. 1997. City-size Dynamics in Urban Systems. In S. van der Leeuw and J. McGlade (eds) *Time, Process and Structured Transformation in Archaeology*: 97-117. London: Routledge.

Raab, H.A. 2001. *Rural Settlement in Hellenistic and Roman Crete: The Akrotiri Peninsula* (BAR International Series 984). Oxford: Archaeopress.

Rethemiotakis, G. 1984. Ανασκαφική Έρευνα στη Λύττο. *Λύκτος 1*: 49-65.

Rouanet-Liesenfelt, A.-M. 1992. Les plantes médicinales de Crête à l'époque romaine. *Cretan Studies* 3: 173-190.

Sanders, I. F. 1982. *Roman Crete. An Archaeological Survey and Gazetteer of Late Hellenistic, Roman and Early Byzantine Crete*. Warminster: Aris and Phillips.

Spanakis, St. 1968. Το θέατρο στη Ρωμαϊκή Κρήτη. In: *Πεπραγμένα 2ου Διεθνούς Κρητολογικού Συνεδρίου, Chania, 12-17 April 1966*, vol. B: 142-168. Chania: The Philological Association 'Chrysostomos'.

Spanakis, G. S. 1969. *Η Κρήτη - Τουρισμός - Ιστορία - Αρχαιολογία*. Vol. I and II. Herakleion: Sfakianakis editions.

Spanakis, G. S. 1981. 'Υδρευση του Ηρακλείου, *828-1939*. Herakleion: Technical Chamber of Greece (TEE).

Spratt, T. A. B. 1865. *Travels and Researches in Crete* I and II. London: John van Voorst.

Talbert, R. J. A. (ed.) 2000. *Barrington's Atlas of the Greek and Roman World*. Princeton.

Taramelli, A. 1899. *Ricerche Archeologiche Cretesi* (Monumenti Antichi della Reale Accademia dei Lincei IX). Milano.

Taylor, R. 2000. *Public Needs and Private Pleasures: Water Distribution, the Tiber River and the Urban Development of Ancient Rome*. Studia Archeologica, 109. Rome: L'Erma.

Themelis, P. 2002. *Αρχαία Ελεύθερνα. Ανατολικός Τομέας*. Athens: Ministry of Culture. Archaeological Receipts Fund.

Thomas, R. and Wilson, A. 1994. Water Supply for Roman Farms in Latium and South Etruria. *Papers of the British School at Rome* 62: 139-196.

Tsigonaki, Chr. 2015. Οι ανασκαφές του Πανεπιστημίου Κρήτης στην Ακρόπολη της Αρχαίας Ελεύθερνας (Τομέας ΙΙ, Κεντρικός). In: Π. Καραναστάση, Α. Τζιγκουνάκη, Χ. Τσιγωνάκη (επιμ.) *Αρχαιολογικό Έργο Κρήτης 3, Γ΄ Παγκρήτια Συνάντηση για το Αρχαιολογικό Έργο στη Κρήτη, Ρέθυμνο 5-8 Δεκεμβρίου 2013*: 391-406. Ρέθυμνο.

Tsigonaki, Chr. 2007. Les villes crétoises aux VIIe et VIIIe siècles: l'apport des recherches archéologiques à Eleutherna. *L'Annuario della Scuola Archeologica di Atene e delle Missioni in Oriente* 85, series III.7: 263-297.

Tsigonaki , Chr. and Yangaki, A. 2015. Οι αρχαιολογικές έρευνες του Τομέα ΙΙ στις δεξαμενές της Ελεύθερνας και η μαρτυρία της κεραμικής. In: Π. Καραναστάση, Α. Τζιγκουνάκη, Χ. Τσιγωνάκη (επιμ.) *Αρχαιολογικό Έργο Κρήτης 3, Γ΄ Παγκρήτια Συνάντηση για το Αρχαιολογικό Έργο στη Κρήτη, Ρέθυμνο 5-8 Δεκεμβρίου 2013*: 429-448. Ρέθυμνο.

Van Effenterre, H. and Gondicas, D. 1999. Lyttos ville fantôme? In: J.-N. Corvisier and M. Valdher (eds) *La démographie historique Antique* (Cahiers scientifiques de l'université d'Artois 11): 129-139. Arras: Artois Presses Université.

Van Spitael, A. M. 1981. *Cristoforo Buondelmonti, Descriptio insule Crete ... ed Critique*. Herakleion: Syllogos politistikes anaptyxeos Herakleiou.

Viviers, D. 1994. La cité de Dattalla et l'expansion territoriale de Lyktos en Crète orientale. *Bulletin de correspondance hellénique* 118: 229-259.

Vogeikoff-Brogan, N. 2012. Hellenistic and Roman Periods: Expansion of the Isthmus in an International Era. In: L.V. Watrous, D. Haggis, K. Nowicki, N. Vogeikoff-Brogan and M. Schultz (eds) *An Archaeological Survey of the Gournia Landscape. A Regional List of the Mirabello Bay, Crete*: 81-93. INSTAP: Academic Press.

Watrous, L.V. 1982. *Lasithi: A History of Settlement on a Highland Plain in Crete* (Hesperia Supplement 18). Princeton: American School of Classical Studies at Athens.

Wilson, A. I. 2001. Urban Water Storage, Distribution, and Usage in Roman North Africa. In: A. O. Koloski-Ostrow (ed.) *Water Use and Hydraulics in the Roman city* (Archaeological Institute of America. Colloquia and Conference Papers n.s. 3): 83-96. Boston: Dubuque.

Dr Amanda Kelly
Assistant Professor and Fulbright Ambassador for UCD
School of Archaeology and Beijing
Dublin International College
University College Dublin
Belfield, Dublin 4, Ireland
amanda.kelly@ucd.ie

PART II: NYMPHAEA

Shifting tides: approaches to the public water-displays of Roman Greece

Dylan Kelby Rogers

Abstract

The study of Roman water-displays is constantly evolving, with research being conducted on the terminology of the structures, their typologies, their aesthetic programs, and their impact on the social and physical landscapes surrounding them. The water-displays of Roman Greece have received various treatments over the last four decades, including catalogues, their artistic programs, their ties to identity, and their placement in wider Empire-wide schemes of imperial and civic patronage. While much has been done on the Roman fountains of Greece, it is the aim of this essay to survey the terminology used and the historiography therein, in order to consider water-displays that have not received a great deal of attention, in religious, civic, and entertainment-related contexts. We will then posit several ways to advance the study of public Roman water-displays in Greece.

Keywords: nymphaea, water-displays, aesthetics, terminology.

Definitions

Before beginning our appraisal of fountains in Roman Greece, it will be helpful to offer a few comments on terminology. The term 'nymphaeum' has come widely to indicate a large-scale or 'monumental' fountain. A nymphaeum often evokes the large water-displays seen in the city of Rome itself (e.g., the Severan Septizodium) or in Asia Minor, the latter of which is a region known for its large, showy fountains.[1]

There are implications and difficulties with the term 'nymphaeum.' First, there are religious connotations of the word, stemming from sacred spaces of nymphs, especially in the Greek world.[2] It seems, however, that by the High Roman Empire, large-scale fountains had been divorced from religious associations -in the same way one might consider a natural spring, which might be the object of ritual and cult. Indeed, in the Roman period, fountains were man-made; the structures themselves are built *de novo* or are constructed over a natural water source, both of which are the case in Greece.[3] Second, the 'nymphaeum' often implies a certain level of monumentality. Modern scholarship often deems any large-scale fountain a 'nymphaeum,' even with the often lacking epigraphic evidence that the ancients themselves called the structure a 'nymphaeum.'[4] Given the handful of inscriptions still remaining, it is difficult to apply universally the term 'nymphaeum' for all fountains. In fact, there are at least 21 terms that the ancient Romans used for fountains, which indicates that they themselves were fluid in their word choice for their water-displays.[5] The term 'nymphaeum' became popular in the modern period, especially in the Renaissance, when it was used to describe fountains (often of a grotto-like appearance) in Italian villas.[6]

Given the modern adoption of the term 'nymphaeum,' we should not simply reject the term all together. We can, however, employ other terms, such as 'water-display,' which do not have the same connotations

[1] For more on the number and reasons for the large-scale fountains of Asia Minor, see the recent article of Lendon (2015), which posits that the influx of monumental fountains there after the Flavian period can be attributed to the formal education of young men, which included extolling the importance of water in urban settings.
[2] Walker 1979, 1987; Richard 2012: 4-7.
[3] Agusta-Boularot 2001: 167-168.
[4] There are a handful of inscriptions that describe fountains as 'nymphaea' in the Roman world, including the 1st century AD *nymphaeum* from Metz, France (*CIL* 13.4325), the AD 104/105 Trajanic fountain in Syria (*IGR* III,1273), and the second half of the 2nd-century νυμφαῖον from the Agora of Argos (Argos Inv. E.266). For a complete discussion of the epigraphic evidence, see Settis 1973.
[5] The terms include: *alveus, cantharus, castellum [aquae/divisorium], cisterna, concha, crater, euripus, hydreion/ὑδρεῖον, hydrekdocheion/ὑδρεκδοχεῖον, labrum, lacus, Meta Sudans, munus, nymphaeum/νυμφαῖον, phiala, puteus, saliens, septizodium, silanus*, and *solium*. The meaning of these terms is discussed in detail by Settis (1973) and Rogers (2015: 30-79).
[6] Richard 2012: 2-4.

that the 'nymphaeum' has. A water-display has two functions: to show off water and to collect the water for a secondary use, such as ritualistic washing or refreshment. A water-display, then, is not restricted to a particular size, which means that a wider body of examples can be used in studying Roman fountains – not just the often impressive monumental façades that one associates with nymphaea. The water-displays considered in this essay are from public contexts, although there is a growing scholarship on the study of domestic fountains in Greece and the rest of the Empire.[7]

State of the field

The state of the field of water-displays in Roman Greece fits well into general trends of study of Roman fountains.[8] The first archaeological treatises on Roman fountains tended to focus on typologies, such as Norman Neuerburg's 1965 collection of structures in Italy, which was also one of the first regional gazetteers produced regarding fountains. Neuerburg was followed up by René Ginouvès (1969) and Pierre Aupert (1974), which examined structures in North Africa, while placing them in a larger typological scheme, in addition to the seminal work by Salvatore Settis (1973). Wolfram Letzner's 1990 (and later revised in 1999) catalogue of the fountains of the western half of the Empire, is still an important synthesis of over 450 water-displays of various sizes, along with crucial information about the study of fountains in terms of decoration, terminology, etc. Other regional syntheses of fountains include: Gaul,[9] North Africa,[10] Asia Minor,[11] and the Roman East.[12] Such catalogues have allowed for greater, Empire-wide syntheses, most notably by Pierre Gros (1996), Franz Glaser (2000, focusing on chronological developments and typologies), Brenda Longfellow (2011), and Nicholas Lamare (2014).[13] Other noteworthy recent approaches have included the nature of water-displays' placement within water infrastructure systems[14] or wider geological landscapes.[15]

Turning to the fountains of Greece, there are two early major studies that are still useful today. The first is Susan Walker's 1979 University of London DPhil dissertation, entitled 'The Architectural Development of Roman Nymphaea in Greece.' Walker successfully traces the development of over 50 Roman water-displays in Greece from the late Hellenistic period to the 4th century AD, primarily in the regions of Achaea, Macedonia, and Epirus. Pushing away from a catalogue and a study based on typologies, Walker organizes her examples based on their function and situation in the landscape. Her approach allows her to ask broader questions, such as the nature of the use of these structures, the social implications of the fountains, and more detailed questions about construction techniques (including brickwork and the marble trade of the eastern Empire). In 1987, Walker presented a distillation of the thesis in a chapter in an edited volume about Roman architecture in the Greek world, in which she used the Roman nymphaea of Greece as markers of architectural and artistic production (especially sculpture) and of prosperity, along with comments about the meaning of 'nymphaeum' in Greece. Glaser's 1983 *Antike Brunnenbauten* (KPHNAI) *in Griechenland* presents a catalogue of nearly 100 fountains in Greece from the 7th century BC until the 2nd century AD. The study is organized primarily by typologies, followed by discussions of fountains in the city, sanctuaries, and as decoration on vases. Glaser, focusing primarily on Greek examples, still provides a number of Roman fountains not included in Walker's study – making it an invaluable addition to fountain studies.

In the same vein, Sandrine Agusta-Boularot (2001) presents a discussion and catalogue of 46 fountains in Greece from the Roman conquest until the Flavian period (2nd century BC until 1st century AD). Agusta-Boularot provides an examination of water-displays in Greece through a number of categories, based

[7] On domestic fountains in Roman Greece, see Walker (1979: 229-59), Papaioannou (2007), and Zarmakoupi (2013). The Italian peninsula has received more attention regarding domestic fountains, from Neuerburg (1965) to Dessales (2013), Rogers (2013), and Zarmakoupi (2014: 141-177).

[8] The following cannot be an exhaustive historiography of Roman fountain research, for which one can see: Lendon 2015: 123-127; Richard 2012: 1-12.

[9] Agusta-Boularot 1997, 2004; Lavagne 2012.

[10] Lamare 2014; Tomasello 2005.

[11] Dorl-Klingenschmid 2001; Richard 2012.

[12] Segal 1997.

[13] Lamare 2014: 1.93-127.

[14] De Kleijn 2001; Richard 2007, 2012; Schmölder-Veit 2009.

[15] Weiss 2011. See also the work of Rinne (2011) on similar approaches to the use of water in Baroque-era Rome.

on their location: fountains constructed in the 2nd and 1st centuries BC; fountains constructed before the 2nd century and still in service between the 2nd century BC and 1st century AD; the placement of fountains between the 2nd century BC and the 1st century AD. The result of the study is a cohesive narrative of the transition between traditional Greek fountain forms (whether artificial or natural springs monumentalized) and Roman fountain forms in addition to how the Romans capitalized on the aquatic landscape in the early part of their dominion of Greece, thus paving the path for the large-scale fountains built there in the 2nd century AD.

In addition to these catalogues that document a number of water-displays of Roman Greece, the studies that document single fountains are important for a variety of reasons, including for our understanding of architectural developments, restorations and renovations, artistic programs, and the relationship with the surrounding landscape and water systems. In 1984, Renate Bol produced the seminal work on the Nymphaeum of Herodes Atticus at Olympia. By reexamining the architectural and sculpture remains of the water-display, Bol was able to demonstrate that the monument had two stories, full of depictions of both the families of the imperial house and Herodes himself – providing insight into the nature of Herodes' elevated position in contemporary Greek society. Furthermore, Bol's reconstruction has been widely accepted. The 1995 monograph by Patrick Marchetti and Kostas Kolokotsas on the mid-2nd-century Round Nymphaeum on the northern side of the agora of Argos presents a glimpse at a unique water-display found in Greece. The reconstruction and chronology of the monument, as presented by Marchetti and Kolokotsas, has been questioned by some scholars, especially in terms of phasing and building use.[16]

Shawna Leigh's 1997 revaluation of the Hadrianic Lykabettus Nymphaeum (located in the modern *Plateia Dexamenis,* in Kolonaki) is an important article in our understanding of the aquatic landscape of 2nd-century Athens. Using various clues from archival materials, original inscriptions found in the National Gardens, and the ancient remains (e.g., the bases of the Ionic façade) of the 19th-century reservoir built directly on top of its precursor, Leigh was able to reconstruct the ancient water-display, placing it at the beginning of the water leading down into the city of Athens.

The last decade has also continued to provide new studies on singular monuments. The Fountain House of Arsinoë at Messene was restudied in 2009 by Claus Reinholdt. The monument is a difficult one to understand, given that the architectural blocks and members were constantly reused in subsequent building phases through the Hellenistic and Roman periods, which indicates the importance of this study. Betsey Robinson's 2011 monograph on the Peirene fountain of Corinth delivers an important glimpse into the hydro-geography of the area, the mythology and iconography associated with the spring, and its architectural developments in the Greek and Roman periods, and beyond.

Artistic decoration of water-displays has also been of interest to many scholars. By studying the aesthetic program of fountains, especially through sculpture, we can begin to understand notions of civic and imperial patronage, iconographic trends, and how sculpture can interact with the actual display of water (e.g., statues outfitted with holes for the flow of water). The original study of fountain sculptures in the Greek and Roman world began with the short monograph by Balázs Kapossy (1969), in which he catalogued different statue types found on water-displays. Deena Berg's 1994 dissertation illustrates the artistic development of fountains, from 700 to 30 BC, providing a foundation that subsequent studies on Roman fountains can grow from. Longfellow's 2011 monograph, through the various examples presented, often incorporates discussions of the impact of sculpture on the understanding and meaning of a fountain. For example, she discusses the sculpture associated with the Larissa Nymphaeum of Argos, installed in AD 124, with the inauguration of the Hadrianic aqueduct there.[17] The back niche of the grotto fountain would have included a large-scale bare-chested statue of Hadrian, carrying a sword; water would have poured forth, giving Hadrian an almost divine appearance, especially as the donor of this water.[18]

[16] Piérart 1999.
[17] Longfellow 2011: 117-119.
[18] Cf. Aristodemou 2012: 70-71.

The most complete study to-date of sculpture associated with Roman water-displays in Greece (and Asia Minor) is by Georgia Aristodemou, whose 2012 monograph not only catalogues the known sculptures, but also attempts to contextualize these objects in terms of their iconography and display programs.

The last decade in particular has seen a renewed interest in Roman fountains in Greece, moving past architectural typologies and decoration. Studies are beginning to focus on the implications of the construction of these structures, indicating more about identity tied to the fountains, their patronage, and their wider cultural contexts. Robinson's 2013 article on the fountains of Corinth, uses five water-displays (Peirene, Glauke, the Fountain of Neptune, the South Stoa Fountain, and the North Market Nymphaeum) to explore notions of memory and identity in Roman Corinth. In a city with fountains with ancient mythologies (i.e., Peirene and Glauke), the inhabitants of Corinth, especially as it transformed from a *colonia* into the capital of the province of Achaea, used the water-displays there to tie themselves to an aquatic past, while also associating them with the surrounding natural landscape.

Longfellow's 2011 monograph on Roman 'monumental fountain complexes' explores the role of civic and imperial patronage of these structures. Tracing the development of large-scale fountains in the first three centuries AD, with an emphasis on Rome, Greece, and Asia Minor, Longfellow has demonstrated the important role of hydraulic benefaction, particularly in terms of the role of the Emperor and his role with members of the local élite.[19] Longfellow followed up her monograph in 2012 with an article on the placement of Roman fountains in Greek sanctuaries (i.e., the Hadrianic fountain of the forecourt of the Sanctuary of Demeter and Kore at Eleusis, the Nymphaeum of Herodes Atticus at Olympia, and the fountain of the Letoon at Xanthos in Asia Minor), delving into the impact that those structures had on the surrounding landscape and their implications for their donors. Julian Richard's 2012 monograph attempts to place fountains in their wider hydraulic systems, since understanding the management of water and its relationship with the water's display is lacking in the scholarship. The functionality of water and its display can begin to show how 'monumental fountains' fit into wider socio-cultural, socio-economic, and socio-political contexts. Finally, Dylan Rogers (2015) uses different physical contexts (civic, religious, and entertainment-related) to explore how spaces are shaped by the addition of water-displays, particularly with the framework of the 'archaeology of the senses.' Attempting to imagine how an ancient Roman would have acted and reacted to a fountain is the key to unlocking the reasons for the placement of these structures into the built environment.

In the year 2015, there were also two important conferences held in Athens regarding water throughout the Greek and Roman periods of Greece. First was the 16th International Conference, the *Cura Aquarum in Greece*, sponsored by the Deutsche Wasserhistorische Gesellschaft (DWhG). It provided a forum for an array of scholars studying various aspects of water, especially regarding water infrastructure systems, to present their work. There are currently plans to publish the proceedings. Secondly, the HYDRΩMED Project, based at the University of Aix-Marseille (France), explores water management in the Mediterranean in the 1st millennium BC through a series of conferences. In July 2015, the project presented 'Water Cult and Culture in the Mediterranean World of the 1st Millennium BC,' providing a number of papers that explored the implications of water in cultic settings of Greece and beyond, offering a strong foundation for scholars of the Roman period of Greece to build upon.

Fountains to (re)consider

The history and development of fountains in Greece, as mentioned previously, is well documented. While there was a pre-existing fountain culture in Greece, the first Roman fountain constructed *de novo* in Greece was the Fountain of Neptune in Corinth, dated to the late 1st century BC-early 1st century AD.[20] It was not until the 2nd century AD, with the increased flow of water from the aquatic euergetism of Hadrian and other members of the local élite, that Greek cities were able to sustain larger fountains with open-air

[19] See also the pointed critiques of this monograph by Campagna (2011) and Burrell (2012).

[20] Robinson 2013: 352-357.

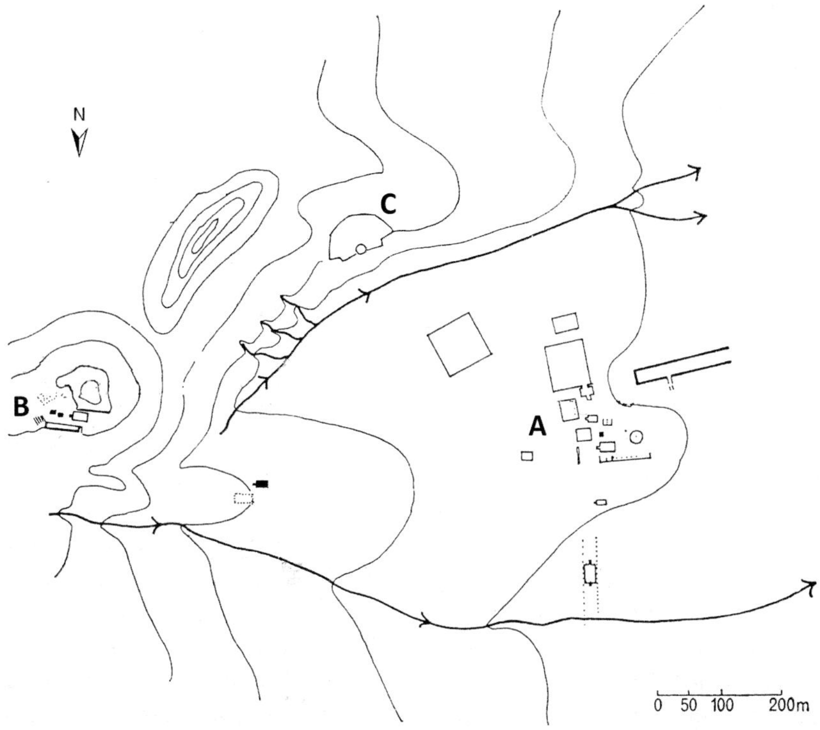

Figure 1. Epidauros. Sanctuary of Asclepius (A), Sanctuary of Apollo Maleatas (B), Theatre (C) (after Melfi 2007: Fig. 2)

basins (unlike their Greek forbearers of closed-forms).[21] While 'great' fountains have been studied in-depth, there remain a number of examples that have received less attention in the secondary scholarship. These examples, which can be categorized by their contexts (i.e., religious, civic, and entertainment-related), are essential to understand better the reasons behind the construction and location of public water-displays in Roman Greece.

Epidauros: sanctuary of Apollo Maleatas

Throughout the Greek and Roman world, religious sites included water for a variety of purposes, often aiding in purification and lustration, especially when placed at the entrances to sanctuaries.[22] Roman sanctuaries that prominently featured water were those tied to water sources (e.g., the *Lacus Iuturnae* in Rome) and healing gods.[23] Healing sites in the Roman world, such as those connected to Apollo, are tied to thermal water and are usually extra-urban.[24] The Sanctuary of Apollo Maleatas at Epidauros, then, fits nicely into an Empire-wide scheme of Apollo sanctuaries associated with water, despite its ancient connections to it surrounding landscape and connections to Asclepius.

The site of Epidauros in the Argolid of the Peloponnese is well known for its healing sanctuary of Asclepius (Figure 1).The sanctuary used water to heal its pilgrims, as water was plentiful in the valley of the sanctuary, in addition to its uses in other ritualistic practices, such as incubation.[25] In fact, the sheer amount of water at the site, both in the natural resources of the valley and in the built environment (e.g., aqueducts, cisterns, basins, bathing complexes), demonstrates how the cult of Asclepius here was believed to work its therapeutic magic.[26] The mysterious powers of the water served to 'symbolize [...] the sacred character of Asclepius [and] served to purify the body and soul'.[27] Thus, water and its display are crucial to the experience of healing at Epidauros.

[21] Lolos 1997: 302-314; Longfellow 2011; Robinson 2013.
[22] *Thesaurus cultus et rituum antiquorum* 2.3a.IV.A, s.v.,'Purificazione, Romana, Mezzi impiegati nelle purificazioni, liquidi e unguenti' (V. Saladino).
[23] Aupert 2012; Scheid 1991.
[24] Cole 1988: 162.
[25] Lambrinoudakis 1994: 226; Melfi 2007: 17-209; Prignitz 2014.
[26] Lambrinoudakis 1994: 226-227.
[27] Trümper 2014: 211.

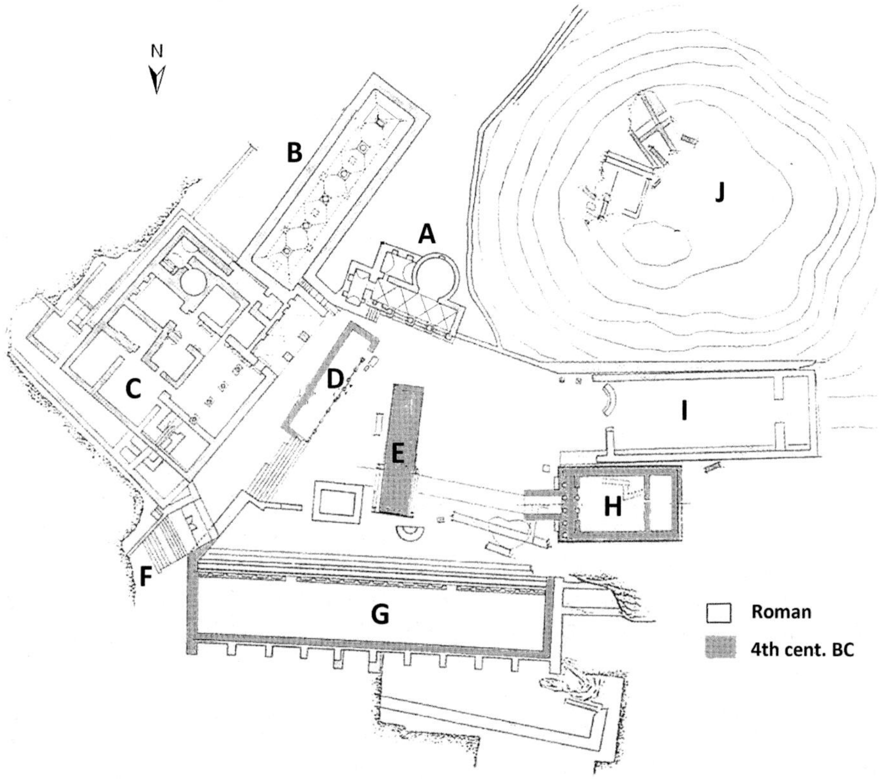

Figure 2. Epidauros. Sanctuary of Apollo Maleatas Plan: Nymphaeum (A), Cistern (B), Skana or Priest's House (C), Mouseion (D), Altar (E), Propylon (F), Stoa (G), Temple (H), Altar of Ashes (I), Mycenaean-era Remains (J) (After Melfi 2007: Fig. 4)

The sanctuary of Apollo Maleatas on the Kynortion Hill, directly east of the Sanctuary of Asclepius, is a religious space whose use extends back to the Bronze Age.[28] Apollo Maleatas is believed to be a father figure of Asclepius, thus tying him to the healing cult associated with the site, and the epithet 'Maleatas' is probably connected to Cape Maleas in Laconia, where the figure of Maleatas is thought to have been a vegetation god.[29] Over the course of nearly a thousand years, the Sanctuary of Apollo Maleatas went through a series of expansions and destructions. When the Asclepius sanctuary started gaining momentum in the 8th century BC, so too did the Sanctuary of Apollo Maleatas. The space sacred to Apollo was built up in the 4th to 3rd centuries BC, suffered destruction in the 1st century BC, but it was, like the Asclepius sanctuary, built up during the Julio-Claudian period, and especially in the 2nd century, by the generosity of Sextus Iulius Maior Antoninus Pythodorus.[30]

Pythodorus plays a prominent role in Epidauros, and he constructed a number of water-related structures. He was a wealthy senator of the middle of the 2nd century AD, from Nysa-on-the-Meander in Asia Minor.[31] We know from Pausanias (2.27.6-7) that Pythodorus heavily restored not only the Asclepius sanctuary, particularly by building baths, but also the Apollo sanctuary, where he installed a large cistern and renovated the priest's house, where roof tiles with his name (*ANTONEIN*) stamped on them were found.[32] Pythodorus, as a wealthy benefactor in the time of the Second Sophistic with connections to Pergamon and with the imperial family in Rome, was acting in a similar manner to his contemporaries, like Herodes Atticus, restoring older cults of the Greek world.[33] Among the buildings of the Apollo Sanctuary, there was a Temple of Apollo, a priest's house, or *skana*, with a bath complex, a *Mouseion*, or *temenos* of the Muses, along with a monumental propylon, which led down to the Asclepius sanctuary (Figure 2). The area is well watered, like the space of Asclepius below, and the large cistern of Pythodorus would have been fed not only with the waters of the Kynortion Hill, but also by rainwater.[34]

[28] For the excavation of the sanctuary, see the various reports and articles by Lambrinoudakis (1984; 1987; 1988; 1989; 1993; 1994; 2002), along with Lambrinoudakis *et al.* (1999). On the Bronze Age origins of the space, see Riethmüller (2005: 152).

[29] Riethmüller 2005: 157.

[30] Lambrinoudakis 1994: 226-228; Melfi 2010; Riethmüller 2005: 156; Trümper 2014: 220-230.

[31] Melfi 2007: 121-123; Melfi 2010: 334.

[32] Melfi 2010: 334.

[33] Melfi 2010: 338.

[34] Pausanias reports that the cistern was fed by rainwater, but modern excavations indicate that the water of the hill was also tapped for the cistern (2.27.6-7). See also Melfi (2007: 120).

Figure 3. Epidauros. Sanctuary of Apollo Maleatas: Nymphaeum (Eastern Room in Foreground). Ephorate of Antiquities of Argolida (Photo Author) ©Ministry of Culture and Sports: Archaeological Receipts Fund

Immediately adjacent to the cistern is a structure identified by excavators as a nymphaeum, which Pythodorus is believed to have installed.[35] The construction of the cistern and the nymphaeum, along with the *skana*, probably occurred at the same time, since the techniques employed are very similar.[36] The form of the nymphaeum is unusual (Figure 3). One enters a colonnade, opening onto a cross-vaulted interior space, which, in effect, restricts access. On the east and south ends are small barrel-vaulted spaces, with fountains on their walls. The south side also contains a circular room, again, with a fountain on the back wall. It has been suggested the water displayed in these rooms was used for lustral activities, especially in connection with the immediately adjacent *Mouseion* and the Temple of Apollo.[37] The circular room, in fact, reminds one of architectural forms that are found in bathing establishments. The architect was playing with common and familiar architectural shapes, but employing them in a novel manner.

The set of structures at the Sanctuary of Apollo Maleatas is unparalleled, particularly locating a nymphaeum next a *Mouseion*. It is not out of the question to see connections between water and the Muses, who were often conflated with the nymphs, as the Muses were companions of Apollo.[38] The well-watered nature of this area indicates that while water was being used for display in the nymphaeum, it also served a variety of purposes in other structures, such as bathing in the priest's house or in the cultic activities of the *Mouseion*. The water-display in the nymphaeum, however, would have created an

[35] There is not much scholarship on this nymphaeum, other than its initial publication by Lambroudakis (1984: 229-230; 1989: 49-50, 54), Leigh (1997: 281-284), and Melfi (2007: 120-121; 2010: 336). Leigh compares the architectural form of the nymphaeum at Apollo Maleatas with that of the Hadrianic nymphaeum on Lykabettos in Athens, stating that there are two aisles, covered by double barrel-vaulted roof. The structure does not appear to have two aisles, however, making her parallel with the Athenian example not completely accurate.
[36] Lambrinoudakis 1993: 37-43; Melfi 2007: 120. Pausanias (2.27.6-7) might also indicate that the nymphaeum was constructed in connection with the cistern in the passage about Antoninus' activities at the site (ἔλυτρον κρήνης: 'a reservoir of the fountain').
[37] Melfi 2007: 120.
[38] Cf. Arnaldi 2006: 59-66.

intimate and inviting space in this small sanctuary, using novel architectural forms. The patronage of Pythodorus would not have gone unnoticed by pilgrims, as Pausanias even reports this to us. The addition of a more 'Roman' space devoted to water-display would have certainly transformed the Sanctuary of Apollo Maleatas, used since the Bronze Age, into a truly Roman cult site, employing Roman architectural technologies, while still celebrating local Greek deities. Furthermore, it is important to note the role of Pythodorus, who was not a member of the local élite, but a visitor from Asia Minor, who put his mark on the local religious landscape of Epidauros. He was concerned with restoring not only older structures associated with water, such as bathing establishments, but also with building new water structures, such as the nymphaeum in the Maleatas *temenos*. While promoting himself, he made sure to exploit water in this space for the benefit of all the pilgrims traveling there.

The forum of Philippi

In various agoras and fora throughout the Empire, different elements of hydraulic infrastructure have been found, especially pipes and drainage channels, of which the latter would have ensured dry spaces.[39] Usually tied to centrally located spaces, fora and agoras tended to be positioned with easy access to urban water distribution networks.[40] Water-displays in these public spaces come in three forms: as subsidiary; entrance-adjacent; or as focal points.[41] The latter form of water-display draws the attention of the pedestrians either to congregate at the structure itself or to focus their gaze on the fountain or an adjacent structure (e.g., temple). Focal points can either stand alone or act as architectural pendants to other important buildings in the forum. Focal points do not have to be planned, as they can be part of the organic development of an urban node, when a public square grows around an area where there is a fountain and a major street.

At Philippi, in northern Greece, two long rectilinear fountains on the northern side of the forum show how a water-display can help to emphasize the central area of the space, acting as focal points. The forum of Philippi (*Colonia Julia Augusta Philippiensium*) was monumentalized in three subsequent phases, from the colony's refounding in 27 BC to the 5th century AD (Figure 4).[42] On the slopes directly to the north of the *Via Egnatia* and the lower forum area, a terrace with three different temples was installed. During the first monumentalizing phase (mid-1st century AD), along the northern edge of the open space of the forum, below the temple terrace, a line of different structures was constructed that included a *rostrum*, two small temples, and two ramps, the latter of which provided access to the *Via Egnatia*.[43] In the second phase (between AD 161-175), the open space of the forum itself was formalized, enclosing all four sides (making an area of roughly 98m by 48m). On the north, the space between the small temples and the ramps was filled by two extremely long fountains (22.00m by 3.10m, nearly 0.80m deep), which were minimally decorated with a single pillar on the interior that had a lion-head spout that poured water into the basin (Figure 5). An extremely fragmentary inscription names one Decimus as responsible for the construction of the fountains (at a cost of 30,000 sesterces).[44]

The fountains in the Philippi forum effectively use their placement in the landscape to focus the attention of the passer-by. Although the fountains themselves are relatively simple basins with solitary pillars of moving water, they filled an area that was previously empty. Nestled on the northern edge of the forum, they provide continuity in the architectural space among ramps, temples, and the *rostrum*. The single visual line then prompted the viewer not only to look at them, but also at what was above, namely the *Via Egnatia* and the temple terrace (Figure 6). By working with other architectural elements and the natural

[39] Akaiturri 2008: 86-97.
[40] Richard 2012: 189.
[41] Rogers 2015: 134-146.
[42] Evangelidis 2010: 259-275; Sève and Weber 2012.
[43] Sève and Weber 2012: 14-16.
[44] *[... L. Decimu]s L(uci) f(ilius) Vol(tinia) Bassus aed(ilis) Philippis testament sibi et L(ucio) Decimio L(uci) f(ilio) Vol(tinia) | [q(uaestori) IIvir(o) Philippis patri(?) e]t C(aio) Decimio L(uci) f(ilio) Vol(tinia) Maxsimo fratri fieri iussit HS ((I)) ((I)) ((I))*. For the inscription, see Collart (1937: 346-347) and Pilhofer (2000: 230). Sève and Weber (2012: 72) disagree with the attribution of the inscription to the fountain, but suggest that it perhaps belongs to some other structure, such as a funerary monument.

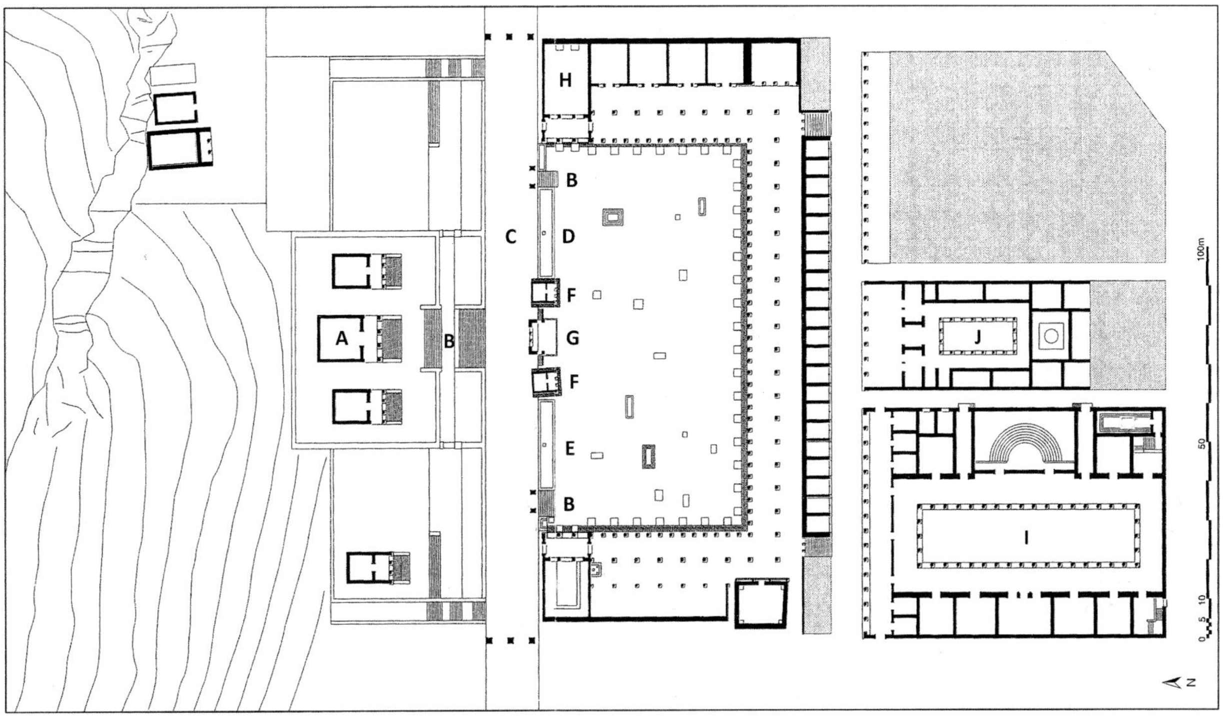

Figure 4. Ancient Philippi. Forum: Temple Terrace (A), Ramps (B), via Egnatia (C), East Fountain (D), West Fountain (E), Temples (F), Rostrum (G), Temple of the Imperial Cult (H), Palaestra (I), Market (J) (after Sève and Weber 2012: fig. 67)

Figure 5. Ancient Philippi. Forum: East Fountain, with Lion Head Waterspout in Basin, Terracing on the North Side to via Egnatia and Temple Terrace. Ephorate of Kavala and Thasos (Photo Author)
©Ministry of Culture and Sports: Archaeological Receipts Fund.

Figure 6. Ancient Philippi. View from North down into Forum from Temple Terrace. Ephorate of Kavala and Thasos (Photo Author) ©Ministry of Culture and Sports: Archaeological Receipts Fund

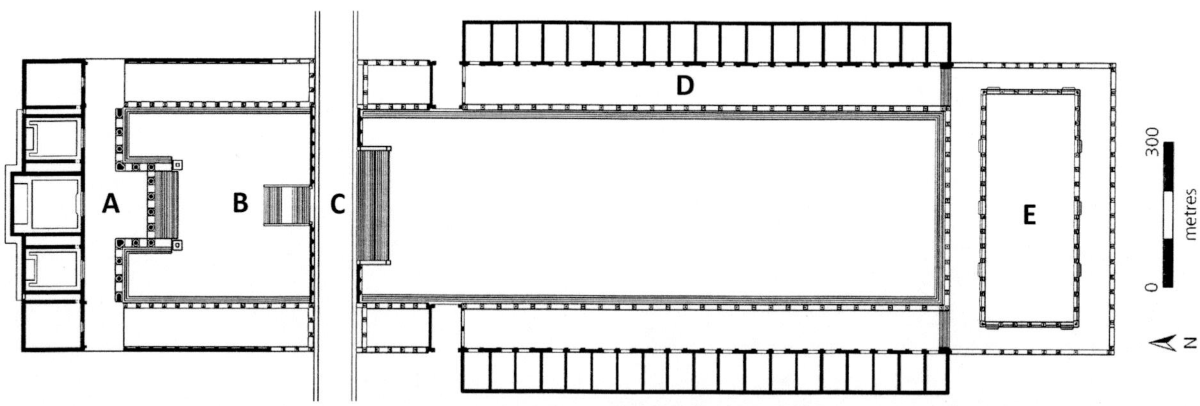

Figure 7. Brescia, Italy. Forum: Temples (A), Ramps (B), Main Street (C), Colonnade (D), Basilica (E) (after Laurence, Esmonde Cleary, Sears 2011: fig. 7.2)

terracing of the site, the architects and patrons effectively monumentalized the northern edge of the Philippi forum.

At the same time, the division of space in the forum of Philippi exemplifies the Roman desire to create an *area sacra* in the forum. Throughout the Roman Empire, fora are known for including *areae sacrae* that are spatially (and sometimes visually) separated, particularly through axial configurations of the space, often with a temple at the end of the forum.[45] In addition to axial arrangements, one finds terracing that can aid in the visibility of structures (evident especially through steps and ramps) and occasionally streets

[45] Evangelidis 2014: 351; Laurence *et al.* 2011: 176-179.

passing through (which increases traffic in a space), which is found at a number of sites throughout the Empire, such as at Brescia (Italy), where a line of temples culminates in an axis that is punctuated by ramps, a main street, a long colonnade, and a basilica (Figure 7).[46] At Philippi, this organization of space is clear, especially with the bifurcation created by the terracing, the path of the Via Egnatia, and the line of monuments, including the fountains, in the forum proper.

Theatrical Waters

Roman entertainment complexes (e.g., theatres, amphitheatres, odeia, stadia, and circuses) used water for a variety of purposes, but the study and understanding of water-displays in and around them is too often overlooked in scholarship. Perhaps such work is hampered by the fact that very rarely in the modern period is water used in the context of theatrical performances, making such a notion foreign to us, except in rare cases, such as Mary Zimmerman's 1996 play, *Metamorphoses*, an adaptation of Ovid, which uses a pool of water that actors interact with as part of its set.[47] Water-displays and fountains, however, were often placed inside or nearby Roman entertainment complexes, demonstrating the entertainment value and utility of water in the Roman theatre.

The archaeological remains of Roman theatres show evidence that the theatres were equipped for the use of water. Hydraulic installations in the theatre could include cisterns for the storage of water, lead and terracotta piping, drainage channels, and fountains. First and foremost, the water that was provided from fountains was certainly used for drinking, ensuring that spectators were sufficiently hydrated while attending performances and were thus comfortable.[48] With drainage channels, especially those lining the perimeter of the orchestra, water could have been used for cleaning the theatre space. Water was evidently also used for cooling. Vitruvius (*De architectura,* 5.9.6) states that in the *postscaenium*, where walking under porticos is healthy for spectators, water can be used to ensure cool spaces. Valerius Maximus (2.4.6) mentions that Pompey the Great was the first to use water channels to cool his Porticus, immediately adjacent to his theatre complex in Rome. Propertius (2.32.11-16) also discusses the same colonnade, with shade from the columns and plane trees, along with water flowing from fountain sculptures.

Finally, there is the phenomenon of liquid *sparsiones* in the Roman theatre.[49] Literally a 'scattering', the liquid *sparsio* was generally perfumed water that was sprayed in order to cool air, cut down on dust, or to water vegetation.[50] Water for *sparsiones* could be held in small fountains or basins within the theatre, or perhaps even in channels that ran down the cavea steps. In fact, Hadrian (SHA *Hadr.* 19.5) is reported to have ordered *sparsiones* of balsam- and crocus-perfumed water to be sent down the cavea steps of a theatre in Rome. In reality, the *sparsiones* were probably dispersed using a siphon pump, akin to the *machina Ctesibica* described by Vitruvius (*De architectura,* 10.7). The mist of the *sparsiones* could have been directed at either the audience or the orchestra itself.[51] The *sparsio*, a great show of pneumatic technology, would have added to the entertaining atmosphere of the Roman theatre, with its coloured water slightly tinting the air. One can imagine the refreshing nature of a *sparsio* in the Roman theatre, not only cooling the area, but also scenting the air.[52]

In order to explore water-displays in the Roman theatre, two different spaces of the theatre are examined. First, there are water-displays on the stage and in the orchestra. Fountains were present on the *proscaenium* of a number of theatres throughout the Roman world, whether in the form of simple jets and basins, or with fountain sculpture. Then, there are fountains found in the vicinity of theatres. There are certain characteristics that aid in placing the water back into theatres, such as drainage channels and

[46] Laurence *et al.* 2011: 176-178.
[47] Zimmerman 2002.
[48] Cf. Rose 2005.
[49] Nibley 1945.
[50] Daremberg-Saglio 4.2, 1419, s.v.'Sparsio' (Ph. Fabia); Fleury 2008; Fuchs 1987: 143; Spano 1913: 144-146.
[51] Fleury 2008: 97.
[52] Cf. James 2004.

water-display accoutrement, which can include basins and sculpture; such elements offer clues to the nature of water-display in the Roman theatre.

When there is evidence for water-displays within the Roman theatre, it can be found on the *proscaenium* or in the orchestra. The *scaenae frons* of the Roman theatre included a parapet at the *proscaenium*, which is often termed the *frons pulpiti*, or the front of the stage, an ideal place to include a water-display.[53] While the *frons pulpiti* can be a plain, undecorated flat surface, many extant examples were recessed, with alternating rectilinear and semi-circular exedras. The recessed *frons pulpiti*, however, is only found in areas where the Roman- or Latin-style theatres were adopted for use.[54] After the dissemination of the Roman-style theatre type out of Rome and throughout the Italian peninsula, it was then found in southern France, southern Spain, North Africa, and some parts of the East, where a more Greek style theatre usually still prevailed.[55] Roman style theatres could be found at sites in the East with close connections with Rome (e.g., Corinth, Ephesus, Miletus). Greek theatres of the Roman period that have drains around their *frontes pulpiti* include: Argos;[56] Elis;[57] Sikyon.[58] Although the state of the evidence does not allow for a reconstruction of any sort of water-display (or even acting as simple rainwater drains), the Greek examples are suggestive of a larger, Empire-wide phenomenon.

In addition to water-displays on and around the actual stage of the Roman theatre, fountains could be found in the *porticus* of the *postscaenium*, around main entrances, in *paradoi* or *aditi*, and, on sight lines with theatres within the urban fabric. The presence and locations of these water-displays demonstrate a number of points. The practicality of the water installations (e.g., collecting water or cooling a space) can help to explain the placement of these structures. Fountains could also demarcate a new space, showing spectators that they were moving from one location to another. For example, when a spectator approached a theatrical complex, he could first stop in the *porticus* of the *postscaenium* before entering the theatre proper, in order to rest, to take water, or to socialize with others in the space. Finally, water-displays interact with the urban armature to create nodes and cohesive urban units. Thus, the importance of fountains in proximity to theatres can be essential for our understanding of the use of theatrical space.

The Roman-style theatre complex included not only the seating, orchestra, and skene building, but also other vital structures. Vitruvius (*De architectura*, 5.3) mentions that when choosing the site for a theatre, the builders must choose a healthy site, thinking about the wellbeing of the future spectators. Further, he states (*De architectura*, 5.9.1, 6) that the colonnaded space behind the scene building, a *porticus* or *postscaenium*, is essential to cool spectators off during hot weather, as a place of refuge during sudden rains, and as a storage space for theatrical equipment. Within the *postscaenium*, the presence of water-displays would have had a cooling effect on the spectators. An early example of this can be seen at the theatre of Ephesus, dated from the Hellenistic into the early Roman periods.[59] Although not in a traditional porticoed area, the small covered fountain is near an entrance to the theatre, along with the adjacent theatre gymnasium, both areas of congregation. In another example, the *porticus* associated with the Theatre of Pompey in Rome has been described as the first public park of Rome, as it would have provided many amenities to the spectator, including refreshing and cooling waters in the courtyard, along with gardens, sculptures, and libraries.[60] While the *porticus* of Pompey was certainly large and impressive compared to other examples, it helps us to imagine how the space of the *postscaenium* functioned.

There are two known examples in mainland Greece of water-displays in the *postscaenium*: Sikyon and Corinth. An early imperial example can be found at Sikyon, dated to the early Empire.[61] The form of the fountain here is an exedra (nearly 6m wide, 5m deep), with four columns at the front of the structure,

53 Di Napoli 2013: 101-102,153-154; Sear 2006: 83-95; Sturgeon 1972: 124-129.
54 Rogers 2015: 358-359.
55 Sear 2006: 96-97.
56 Di Napoli 2013: 52; Sear 2006: 386.
57 Di Napoli 2013: 64; Sear 2006: 396.
58 Fiechter 1931: 22; Sear 2006: 405.
59 Dorl-Klingenschmid 2001: 178-180.
60 Kuttner 1999.
61 Bressan 2009: 228-231.

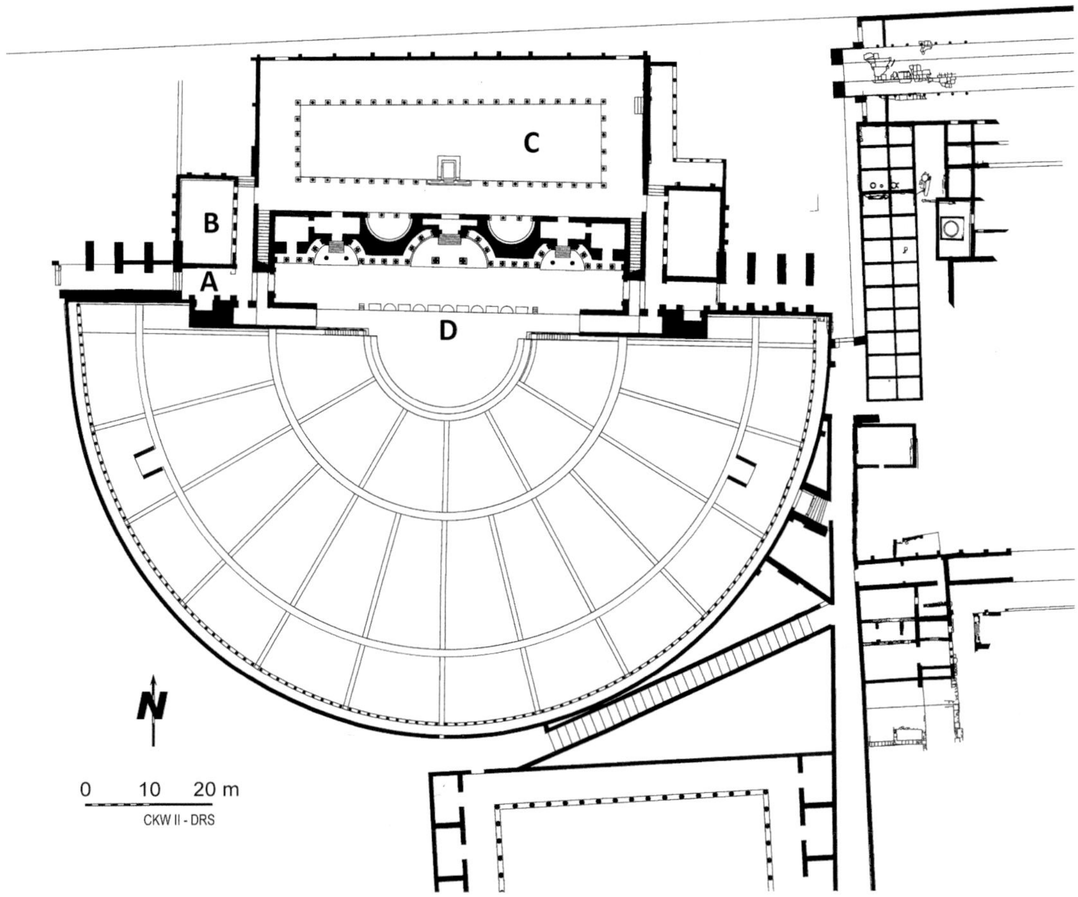

Figure 8. Corinth. Theatre, Plan of the 5th Phase: Fountain (A), Lesser Plaza (B), North Peristyle Court (Postscaenium) (C), *Frons pulpiti* (D) (after Williams 2013: Fig. 3)

and a water connection located in the back of the basin. The structure is located at the north end of the *postscaenium*, immediately adjacent to the portico. The columns of the fountain, then, would have mimicked the porticoed area, creating a cohesive façade at the back of the stage building of the theatre. The architectural ensemble of this area of the theatre would have created an important way station for spectators, in addition to providing them with water for drinking or other purposes before and after performances.

At the site of Corinth, a fountain was located in the porticoed area to the north of the stage building. Recent excavations by Charles Williams have begun to clarify the theatre's chronology (assigning to it 11 phases of activity), which aid in understanding when the fountain was installed and when it went out of use (Figure 8).[62] In the second phase (*c.* AD 22-77), the North Peristyle Court was installed.[63] By the fifth phase, dated to the Hadrianic period, subsidiary rooms on the east and west of the peristyle court were added. On the west side, the so-called Lesser Plaza was installed, with a fountain in the southwest corner.[64] Not much is known about the actual structure of the fountain, although we have remains of its pebbled floor surface and some hydraulic installations, because in the eighth phase (the 4th century AD), the fountain was dismantled.[65] It is clear, however, that the water-display was incorporated into the edifices that made up the theatre complex, especially in the North Peristyle Court area. The area must have been a comfortable place to congregate, not only with the refreshing waters associated with a fountain, but also the colonnaded space and sculpture (e.g., a large Amazon, a male figure wearing a hip mantle, a

[62] Bressan 2009: 158-172; Sear 2006: 392-393; Williams 2013.
[63] Williams 2013: 490.
[64] Robinson 2001: 293-307; Williams 2013: 538.
[65] Williams 2013: 497.

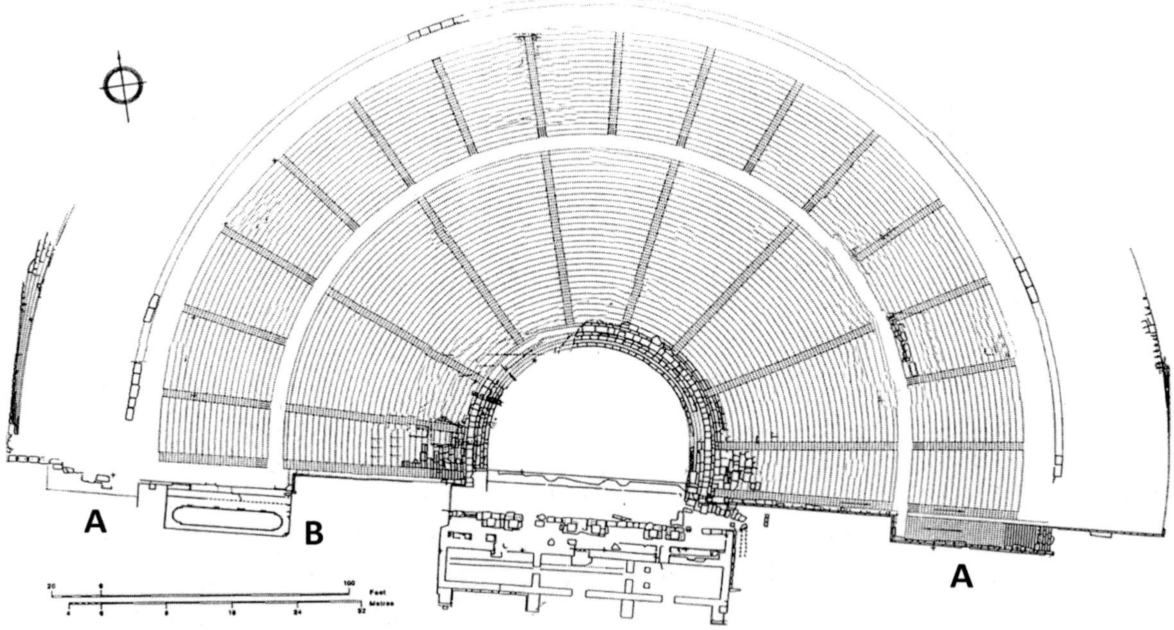

Figure 9. Sparta. Theatre, Plan: Paradoi (A), Fountain (B) (after Di Napoli 2013: fig. 38).

bronze statue of a boy actor, etc.).[66] Further, it appears that one could pass from the peristyle, through the *paradoi*, to enter the seating area of the theatre. Again, we can imagine spectators congregating and using this space, enjoying the amenities provided for them before, during, and after theatrical productions.

The next category to consider is fountains found in *paradoi* or *aditi*, the corridors that led to the orchestra space of the theatre. Of the four known examples, all come from the eastern part of the Mediterranean. Three examples come from the Peloponnese. Argos has a fountain structure (perhaps related to the 2nd century AD rebuilding of the stage building) in its northern *parados*, although not much is known about the structure.[67] At Elis, on the eastern *parados*, a small (roughly 2m by 2m) square fountain, with two engaged columns at the entrance to the basin, has been found.[68] Sparta preserves a long rectangular framed basin (15.10m by 4.00m, with an interior basin of 13.20m by 2.20m) on the west *parados* of the theatre (Figure 9).[69] Because the fountain uses building material of the theatre from the 1st century AD, along with the use of an inscribed statue base with a *terminus post quem* of first half of the 3rd century, it is believed that the structure was installed after the Herulian invasion of 267.[70] The basin would have been clad in marble veneer, and decorated with statues of a seated lion (with a water-spout), a running boar, and a roughly finished Herakles on the back ledge of the fountain (Figure 10). The water that supplied the fountain probably arrived from an aqueduct that was situated above the theatre. Five spouts on the front of the basin would have poured water into accompanying troughs, which would have allowed for collection by passers-by, or even by animals.[71]

Finally, at the site of Nea Paphos on Cyprus, while not on the mainland of Greece, another similar rectangular-form fountain has recently been excavated on the eastern *parados*.[72] The structure, dated to the 1st century AD, measures 20m long by 5m wide, and there is evidence for a niche above the wall

[66] Sturgeon 2004: 15-16.
[67] Aristodemou 2011: 178; Bressan 2009: 76-85; Di Napoli 2013: 51-54; Glaser 1983: 40, 73; Sear 2006: 386-387.
[68] Aristodemou 2011: 178; Bressan 2009: 134-140; Di Napoli 2013: 63-64; Glaser 1983: 39-41; Sear 2006: 396.
[69] Aristodemou 2012: 299-300; Bressan 2009: 233-240; Di Napoli 2013: 86-93; Sear 2006: 405-406; Walker 1979: 211-217.
[70] Cartledge and Spawforth 1989: 122; Walker 1979: 214-215.
[71] Walker 1979: 212.
[72] Barker 2010a; 2010b; 2013.

Figure 10. Sparta. Theatre, Western Parados: Fountain Basin. Ephorate of Laconia (Photo Author) ©Ministry of Culture and Sports: Archaeological Receipts Fund

with sculpture, along with marble architectural members that were later dumped inside the basin.[73] The interior of the basin was covered with a star-pattern mosaic floor and was lined with chamfered plaster. The fountain was placed at a strategic point of not only at the *parados* of the theatre, but also along a main thoroughfare in the city, indicating that this area would have been heavily trafficked.

It is suggestive that examples of water-displays in the *paradoi* have only been found in theatres in the eastern half of the Mediterranean. This can perhaps be explained in terms of theatre usage. Greek theatres, on the one hand, took advantage of the surrounding landscape and were usually situated in a hillside. The entrance to the structure was through the (uncovered) *paradoi*, often a highly trafficked area.[74] In the Roman theatre, on the other hand, because of the technological advances in building materials and techniques, hillsides were not needed. This shift allowed for the entrance of the theatre to be moved to the space below the *cavea*, and for the *paradoi* then to be covered and vaulted, in effect, closing up the stage building. What we see in these four examples are Greek-style theatres, originally built into the surrounding hillsides but subsequently adapted to become Roman-style theatres, allowing for easy access to the water at the entrances of the theatres. Indeed, the eastern Mediterranean, even after the Roman conquest, continued to use more Greek-style theatres.[75] In fact, there are very few theatres constructed *de novo* during the Roman period in Greece.[76] In these four examples, we have Greek-style uncovered *paradoi*, which would have served as entrances to the theatre. With the high traffic, builders were eager to

[73] Barker 2010b; 2013: 18.
[74] Bieber 1961: 189; Di Napoli 2013: 99-100.
[75] Sear 2006: 113-115.
[76] Di Napoli 2013: 102.

provide easy access to water for all the spectators. Because the *paradoi* were not open in the West, there we more often see fountains in the *postscaenium*, where the large crowds could congregate.

Where to now?

The last section of this essay has demonstrated the importance of singling out different contexts of Roman water-display, helping to show the reasons behind their construction and novel locations for water-displays. Placing fountains in their original built complexes, especially by examining their functionality as providers of water for consumption and delight, is crucial as a means for understanding them better. In examining these few examples from Roman Greece, it is easy to discern how they interacted in these various contexts. And while the examples are extremely important for comprehending water-usage in Greece, when the water-displays are placed into a wider Empire-wide scheme, they fit nicely into trends that are discernible all throughout the first three centuries of the Roman Empire. While there are certainly numerous variables that make the water-displays of Roman Greece unique, such as specific regional trends (e.g., Roman fountains in the *paradoi* of Greek-style theatres in Greece), these structures exemplify patterns that are found throughout the Empire – thus stressing the importance of water-displays more generally.

The goal of using three different contexts of lesser studied water-displays in Greece also stresses the importance of each of these structures. Just because these fountains were not the monumental nymphaea that the Romans are especially known for, does not mean that they are any less vital for a better understanding of Roman life. By re-examining the evidence, we can replace water in religious sanctuaries, in the middle of a forum, and throughout various parts of the Roman theatre. While these examples might not have the grand façades of large-scale fountains, they still *alter* and *impact* their surrounding spaces, particularly for those ancient Romans using and enjoying those structures. Indeed, such contextualization—in addition to replacing the ancient Roman on the ground—allows us to apply new theoretical approaches to Roman water-displays, such as the 'archaeology of the senses,' in order to understand better these fountains throughout the Empire.

Acknowledgments

This essay is based in part on the author's doctoral dissertation, 'Water-Display and Meaning in the High Roman Empire' (2015, University of Virginia). The author wishes to thank Prof. Tyler Jo Smith for discussing and reading the essay, Ms. Ioanna Damanaki (American School of Classical Studies at Athens) for obtaining the permissions for the figures included, and especially the chief editor of this volume, Dr Georgia Aristodemou, for soliciting this contribution.

Bibliography

Abbreviations used

CIL *Corpus Inscriptionum Latinarum* (Berolini 1863-)

IGR III *Inscriptiones graecae ad res Romanas pertinentes*, ed. René Cagnat et al. 3 vols. Paris 1901-1927. Vol. 3, fasc. 1-6, with Jules F. Toutain, Pierre Jouguet and Georges Lafaye. Paris 1902-1906. Reprint: Chicago, Ares 1975.

....

Agusta-Boularot, S. 1997. La fontaine, la ville et le Prince. Recherches sur les fontaines monumentales et leur fonction dans l'urbanisme impérial, de l'avénement d'Auguste au règne des Sévères. Unpublished PhD dissertation, Université de Provence, Aix-Marseille 1.

Agusta-Boularot, S. 2001. Fontaines et fontaines monumentales en Grèce de la conquête romaine à l'époque Flavienne: Permanence ou renouveau architectural? In: J.-Y. Marc and J.-C. Moretti (eds)

Constructions publiques et programmes édilitaires en Grèce entre le IIe siècle av. J.-C. et le Ier siècle ap. J.-C.: 167-236. Paris: Diffusion E. de Boccard.

Agusta-Boularot, S. 2004. Les fontaines urbaines en Gaule. In: S. Agusta-Boularot and É. Follain (eds) *Fontaines et Nymphées en Gaule Romaine: Fontaines publiques et privées, les nymphées dans les thermes*: 4-11. Dijon: Éditions Faton.

Akaiturri, A.E. 2008. *Los foros Romanos republicanos en la Italia centro-meridional tirrena: Origen y evolución formal*. Madrid: Consejo Superior de Investigaciones Científicas.

Aristodemou, G. 2011. Theatre Façades and the Façade Nymphaeum. The Link Between. *Bulletin de Correspondance Hellénique* 135: 163-197.

Aristodemou, G. 2012. *Ο Γλυπτός Διάκοσμος Νυμφαίων και Κρηνών, στο Ανατολικό Τμήμα της Ρωμαϊκής Αυτοκρατορίας* (*Sculptured Decoration of Monumental Nymphaea and Fountains at the Eastern Provinces of the Roman Empire*). Thessaloniki: Kornelia Sfakianakis Editions.

Arnaldi, A. 2006. La valenza 'salutare' del culto delle *Nymphae* nell'Italia Romana. In: L. Gasperini (ed.) VSVS VENERATIOQVE FONTIVM: *Fruizione e culto delle acque salutari nell'Italia Romana*: 55-84. Tivoli: Editrice Tipigraf.

Aupert, P. 1974. *Le nymphée de Tipasa et les nymphées et 'septizonia' nord-africains*. Rome: École Française de Rome.

Aupert, P. 2012. Eau et religion. In: J.-P. Bost (ed.) *L'eau: usages, risques et représentations dans le Sud-Ouest de la Gaule et le Nord de la Péninsule ibérique de la fin de l'Age du fer à l'Antiquité tardive (IIe s. a.C.-VIe s. p.C.)*: 293-320. Pessac: Fédération Aquitania Publications.

Barker, C. D. 2010a. Fabrika Hill: Excavating a Hellenistic-Roman Theatre. *Past Horizons-Adventures in Archaeology* 12: 28-35.

Barker, C. D. 2010b. 'Paphos 2010 Season Press Release'. Accessed from http://www.paphostheatre.com/2010-press-release.html

Barker, C. D 2013. Nea Paphos Theatre Excavations 2011. *Australian Archaeological Institute at Athens Bulletin* 9: 18-9.

Berg, D. 1994. *Fountains and Artistic Water Displays in Classical Antiquity: Origins and Development from 700 to 30 B.C.* Unpublished PhD dissertation, University of Texas at Austin.

Bieber, M. 1961. *The History of the Greek and Roman Theater*. Princeton: Princeton University Press.

Bol, R. 1984. *Das Statuenprogramm des Herodes-Atticus-Nymphäums*. Berlin: Walter de Gruyter.

Bressan, M. 2009. *Il teatro in Attica e Peloponneso tra età greca ed età Romana: Morfologie, politiche edilizie e contesti culturali*. Rome: Edizioni Quasar.

Burrell, B. 2012. Book Review of *Roman imperialism and Civic Patronage: Form, Meaning, and Ideology in Monumental Fountain Complexes*. *American Journal of Archaeology* 116.2 (online).

Campagna, L. 2011. Le fontane nella città dell'impero Romano: 'Monuments Indelibly Associated with the Emperor'? *Journal of Roman Archaeology* 24: 648-656.

Cartledge, P. and Spawforth, A. 1989. *Hellenistic and Roman Sparta: A Tale of Two Cities*. London: Routledge.

Collart, P. 1937. *Philippes: Ville de Macédoine, depuis ses origines jusqu'à la fin de l'époque romaine*. 2 vol. Paris: Diffusion E. de Boccard.

Cole, S.G. 1988. The Uses of Water in Greek Sanctuaries. In: R. Hägg, N. Marinatos, and G. C. Nordquist (eds) *Early Greek Cult Practice*: 161-15. Stockholm: Paul Åstöms Förlag.

De Kleijn, G. 2001. *The Water Supply of Ancient Rome: City Area, Water, and Population*. Amsterdam: J. C. Gieben.

Dessales, H. 2013. *Le partage de l'eau: Fontaines et distribution hydraulique dans l'habitat urbain de l'Italie romaine*. Rome: École Française de Rome.

Di Napoli, V. 2013. *Teatri della Grecia Romana: Forma, decorazione, funzioni. La provincia d'Acaia*, Meletemata 67. The National Hellenic Research Foundation/ Research Centre for Greek and Roman Antiquity (Ethnikon Hidryma Ereunōn/ Kentron Hellēnikēs kai Rōmaïkēs Archaiotētos [EIE/KEPA]). Athens-Paris: Diffusion De Boccard.

Dorl-Klingenschmid, C. 2001. *Prunkbrunnen in kleinasiatischen Städten: Funktion im Kontext*. München: Verlag Dr. Friedrich Pfeil.

Evangelidis, V. 2010. *Η Αγορά των πόλεων της Ελλάδας από τη ρωμαϊκή κατάκτηση ως τον 3ο αι. μ.Χ.* (*The Agoras of the Cities of Greece from the Roman Conquest to the 3rd Century* AD). Thessaloniki: University Studio Press.

Evangelidis, V. 2014. Agoras and Fora: Developments in the Central Public Space of the Cities of Greece during the Roman Period. *Annual of the British School at Athens* 109.1: 335-356.

Fiechter, E. 1931. *Das Theater in Sikyon*. Stuttgart: Verlag von W. Kohlhammer.

Fleury, P. 2008. Les *sparsiones* liquides dans les spectacles romains. *Revue des études latines* 86: 97-112.

Fuchs, M. 1987. *Untersuchungen zur Ausstattung römischer Theater in Italien und den Westprovinzen des Imperium Romanum*. Mainz am Rhein: Philipp von Zabern.

Ginouvès, R. 1969. L'Architecture. In J. Des Gagniers *et al.* (eds) *Laodicée du Lycos: Le Nymphée, Campagnes 1961-1963*: 13-185. Paris: Diffusion E. de Boccard.

Glaser, F. 1983. *Antike Brunnenbauten (KRHNAI) in Griechenland*. Wien, Verlag der Österreichischen Akademie der Wissenshaften.

Glaser, F. 2000. Fountains and Nymphaea. In: Ö. Wikander (ed.) *Handbook of Ancient Water Technology*: 413-451. Leiden: Brill.

Gros, P. 1996. *L'architecture romaine du début du IIIe siècle av. J.-C. à la fin du Haut-Empire, Vol. 1, Les monuments publics*. Paris: Picard éditeur.

James, L. 2004. Senses and Sensibility in Byzantium. *Art History* 27.4: 522-537.

Kapossy, B. 1969. *Brunnenfiguren der hellenistischen und römischen Zeit*. Zürich: Juris Druck & Verlag.

Kuttner, A. L. 1999. Culture and History at Pompey's Museum. *Transactions of the American Philological Association* 129: 343-373.

Lamare, N. 2014. *Les fontaines monumentales en Afrique romaine*. Unpublished PhD dissertation, Université Paris-Sorbonne.

Lambrinoudakis, V. 1984. Ἀνασκφή Ἐπιδαύρου: Ἐργασίες συντήρης στό ἱερό τοῦ Ἀπόλλωνος Μαλεάτα. *Praktika tes en Athenais Archaiologikes Etaireias*: 229-232.

Lambrinoudakis, V. 1987. Ἱερό Ἀπολλωνος Μαλεάτα Ἐπιδαύρου. *Praktika tes en Athenais Archaiologikes Etaireias*: 52-65.

Lambrinoudakis, V. 1988. Ἱερό Ἀπολλωνος Μαλεάτα Ἐπιδαύρου. *Praktika tes en Athenais Archaiologikes Etaireias*: 21-29.

Lambrinoudakis, V. 1989. Ἱερό Ἀπολλωνος Μαλεάτα Ἐπιδαύρου. *Praktika tes en Athenais Archaiologikes Etaireias*: 43-56.

Lambrinoudakis, V. 1993. Ἱερό Ἀπολλωνος Μαλεάτα Ἐπιδαύρου. *Praktika tes en Athenais Archaiologikes Etaireias*: 37-43.

Lambrinoudakis, V. 1994. L'eau médicale à Épidaure. In R. Ginouvès *et al.* (eds) *L'eau, la santé et la maladie dans le Monde Grec*: 225-236. Paris: Diffusion De Boccard.

Lambrinoudakis, V. 2002. Conservation and Research: New Evidence on a Long-Living Cult. The Sanctuary of Apollo Maleatas and Asklepios at Epidauros. In M. Stamatopoulou and M. Yeroulanou (eds) *Excavating Classical Culture: Recent Archaeological Discoveries in Greece*: 213-224. Oxford: Archaeopress.

Lambrinoudakis, V. (ed.) 1999. *Το Ασκληπιείο της Επιδαύρου: η έδρα του θεού γιατρού της αρχαιότητας: η συντήρηση των μνημείων του*. Athens: Prefecture of Peloponnese.

Laurence, R., Esmonde Cleary, S., and Sears, G. 2011. *The City in the Roman West, c. 250 BC- c. AD 250*. Cambridge: Cambridge University Press.

Lavagne, H. 2012. Quelques fontaines publiques et nymphées en Gaule romaine. In J. Jouanna *et al.* (eds) *L'eau en Méditerranée de l'Antiquité au Moyen Âge*: 117-139. Paris: Diffusion E. de Boccard.

Leigh, S. 1997. The 'Reservoir' of Hadrian in Athens. *Journal of Roman Archaeology* 10: 279-290.

Lendon, J. E. 2015. Rhetoric and Nymphaea in the Roman Empire. *Chiron* 45: 123-149.

Letzner, W. 1999. *Römische Brunnen und Nymphaea in der westlichen Reichshälfte*. 2nd Rev. Ed. Münster: Charybdis.

Lolos, Y. 1997. The Hadrianic Aqueduct of Corinth (With an Appendix on the Roman Aqueducts in Greece). *Hesperia* 66.2: 271-314.

Longfellow, B. 2011. *Roman imperialism and Civic Patronage: Form, Meaning, and Ideology in Monumental Fountain Complexes*. Cambridge: Cambridge University Press.

Longfellow, B. 2012. Roman Fountains in Greek Sanctuaries. *American Journal of Archaeology* 116: 133-155.

Marchetti, P. and Kolokotsas, K. 1995. *Le nymphée de l'agora d'Argos: Fouille, étude architecturale et historique*. Paris: Diffusion E. de Boccard.

Melfi, M. 2007. *I santuari di Asclepio in Grecia. I*. Rome: 'L'Erma' di Bretschneider.

Melfi, M. 2010. Rebuilding the Myth of Asklepios at the Sanctuary of Epidauros in the Roman Period. In: A. D. Rizakis and C. E. Lepenioti (eds) *Roman Peloponnese III: Society, Economy and Culture under the Roman Empire: Continuity and Innovation*: 329-339. Athens/Paris: Diffusion E. de Boccard.

Neuerburg, N. 1965. *L'architettura delle fontane e dei ninfei nell'Italia antica*. Naples: Gaetano Macchiaroli Editore.

Nibley, H. 1945. 'Sparsiones'. *Classical Journal* 40.9: 515-543.

Papaioannou, M. 2007. The Roman *domus* in the Greek World. In: R. Westgate *et al.* (eds) *Building Communities: Houses, Settlement and Society in the Aegean and Beyond*: 351-361. London: British School at Athens.

Piérart, M. 1999. Les puits de Danaos et les fontaines d'Hadrien. In: J. Renard (ed.) *Le Péloponnèse: Archéologie et Histoire*: 243-268. Rennes: Presses Universitaires de Rennes.

Pilhofer, P. 2000. *Philippi, Band II: Katalog der Inschriften von Philippi*. Tübingen: Mohr Siebeck.

Prignitz, S. 2014. *Bauurkunden und Bauprogramm von Epidauros (400-350): Asklepiostempel, Tholos, Kultbild, Brunnenhaus*. München: Verlag C.H. Beck.

Reinholdt, C. 2009. *Das Brunnenhaus der Arsinoë in Messene: Nutzarchitektur, Repräsentationsbaukunst und Hydrotechnologie im Rahmen hellenistisch-römischer Wasserversorgung*. Wien: Phoibus Verlag.

Riethmüller, J. W. 2005. *Asklepios: Heiligtümer und Kulte*. Heidelberg: Verlag Archäologie und Geschichte.

Richard, J. 2007. Roman Monumental Fountains in the Levant: Water Supply vs. Urban Aesthetics? In: C. Ohlig (ed.) *Cura Aquarum in Jordanien*: 263-284. Norderstedt: Deutsche Wasserhistorische Gesellschaft.

Richard, J. 2012. *Water for the City, Fountains for the People: Monumental Fountains in the Roman East, An Archaeological Study of Water Management*. Turnhout: Brepols.

Rinne, K. W. 2011. *The Waters of Rome: Aqueducts, Fountains, and the Birth of the Baroque City*. New Haven: Yale University Press.

Robinson, B. A. 2001. *Fountains and the Culture of Water at Roman Corinth*. Unpublished PhD dissertation, University of Pennsylvania.

Robinson, B. A. 2011. *Histories of Peirene: A Corinthian Fountain in Three Millennia*. Princeton: American School of Classical Studies at Athens.

Robinson, B. A. 2013. Playing in the Sun: Hydraulic Architecture and Water Displays in imperial Corinth. *Hesperia* 82: 341-384.

Rogers, D. K. 2013. Il piacere d'acqua: I ninfei domestici di Pompeii. In: V. Gheller (ed.) *Richerche a Confronto: Dialoghi di Antichità Classica e del Vicino Oriente*: 154-169. Montorso Vicentino: Edizioni Saecula.

Rogers, D. K. 2015. *Water-Display and Meaning in the High Roman Empire*. Unpublished PhD dissertation, University of Virginia.

Rose, P. 2005. Spectators and Spectator Comfort in Roman Entertainment Buildings: A Study in Functional Design. *Papers of the British School at Rome* 73: 99-130.

Scheid, J. 1991. Sanctuaries et thermes sous l'empire. In *Les thermes Romains*: 205-214. Rome: École Française de Rome.

Schmölder-Veit, A. 2009. *Brunnen in den Städten des westlichen Römischen Reichs*. Wiesbaden: Dr. Ludwig Reichert Verlag.

Sear, F. B. 2006. *Roman Theatres: An Architectural Study*. Oxford: Oxford University Press.

Segal, A. 1997. *From Function to Monument: Urban Landscapes of Roman Palestine, Syria, and Provincia Arabia*. Oxford: Oxbow.

Settis, S. 1973. 'Esedra' e 'Ninfeo' nella terminologia architettonica del mondo Romano: Dell'età repubblicana alla tarda antichità. *Aufstieg und der Niedergang der römischen Welt* 1.4: 661-745.

Sève, M., and P. Weber. 2012. *Guide du forum de Philippes*. Athens/Paris: École française d'Athènes.

Spano, G. 1913. Il teatro delle fontane in Pompeii. *Rendiconti dell'Accademia di archeologia, lettere e belle arti, Napoli* 2.2: 109-148.

Sturgeon, M. 1972. *Corinth, IX. 2: Sculpture: The Reliefs from the Theater*. Princeton: American School of Classical Studies at Athens.

Sturgeon, M. 2004. *Corinth IX. 3: Sculpture: The Assemblage from the Theater*. Princeton: American School of Classical Studies at Athens.

Tomasello, F. 2005. *Fontane e ninfei minori di Leptis Magna*. Rome: 'L'Erma' di Bretschneider.

Trümper, M. 2014. Bathing in the Sanctuaries of Asklepios and Apollo Maleatas at Epidauros. In: A. Avramidou and D. Demetriou (eds) *Approaching the Ancient Artifact: Representation, Narrative, and Function*: 211-231. Berlin: De Gruyter.

Walker, S. 1979. The Architectural Development of Roman Nymphaea in Greece. Unpublished DPhil dissertation, University of London.

Walker, S. 1987. Roman Nymphaea in the Greek World. In: S. Macready and F. H. Thompson (eds) *Roman Architecture in the Greek World*: 60-71. London: Thames and Hudson.

Weiss, C.F. 2011. Living Fluidly: Uses and Meanings of Water in Asia Minor (Second Century BCE-Second Century CE). Unpublished PhD dissertation, Brown University.

Williams, C.K. 2013. Corinth, 2011: Investigation of the West Hall of the Theater. *Hesperia* 82.3: 487-549.
Zarmakoupi, M. 2013. The Villa Culture of Roman Greece. In: A. D. Rizakis and I. P. Touratsoglou (eds) *Villae Rusticae: Family and Market-Oriented Farms in Greece under Roman Rule*: 752-761. Athens: Diffusion E. de Boccard.
Zarmakoupi, M. 2014. *Designing for Luxury on the Bay of Naples: Villas and Landscapes (c. 100 BCE-79 CE)*. Oxford: Oxford University Press.
Zimmerman, M. 2002. *Metamorphoses: A Play*. Evanston: Northwestern University Press.

Dr Dylan Kelby Rogers
Assistant Director
American School of Classical Studies at Athens
dylan.rogers@ascsa.edu.gr

Fountain figures from the Greek provinces: monumentality in fountain structures of Roman Greece as revealed through their sculptural display programs and their patrons

Georgia Aristodemou

Abstract

Monumental civic fountains and nymphaea were financed by Emperors and wealthy patrons in cities throughout the ancient Roman Empire. These imposing edifices stood one to three stories high and contained elaborate sculptural programs that often interacted with flowing water to create innovative kinetic displays. Typically fed by aqueducts and positioned in heavily trafficked areas of the city, these extravagant fountains showcased a variety of sculptures, from donor portraits to divinities to local heroes, carefully arranged to highlight the prestige of the local community and benefactor.

The study of fountain sculptures, that is sculptures with a visible water outlet ascertaining a spouting function or sculptures that regardless their function derive from a nymphaeum or an elaborated fountain, have only the past few years attracted the scholars' attention. This paper discusses the presence and function of sculptures within the fountain structures based on their direct or indirect connection with water. Monumental nymphaea are centrally located in major cities of the Roman East and are usually financed by the Emperor, the city, or local elite members. Therefore, the presence of Imperial and private portraits underlines the imperial power and the prestige of local Elite. Evidently, sculptural display programs are used either to glorify the Emperor and his family, or honoring the local gods, or in some cases, underline the dynamic local Elite. The significance of choices on a local, regional, imperial level led us to study sculptural display programs of monumental nymphaea as a reflection of the patrons and the city's status and power.

Monumental nymphaea are products of historical and social conditions. This study, apart from highlighting certain aspects about the sculptural display programs of monumental fountains in the Greek provinces during the Roman Period, also tries to reconstruct the social, political as well as the cultural role that these public buildings held in the Greek Provinces.

Keywords: Nymphaea, fountain sculpture, iconography, patronage, Greek Provinces

Monumental fountain structures, built by emperors and wealthy patrons in cities throughout the Roman Empire, have always been striking features of Roman architecture. They were designed as elaborate multi-storied edifices with a columnar façade, often supplemented with projecting lateral wings.[1] These imposing edifices were centrally located and functioned as landmarks and nodal points in major cities of the Roman East.[2] They housed sculptural assemblages that often interacted with the flowing water to create innovative displays, carefully arranged to highlight the prestige of the benefactor and the local community.[3] J. Lendon comprehends the emergence of nymphaea (initially in the western Asia Minor and, later on, in Syria and the Levantine provinces) as the result of a 'growing fixation upon local water-supply', associated with the Second Sophistic and its rhetoric practices upon praising cities as if they were persons.[4] Within this framework and while discussing Libanius' appraisal speech (*Or.*11, *Antiochikos*) about his home city, Antioch-on-the-Orontes, Lendon reaches to the conclusion that abundant natural water resources raised significantly the status of a city – or, in other words, 'the excellence of a city's position was illustrated by how well watered it was'.[5]

[1] For a discussion on the architectonic forms and typology of Roman nymphaea, see Neuerburg 1960 and 1965; Letzner 1990: 17-216; Gros 1996: 418-444; Ginouvès 1998: 92-100; Agusta-Boularot 2001; Dorl-Klingenschmid 2001: 21-61; Longfellow 2005: 334-342. Longfellow 2011: 1-5; Aristodemou 2012: 28-41; Richard 2012; Lendon 2015: 123-124. On a comparative discussion between nymphaea and theatres, cf. Parra 1976: 89-118; Burrell 2006: 437-469; Aristodemou 2011b: 163-197.

[2] Bauer 1996: 392; Dorl-Klingenschmid 2001:149-150; Uğurlu 2009: 42-45; Aristodemou 2014.

[3] Berg 1994. Also, Lendon 2015: 124.

[4] Lendon 2015: 131-135.

[5] Lendon 2015: 135-139.

This paper approaches these monuments through their sculptural decoration. Contrary to the numerous studies focusing on the architectural forms of Roman nymphaea, their sculptural decoration never actually attracted the interest of the scholarly research. The only two studies on the subject were those of E. Curtius, *Die Plastik der Hellenen an Quellen und Brunnen* (1876) and E. Lange, *Die Entwicklung der antiken Brunnenplastik* (1920), until a more updated monograph was published in 1969 by B. Kapossy (*Brunnenfiguren der hellenistischen und römischen Zeit*). His descriptions were accurate but rather short (thus not covering the subject in depth), whereas the included photographic material appeared rather insufficient. B. Kapossy studied the material by categorizing it in groups according to their depicted subjects (e.g. the *Thiasoi* of Aphrodite, Dionysus, Apollo, etc). He posed some key questions on how these figures were technically related to the monument and the water, as well as on which type of monuments these fountain figures were placed at. He also dedicated part of his study on the predecessors of the fountain figures and on relevant information from ancient authors. Though his study has weaknesses at some points, it was certainly fundamental for the following scholarly research. It took several years for new studies on the sculptural decoration of Roman nymphaea to appear; in 2011 B. Longfellow published her study on *Roman Imperialism and Civic Patronage: Form, Meaning and Ideology in Monumental Fountain Complexes*, where she discussed architectural form and sculptural assemblages on Roman nymphaea as a medium for exercising both Imperial and private propaganda, while in 2012, a new monograph (G. Aristodemou, *The sculptural Decoration of nymphaea and fountain on the Eastern provinces of the Roman Empire*) was released (in Greek) examining both the sculptural decoration of nymphaea of the Eastern Provinces and the concept of fountain figures in depth.

In continuance to the above we shall here discuss issues, such as the definition of the term fountain figure, its position and function in the monument, the selection and in some cases the interpretation of sculptural display programs.

Fountain figures: a definition

1. A sculpted figure is characterized as a fountain figure when it carries visible technical features, such as a water pipe channel or a water outlet - in order to function as a spout; in addition to this, its provenance from a nymphaeum building is archaeologically confirmed.[6] A reclining river god from the Hydrekdocheion of C. Laecanius Bassus in Ephesus,[7] a seated female figure holding a spouting bird from the Miletus Nymphaeum[8] or an Aphrodite statue of the Pudica type from the

Figure 1. Thessaloniki, Archaeological Museum, Inv. No 6680. Sleeping Silenus. Photo: G. Aristodemou

[6] Aristodemou 2012: 88-93. Previous studies on fountain figures: Curtius 1876; Lange 1920; Kapossy 1969; Kent Hill 1974: 107-108; Kent Hill 1981: 84; Söldner 1986: 295-298; Glaser 1987: 103-113; Fuchs 1987: 141-143; Letzner 1990: 258-262.

[7] Ephesus, Arch. Museum, Inv. No.1/36/72: Aurenhammer 1990: 104, no 85, pl.60b; Klementa 1993: 146, no V1.

[8] Constantinople, Arch. Museum, Inv. No. 2006: Hülsen 1919: 55, no 1, fig.1; LIMC II (1984) 94 no 891, s.v. Aphrodite (A. Delivorias).

Figure 2. Thessaloniki, Archaeological Museum, Inv. No 895. Youth male with water vessel. Photo: G. Aristodemou

2. Philadepheia (Amman) Nymphaeum[9] are only three of the numerous examples from the Roman East. Examples from the Greek provinces, such as for example a nymph holding a shell from the Gortyn nymphaeum[10], are considerably fewer, usually and probably due to the lack of relevant data from excavations.
3. Any sculpture with a visible water outlet that ascertains a spouting function is also considered a fountain figure - even without having a confirmed provenance. The majority of these sculptures are found during rescue excavations - or they are random findings; they lack exact find spot data, therefore only speculations can be made about their original exhibition. Such sculptures are numerus in the Greek provinces and the most distinctive among them are: a sleeping Silenus leaning on a wine amphora and a standing youth holding a water vessel from Thessaloniki (Figures 1 and 2)[11] - the vessels in both statues are pierced so as to function as water spouts. A sleeping river god from Thasos Island (Figure 3).[12] Four examples from Athens: i) a reclining Silenus sleeping upon a spouting rocky base;[13] ii) a sculptural complex of Theseus with the Minotaur- of which,

[9] Amman, Arch. Museum, without Inv. No.: Weber 2002: 512 no D14, pl.155C-E; Riedl 2003: 264, 484 no PH.1.

[10] Ghedini 1985: 147-153 no 28, fig. 59; Aristodemou 2012: 367 no. 339, pl. 47.5.

[11] 1) *Thessaloniki*, Arch. Museum, Inv. No. 6680: Stefanidou-Tiveriou 1995: 67-75, pl. 30-34. Despinis *et al.* 1997, I: 124-126, no 97, figs. 269-272 (Th. Stefanidou-Tiveriou); Aristodemou 2012: 301 no 79, pl. 13.1. 2) *Thessaloniki*, Arch. Museum, Inv. No. 895: Despinis *et al.* 1997, I: 126-128, no 98, figs. 273-277 (Th. Stefanidou-Tiveriou); Aristodemou 2012: 301 no 80, pl. 13.2.

[12] *Thasos*, Arch. Museum, Inv. No. Λ100: Klementa 1990: 153 no V 6; Aristodemou 2011a: 150 note 10, fig. 1; Aristodemou 2012: 301 no 77, pl. 12.6.

[13] *Athens*, Museum of Ancient Agora, Inv. No. S 2175: Ajootian 1993: 251 note 3, 255 fig. 8; Klementa 1993: 151 no V 5; Aristodemou 2012: 285, no 3, pl. 1.2.

Figure 3. Thasos, Archaeological Museum, Inv. No Λ100. Reclining River God. (Photo: after Aristodemou 2012: pl. 12.6)

Figure 4. Corinth, Archaeological Museum, Inv. No S 2063. Nymph holding a shell. Photo: G. Aristodemou

the latter had a pierced spouting mouth;[14] iii) a Satyr figure, part of a complex identified as the *Invitation to a Dance*, leaning upon a perforated tree trunk;[15] iv) a reclining Silenus sleeping upon a wine vessel.[16] A seated child figure from Isthmia.[17] A Nymph holding a water spouting shell from Corinth (Figure 4)[18] and a reclining Silenus from Sicyon.[19] An unpublished sleeping male figure, recorded as a sleeping hunter, is known from Patra (Figure 5).[20] A seated Nymph (or Muse?) from Messene,[21] and two Satyr figures (one reclining and one sleeping) from Sparta. In both figures, the wine skin and the wine vessel are pierced so as to function as water spouts.[22] Two small frog figures with pierced bodies originate from Cos island.[23] A reclining Satyr from Lesvos Island sleeps upon a water spouting

[14] *Athens*, National Museum, Inv. No. 1664-1664a: LIMC VI (1992) 579 nr. 66, s.v. Minotauros (S. Woodford); Kaltsas 2001: 96 no 169 with fig.; Aristodemou 2012: 285 no 4, pl. 1.3.

[15] *Athens*, Kerameikos. Muthmann 1931: 87-89, pl. 3940; Muthmann 1951: 73 note 2; Brinkerhoff 1965: 33, no 22; Geominy 1999: 149, pls. 38.1, 3, 39.4; Aristodemou 2012: 286 no 5, pl. 1.4

[16] *Athens*, National Museum, Inv. No. 7053: Stefanidou-Tiveriou 1995: 67-70, pl. 30β-33β; Aristodemou 2012: 286 no 6, pl. 2.1.

[17] *Isthmia*, Arch. Museum, Inv. No. IS 136: Sturgeon 1987: 116 no 26, pl. 54c,f; Aristodemou 2012: 289 no 18, pl. 3.4.

[18] *Corinth*, Arch. Museum, Inv. No. S 2063: Scranton 1951: 34, pl. 15.2; Soles 1983: 85-88 no 4, pl. 21, fig. 19; Robinson 2001: 368 fig. 8.6; Aristodemou 2012: 292 no 34, pl. 6.1.

[19] *Sicyon*, Arch. Museum. Meunier and Launay 1955: 121, pl. 36; Karageorghis 1964: 33 note 4; Aristodemou 2012: 299 no 66, pl. 11.4.

[20] Kolonas and Stavropoulou-Gatsi 2017: 62-63 with a short description and no photo.

[21] *Messene, Mavromati* (Archaeological Museum Depot), Inv. No. 76: Xagorari-Gleißner 2002: 75-80, figs.1-5; Aristodemou 2012: 293 no 38, pl. 6.5.

[22] *Sparta*, Arch. Museum, Inv. No. 22 and 727: Tod and Wace 1906: 138 no 22 and no 727; Ajootian 1993: 251-253 with figs. 1a-b and 2a-b; Aristodemou 2012: 300 no 73, pl. 12.2 and no 74, pl. 12.3.

[23] *Cos*, Arch. Museum, without Inv. No.: Laurenzi 1956: 139, nos 187-188, with figs.; Aristodemou 2012: 336, nos. 204 and 205, pls. 32.6-7.

Figure 5. Patras, Archaeological Museum. Unknown Inv. No. Sleeping hunter (?). Photo: G. Aristodemou

Figure 6. Herakleion, Archaeological Museum, Inv. No 46. Silen holding a wine skin. Photo: G. Aristodemou

Figure 7. Kisamos, Archaeological Museum, Inv. No 694. Silen holding a wine skin. Photo: G. Aristodemou

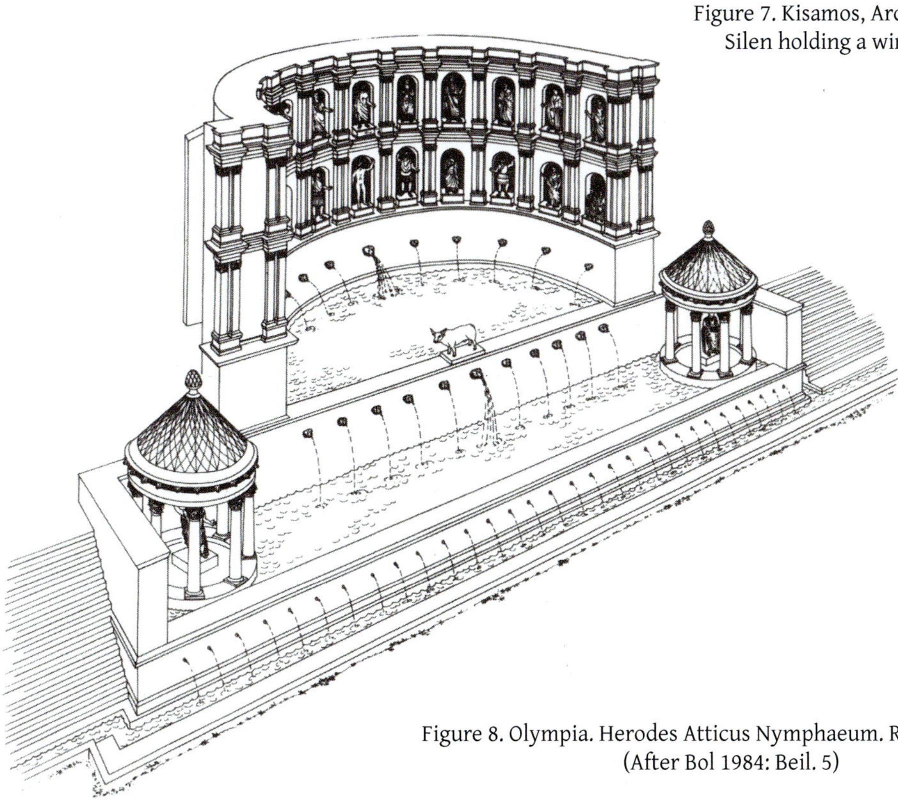

Figure 8. Olympia. Herodes Atticus Nymphaeum. Reconstruction. (After Bol 1984: Beil. 5)

wine vessel.[24] A feline (panther?) figure with a pierced mouth,[25] and a sleeping Satyr (the rocky base upon which he reclines preserves the pipe hole),[26] are known from Rhodes. A reclining river god comes from Samos Island.[27] A Satyr holding a spouting wine skin is known from Knossos (Figure 6),[28] while a standing Satyr stepping on a spouting rocky base (Figure 7) and a standing Silenus leaning on a spouting wine skin come from Kisamos (Crete).[29]

4. A third category includes statues (not necessarily preserving a spouting hole) found within the water basin debris during a nymphaeum's excavation – or statues placed at the upper floors of a multi storey nymphaeum (where they almost never functioned as spouts). These are also characterized as fountain figures.[30] Two typical examples from the Greek provinces are the group of statues found within the basins of the Herodes Atticus Nymphaeum in Olympia (Figure 8)[31] and the nymphaeum F25 of the Praetorium complex in Gortyn.[32] It is also attested that a group of the portrait statues of the Augustus' family decorated the façade of the Arsinöe Fountain in Messene (though nothing has been preserved).[33] So far, no concrete evidence exists regarding the sculptural assemblages originally placed at the twin Nikopolis' nymphaea,[34] or the Southeastern Nymphaeum at the Athenian Agora.[35]

Fountain figures: technical features

Fountain figures can be categorised in two main groups, based on their involvement to the water flow. The main differentiation lies on whether the figure remains intact, or has been perforated so as to function as a spout:

1. The first group includes statues that remain intact, that is, there is *no direct intervention* to the figure itself; water flows from a pipe channel drilled at the plinth or base of the standing, seated or reclining figure.[36] Typical examples from the Greek provinces include figures such as: standing Silens and Satyrs,[37] seated Pans playing the syrinx,[38] seated Nymphs,[39] reclining river gods,[40] reclining Silens and Satyrs,[41] or other sleeping idealistic and mythological figures, such as young hunters or Erotes (Figures 5 and 9).[42] In such figures the plinth is shaped like a natural rock or imitates the sea surface. In cases of standing figures water streams out from a hole at the statue's support, most frequently formed as a tree trunk, an animal or a fish. In cases of human or mythological figures (e.g. children, Nymphs, Silens, Satyrs, Erotes) holding animals, birds and fishes, water vessels, wine-skins, or sea-shells, water flows from a tube running through the animal or the object held by the standing, seated or reclining figure.[43] In the

[24] *Mytilene*, Arch. Museum, Inv. No. 229: Bakalakis 1966: 21, pls. 1-3; Ajootian 1993: 251 fig. 3; Stefanidou-Tiveriou 1995: 67-70, pls. 30β-33β; Archontidou and Acheilara 1999: 98 with fig.; Aristodemou 2012: 336 no 206, pl. 32.8.

[25] *Rhodes*, Arch. Museum, Inv. No. E 153 (82): Gualandi 1976: 224 no 243, fig. 300; Aristodemou 2012: 337 no 207, pl. 33.3.

[26] *Rhodes*, Arch. Museum, Inv. No. 1160: Merker 1973: 29 no 55, pl. 16-17, figs. 37-39; Söldner 1986: 295 note 1105; Gualandi 1976: 202 note 2; 223 note 6; Aristodemou 2012: 337 no 209, pl. 33.2.

[27] *Samos*, Arch. Museum, Inv. No 582: Horn 1972: 110 no 79, pl. 57; Klementa 1993: 153 no V 7; Aristodemou 2012: 337, no 214, pl. 34.1.

[28] *Herakleion*, Arch. Museum, Inv. No 46: Unpublished. Aristodemou 2012: 369 no 354, pl. 49.2.

[29] 1) *Kisamos*, Arch. Museum, Inv. No 694: Aristodemou 2011a: 150 note 15, fig. 3. 2) *Kisamos*, Arch. Museum, Inv. No L3210: Markoulaki 2009: 340 fig.1, note 14. On the above, cf. Aristodemou 2012: 369 nos 355-356, pl. 49.3 and 49.4.

[30] Kapossy 1969: 10, 57; Kent Hill 1981: 84; Dorl-Klingenschmid 2001: 92, 94; Aristodemou 2012: 89.

[31] Settis 1973; Bol 1984; Walker 1979: 182-196, figs. 61-66, no 4.2.3; Fittschen 1999: 127-128 no 46; Aristodemou 2012: 294.

[32] Ortega 1986-1987; On the Nymphaeum sculptural assemblage, Ghedini 1985: 63-248; Longfellow 2011: 138-139 fig. 48 and 206-207 fig. 70; Aristodemou 2012: 366.

[33] *Messene*, Arch. Museum, Inv. No 7611 (a+b): Cited at Themelis 1995: 56-57. See, M. Trabuco dell Torreta's paper in present volume and Aristodemou 2018 (forthcoming).

[34] Longfellow 2005: 174; Longfellow 2011: 131-134. Recently, Zachos 2015: 110. See, Zachos and Leontaris' paper in present volume.

[35] Thompson 1955: 57-59; Walker 1979: 60-71; Glaser 1983: 109 nr 74; Longfellow 2011: 124-130; Leigh 2000: 122-130; Aristodemou 2012: 76, 255, 285.

[36] Kapossy 1969: 55; Kent Hill 1981: 84-85; Aristodemou 2012: 90.

[37] 1) Standing Satyr once holding a flute, *Kisamos*, Arch. Museum, Inv. No. 694: see above, note 29. 2) Standing Silenus holding a wine skin, *Kisamos*, Arch. Museum, Inv. No L3210: see above, note 29. 3) Satyr (?) holding a wine skin on his thigh, *Herakleion*, Arch. Museum, Inv. No 46: see above, note 28.

[38] *Corinth*, Arch. Museum, Inv. No. S2385: Robinson 2001: 374-377, fig.8.8; Aristodemou 2011a: 150; Aristodemou 2012: 90-91, 293 no 36, pl. 6.3.

[39] *Messene*, Arch. Museum Depot, Inv. No 76: see above, note 21.

[40] *Thasos*, Arch. Museum, Inv. No. Λ100: Thouvenot 1974/75 : 96, fig. 2; Klementa 1990: 153, no V 6 ; Aristodemou 2011a: 150 note 10, fig. 1; Aristodemou 2012: 301 no 77, pl. 12.6.

[41] 1) *Athens*, Museum of Ancient Agora, Inv. No S 2175: see above, note 13; 2) *Sparta*, Arch. Museum, Inv. No. 727: see above, note 22. 3) *Sikyon*, Arch. Museum, see above, note 19. 4) *Thessaloniki*, Arch. Museum, Inv. No. 6680: see above, note 11; 5) *Hadrianoupolis*, Hermitage Museum, Inv. No. A106: Waldhauer 1928: 40 no 22, fig. 10; Aristodemou 2012: 303, no 85, pl. 13.5. 6) *Lesvos*, Arch. Museum Inv. No. 229: see above, note 24.

[42] On hunters, cf. above, note 20 (*Patra*). On Erotes, 1) *Delphi*, Arch. Museum, without Inv. No.: Marcadé 1993: 475-484, fig.1-3; 2) *Delphi*, Arch. Museum, Inv. No. 2616: Marcadé 1993: 478-484, figs. 4-6; Also Aristodemou 2011a: 150; Aristodemou 2012: 90-91, 288 no 13, pl. 3.1 and no 14, pl. 3.2.

[43] Aristodemou 2012: 90-91. Cf. 1) *Crouching Silen*, Athens, National Museum, Inv. No.7053: see above, note 16. 2) *Satyr holding a wine skin*,

Figure 9. Delphi, Archaeological Museum, No Inv. No. Sleeping Eros. Photo: (Photo: after Aristodemou 2012: pl. 3.1.2)

complexes of the struggle between Heracles and the Nemean Lion water runs through the beast's muzzle (Figure 10).[44]

2. A second group includes figures that are *organically involved with the flow of water*; their torso is perforated by a pipe (cf. the child figure from Antioch-on-the-Orontes),[45] and water spouts from a natural body opening, such as the mouth, the breast or the genitals.[46] Nymphs holding shells constitute a significant group of statues whose torso is horizontally penetrated by a pipe at the lower abdominal.[47] As a result, the water spouts at a large shell held by the figure at the level of the pudenda (Figure 4). Animal figures, such as lions, dogs, dolphins, etc., whose muzzles (or legs) are penetrated by

Figure 10. Isthmia, Archaeological Museum, Inv. No IS 405. Herakles and the Nemean Lion. (Photo: after Aristodemou 2012: pl.3.3.5)

Herakleion, Arch. Museum, Inv. No. 46: see above, note 28.

[44] Aristodemou 2012: 91-92. Cf. 1) *Athens*, random find (BE 27): Papapostolou 1967: Χρονικά, Β' 1, 36 no 11. 2) *Argos*, Arch. Museum, Inv. No. 737: Salowey 1995: 378 no S11. 3) *Argos*, Arch. Museum Inv. No. Π225: Marcadé 1980: 158 note 60; 4) *Isthmia*, Arch. Museum, Inv. No. IS 405: Sturgeon 1987: 117 no 30, pl.52, 53a. 5) *Corinth*, Arch. Museum, Inv. No. S 2828: Robinson 2001: 373 sq. On the above, cf. Aristodemou 2012: 286 no 7 (*Athens*), 288 nos 11-12, pl. 2.4 (*Argos*), 289 no 17, pl. 3.3 (*Isthmia*), 293 no 37, pl. 6.4 (*Corinth*).

[45] Art Museum, Prinston University, Inv. No. 2000-38: Padgett 2001: 224-225 no 74.

[46] Kapossy 1969: 54; Kent Hill 1981: 90-91; Aristodemou 2012: 92.

[47] Kapossy 1969: 54-55; Aristodemou 2012: 92, 149. Some characteristic examples come from *Corinth*: i) Arch. Museum, Inv. No. 1024: Soles 1983: 90-91 no 7, pl. 22 fig. 24; Robinson 2001: 100, fig. 2.10, 2.11; Robinson 2011: 215-216, fig. 116. ii) Arch. Museum, Inv. No. S 2063: see above, note 18; *Sparta:* Arch. Museum, Inv. No. 99 and 836: Tod and Wace 1906: 148, no 99; *Gortyn*, Nymphaeum F25, (*today lost*): Ghedini 1985: 147-153 no 28, fig. 59; on the above, Aristodemou 2012: 290 no 22, pl.4.4 (*Corinth*), 300 nos 72 and 75, pl. 12.2, 12.4 (*Sparta*), 367 no 339, pl. 47.5 (*Gortyn*).

a water pipe are also included to this group.[48] In those cases too, the water pipe runs through their body and terminates at the water outlet.

3. Other sculptures have been converted into fountain figures at a later phase.[49] Therefore, instead of being internally pierced, their surfaces were carved with grooves that carried the water pipe. In such cases water often spouted awkwardly. A characteristic example comes from Asia Minor and regards an entire statuary group depicting the blinding of Polyphemus, which was reused at the Domitian Fountain in Ephesus after being transferred there from an earlier monument.[50] Its composing figures were converted so as to have a spouting function and to decorate the Domitian's fountain at a later date, having as a result the awkward spouting of water from the newly drilled holes.[51] So far, only few examples are known from the inland or island Greece, namely a child figure holding a rabbit from Isthmia, the sleeping river god from Thasos, the unpublished sleeping and armoured hunter from Patra (Figures 3 and 5).[52]

Reconstruction

Reconstructing the original position of a sculpture in a monument's façade is a challenging process. A general difficutly is that most monumental nymphaea preserve only a small part of their original sculptural assemblage. Moreover and despite the fact that find-spot data is preserved for many examples, the original settings are often difficult to establish, as many of these where found in late or poorly understood contexts or they lack technical features, e.g. pipe holes, that would help restore the figure's position at the monument.[53] Sometimes the monument itself presents us with helpful information. For example, the niches of the twin Nymphaea in Nikopolis (Epirus) preserve the square shaped holes that indicate the exact spots where the fountain statues once stood.[54] Although there is no evidence of the monument's sculptural assemblage, the architectural remains provide proof of their existence.

Three equally important and indicative factors are: the sculpture's *finding spot*, its *iconography* and its *function*.[55] According to B. Kapossy, the higher a figure stands in the hierarchy of the Olympian pantheon the lesser is connected with the water flow.[56] This hierarchy is possible to be detected since the most important figures tend to be placed at the upper floors of a nymphaeum. Furthermore, it is usually considered that sculpture depicting water related subjects are normally placed in contact with the water of the basin below. Water spouting statues are usually placed at the lower floor, so that the water from the shell, the vessel or the wine-skin held by the figure runs directly into the basin below.[57] The same also holds for the seated and reclining figures. In this way the figure correlates to the monument visually, organically as well as conceptually.

A statue's *preservation state* is also evidential, since a well preserved statue assumingly stood at a lower storey and, respectively, a statue broken in many pieces had most possibly fallen from above. The Herodes Atticus Nymphaeum in Olympia stands as an indicative example (Figure 8).[58] It is a two-storied lavishing semicircular monument and the restoration of the statues' original position was possible due to the fact

[48] *Corinth, Peirene*: 1) Scylla group, *not preserved*: Robinson 2001: 329-356 and Robinson 2011: 233-250; 2) Pegasus and Bellerophon complex, *not preserved*, Robinson 2001: 357-361: Robinson 2011: 35-64 and Robinson 2017: 183-184, 187-188. Also, Aristodemou 2012: 290-291, nos 27 and 28; *Sparta*, Arch. Museum, Inv. No. M 5168: Woodward 1926-1927: 32-33, no 7, fig. 12; Aristodemou 2012: 299-300, nos 70 and 75, pl. 11.5 and 12.4; *Cos*, Arch. Museum: see above, note 23; *Rhodes*, Arch.Museum, Inv. No. E153 (82): see above, note 25.

[49] Aristodemou 2012: 92-93.

[50] Ephesus, Arch. Museum. Inv. No. 1093, 1557, 1558, 1559, 1560, 1561, 1562, 14/38/72: Aurenhammer 1990: 168-177 nos 147.1-19, pls.102-109; Andreae 1999a: 162-177; Andreae 1999b: 531-533; Longfellow 2005: 97-102 and Longfellow 2011: 62-76.

[51] Longfellow 2009 on the reuse of statues in fountains and Auinger and Rathmayr 2007 generally on the reuse of sculpture and architectural elements in public monuments. On how the Fountain of Domitian provides visual connection with the classical past and monumentalizes the governor's connections with the Emperor in Rome, Longfellow 2011: 73-74.

[52] *Isthmia*, Arch. Museum, Inv. No. IS 136: see above, note 17. *Thasos*, Arch. Museum, Inv. No. Λ. 100: see above, note 12. *Patra*, Arch. Museum, see above note 20.

[53] Aristodemou 2012: 3, 93-95.

[54] Walker 1979: 140; Zachos and Georgiou 1997: 576-600; Chrysostomou and Kephallonitou 2001: 29-33, figs. 22-25; Longfellow 2011: 131-134; Zachos 2015: 107-110.On the latest studies on the Nikopolis Nymphaea, see Zachos and Leontaris' paper in present volume.

[55] Letzner 1990: 260; Aristodemou 2012: 95.

[56] Kapossy 1969: 83.

[57] Fleischer 1982: 124; Aristodemou 2012: 94.

[58] The monument was thoroughly published by R. Bol (1984).

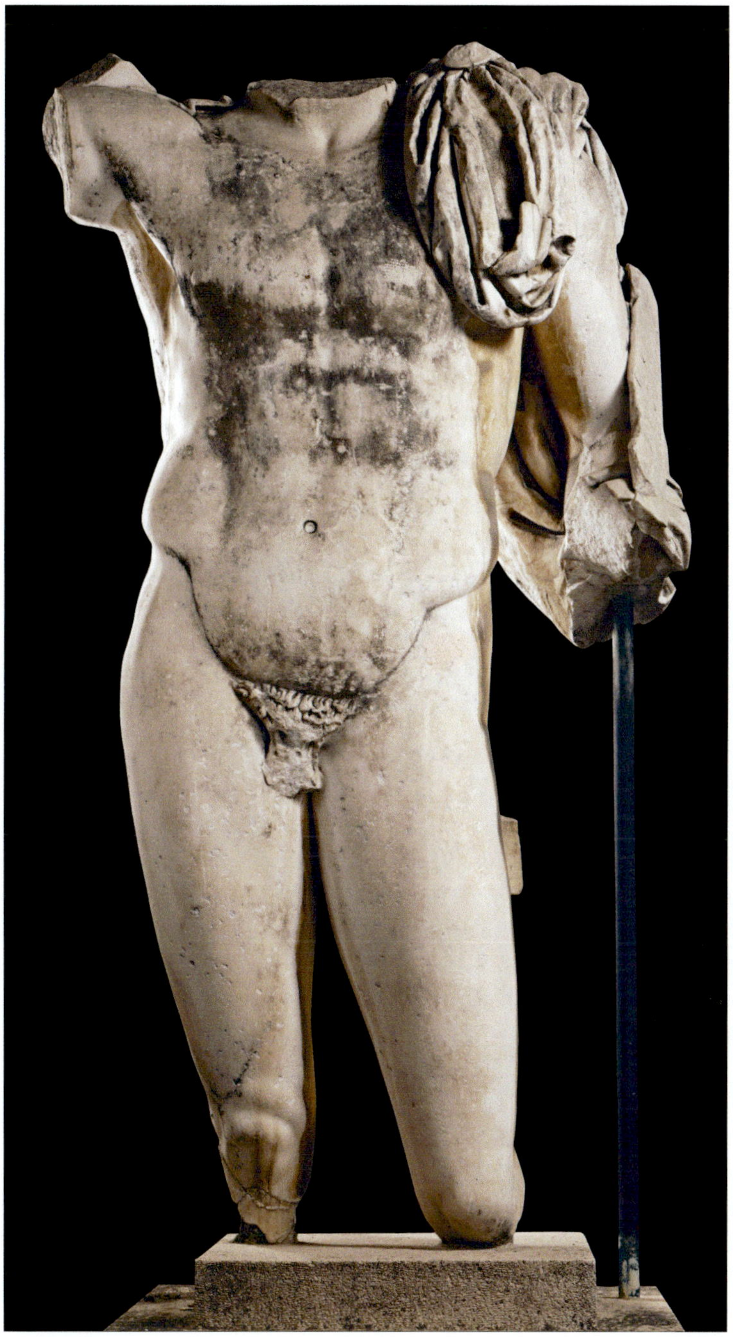

Figure 11. Argos, Archaeological Museum, Inv. No 99. Hadrian's statue.
Photo: Courtesy EfA/ Ph. Collet.

that almost all of its sculptural assemblage was found within the water basin. Based on the archaeological data, the excavators restored each storey with 11 figures.[59] The exedra form enables the statues to deploy symmetrically within the niches of each floor according to a hierarchical and pre-scheduled display programme. The most important figures are placed at the center, while figures of secondary importance are placed towards the edges of the semicircle. The Argos Nymphaeum housed a colossal statue of the Emperor Hadrian (Figure 11). The fact that only a fragmented part of the torso is today preserved, verifies the fact that the statue was placed at a certain height, of which it fell sometime in late Antiquity.[60]

[59] Bol 1984: Beilage 4.

[60] On the Argos Nymphaeum, see below note 129. On Hadrian's statue, Marcadé and Raftopoulou 1963: 49 no 53, figs. 16-18; Maderna 1988: 220 (UD 8), pl. 25.1; Aristodemou 2012: 287 no 9, pl. 2.3 with all previous literature.

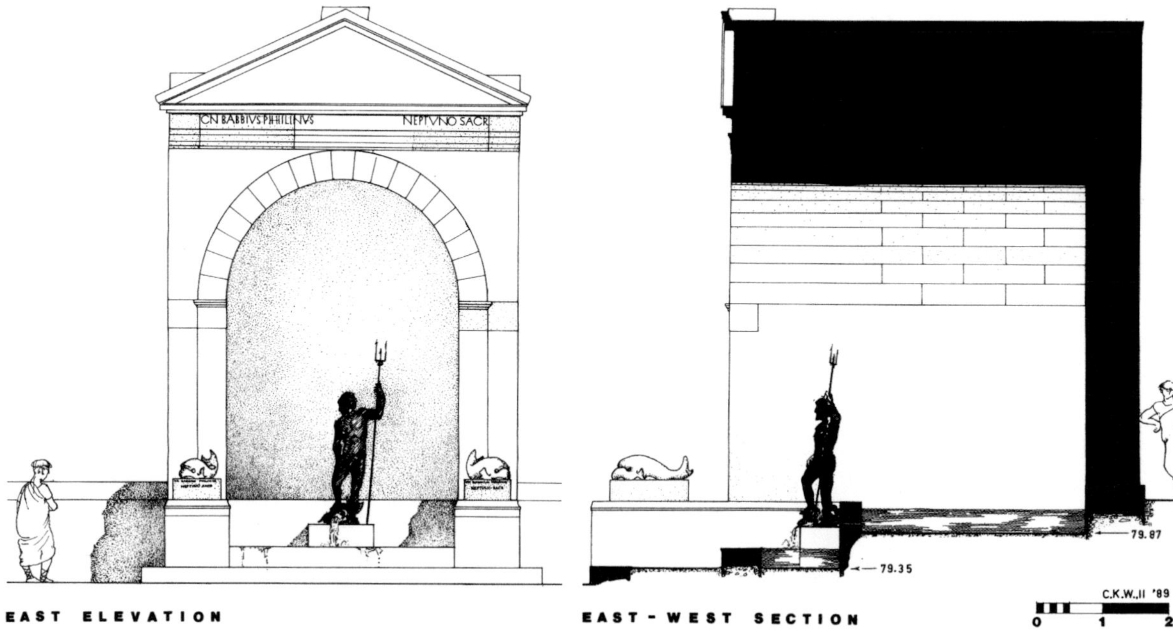

Figure 12. Corinth, Poseidon's Fountain Restored Elevation (Drawing: 1989. 123 001. Artist Williams II, C. K. Courtesy of ASCSA)

Iconography

The sculptured decoration of a monumental nymphaeum derives from a large repertoire of decorative subjects, already established in the majority of public Roman edifices – however adjusted to the particularities of a water monument and having taken into account its architectural form, function and location.[61] Since this paper discusses fountain sculpture strictly from the Greek provinces (inland and island Greece), the material discussed here appears slightly diversified from what is known from the Asia Minor and the Levantine provinces.

1. *Idealistic and mythological figures.* This category includes Olympian and local Gods, minor deities and mythological figures related to water and nature.[62] In the Greek provinces the situation is as such: the *Marine Thiasos* is the predominant subject in fountains and nymphaea.[63] It is represented by a bronze Poseidon statue[64] (Figure 12) and a series of river god figures – once in bull form,[65] but mainly in human form and at a reclining pose.[66] Tritons are also a preferable subject.[67] Dolphins are rarely present as free standing sculpture,[68] since they usually supplement a larger and more significant divine figure, such as Aphrodite and Poseidon, or Erotes.[69] Mythological sea monsters, such as the Scylla are also known as a nymphaeum´s decoration.[70] Erotes, is another large group

[61] Lendon 2015: 134 note 40 (with literature). For the sculptural decoration of theaters, Schwingenstein 1977; Fuchs 1987; Can Özren 1996; for sculptured decoration in Roman thermae, Manderscheid 1981; for sculpture deriving from Italian Roman villas, Neudecker 1988. On the *Purpose and Setting of Graeco-Roman Art in Italy and the Greek Imperial East*, cf. Vermeule 1977; for the *Setting of Greek Sculptures and Garden Sculpture*, cf. Ridgway 1971; 1981 and Kent Hill 1981.

[62] Kapossy 1969: 12-40; Aristodemou 2012: 104-232; on water deities, Becatti 1970/71: 17-58; Lattimore 1976; Amedick 2007. Most recently, on myths depicted on nymphaea of the Roman East, Robinson 2017: 178-203.

[63] For a brief description of the *Hydrekdocheion* of C. Laecanius Bassus in Ephesus as an indicative example of the marine thiasos displayed on nymphaea, Robinson 2017: 197-199 (with previous literature).

[64] *Corinth*, Poseidon's Fountain, 1st century AD (now lost): Robinson 2001: 244-263; Aristodemou 2012: 101, 292 no 29, pl. 5.4 (with all previous literature).

[65] *Olympia*, Arch. Museum, Inv. No. 373: Bol 1984: 109-110, no ι.A.1 note 371, Beil. 5, pl. 2-3; Aristodemou 2012: 294, no 40, pl. 7.1.

[66] *Thasos*, Arch. Museum, Inv. No. Λ100: See above, note 12; *Rhodes*, Arch. Museum, Inv. No. 5288: Aristodemou 2012: 337, no 212, pl. 33.5; *Samos*, Arch. Museum, Inv. No. 582: see above, note 27; *Athens*, National Museum, Inv. No. 1879 (relief): Aristodemou 2012: 286, no 8, pl.2.2 (with previous literature).

[67] *Sparta*, Arch. Museum, Inv. No. M 5196 and 136: Aristodemou 2012: 299, nos 67-68.

[68] *Corinth*, Arch. Museum, Inv. No. S316+316a and I-794: Aristodemou 2012: 292 nos 30-31, pl.5.5.

[69] *Thasos*, Arch. Museum, Inv. No. Λ 2876: Aristodemou 2012: 301, no 78, pl. 12.7; *Sparta*, Arch. Museum, Inv. No. 836 (spring fountain carrying Erotes), see above, note 47.

[70] See above, note 48.

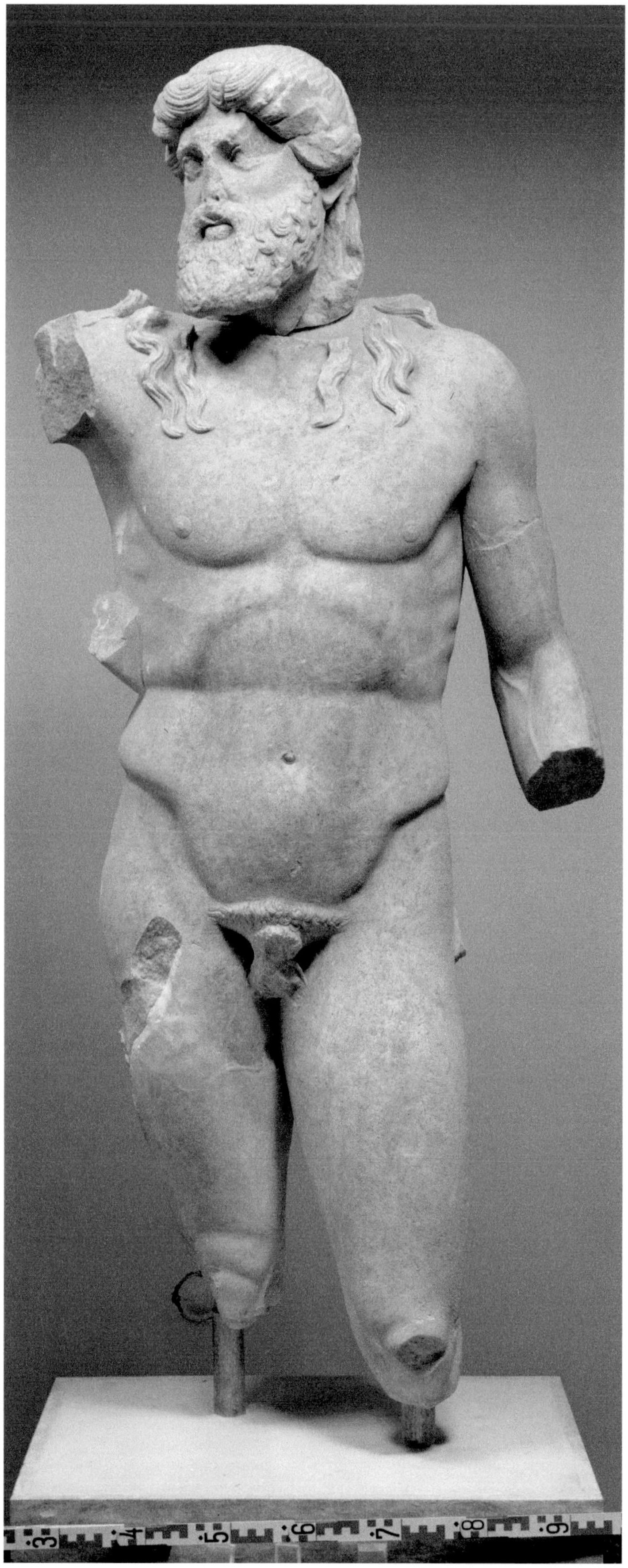

Figure 13. Olympia, Archaeological Museum, Inv. No. Λ170. Zeus' statue. Photo: G. Aristodemou

of figures decorating nymphaea and they are mainly depicted in the sleeping Eros statue type (cf. the New-York statue type).[71] Aphrodite is herself a major sculptural subject for nymphaea.[72] Her *Thiasos*, namely female water deities and Nymphs, is also frequently represented, either as round sculpture (depicted standing,[73] in the Venus Marina type,[74] or seated on rocky bases)[75] – or as relief figures.[76] Nymphs with spouting shells are also a distinctive group.[77] Surprisingly enough, Dionysus is almost absent from nymphaea of the inland and island Greece (though his presence is dominant at the Asia Minor nymphaea).[78] His *Thiasos* is more often recognized than he is and figures such as Silens, Satyrs, Maenads, or Pan usually have a spouting function. Reclining Satyrs and Silens are the most frequently depicted. In majority they are represented elderly and bearded, with a flabby belly – except from one example from Rhodes, depicted youthful and robust.[79] They are represented either reclining on a large wine amphora, whose opening has a spouting function (Athens, Sicyon, Thessaloniki, Lesvos), or sleeping upon a wine skin (Sparta and Rhodes). Most figures are depicted in nudity, apart from the one from Sicyon wearing the himation around the hips. The sculptural complex *Invitation to a Dance* is recognized in Athens, however without a certified provenance.[80] Satyrs and Silens in standing poses are known from Crete; they wear hides and carry objects, such as musical instruments or amphoras, with a spouting function.[81] Pan is frequently recognized, twice as a free standing statue and once on a relief (at the *aposkopein* type).[82] Apollo and Artemis, though frequently present at the Asia Minor nymphaea, do not appear at nymphaea in the Greek provinces. Figures of Muses are rather difficult to be identified, especially when their attributes are missing.[83] Zeus is rarely presented in nymphaea, however when he appears he always holds a prominent position in the monument. The most important example in Greece is his double depiction at the Herodes Atticus Nymphaeum in Olympia (Figure 13).[84] Athena is depicted once,[85] Demeter is a rare deity to be found, and Hermes is rather difficult to be recognized when attributes are missing.[86] Asclepios is not a usual presence in the Greek nymphaea – in those rare cases, he is recorded mainly in nymphaea located within sanctuaries;[87] Hygeia is recognized once.[88] Nemesis and oriental gods though largely attested in the Asia Minor nymphaea, they are not yet identified in the Greek nymphaea. On the other hand, Herakles is a highly important figure for water structures and he is frequently represented in Greek nymphaea, especially in the sculptural complex depicting

[71] *Delphi*, Arch. Museum, Inv. No. 2616: see above, note 42; *Gortyn*, Nymphaeum F25, Louvre Magazin, Inv. No. MNC 1295 (MA 2237): Söldner 1986: 107, 211, 233, 646 no 92, figs. 95-96; Aristodemou 2012: 368 no 352, pl.48.6.

[72] *Athens*, possibly SE Nymphaeum, Ancient Agora Museum, Inv. No. S 1654 (Louvre-Naples type): Brinke 1996: 34 no R23, fig. 16, pl. 26; *Corinth*, Peirene, Arch. Museum, Inv. No. S 54 (Pudica type): Robinson 2011: 215-216 note 44, fig. 115; *Gortyn*, Nymphaeum F25, Herakleion, Arch. Museum, Inv. No. 325 (Louvre-Naples type): Ghedini 1985: 110-115 no 23, fig. 37; On the above, Aristodemou 2012: 285, no 1, pl.1.1 (Athens), 290, no 21, pl.4.3 (Corinth), 367, no 338, pl. 47.4 (Gortyn).

[73] *Delos*, Arch. Museum, Inv. No. A 4289: Marcadé 1996: 126 no 53; Aristodemou 2012: 300, no 76, pl. 12.5.

[74] *Gortyn*, Castellum Aquae, without Inv. No.: Ghedini 1985: 153-155 no 29, fig. 63; Aristodemou 2012: 367, no 340, pl. 47.6.

[75] *Messene*, Arch. Museum depot, Inv. No. 76: see above note 21. *Rhodes*, Arch. Museum, Inv. No. 13614 and Inv. No. 5288: Konstantinopoulos 1997: 72 (Vι.20) and 102-103 no 190, fig.168; Aristodemou 2012: nos 210 and 211, pl. 33.4-5.

[76] *Athens*, National Museum, Inv. No. 1879: See above, note 66.

[77] *Corinth*, Arch. Museum, Inv. No. 1024 and S2063 (see above, notes 18;47); *Sparta*, Arch. Museum, Inv. No. 99 and 836: see above, note 47; *Gortyn*, Nymphaeum F25 (now lost): see above, note 47.

[78] Aristodemou 2012: 166.

[79] *Athens*, Ancient Agora Museum, Inv. No. S2175: Klementa 1993: 151 no V 5; *Athens*, National Museum, Inv. No. 7053: see above, note 16; *Sikyon*, Arch. Museum, without Inv. No.: see above, note 19; *Sparta*, Arch. Museum, Inv. No. 727: see above, note 22; *Thessaloniki*, Arch. Museum, Inv. No. 6680: see above, note 11; *Lesvos, Mytilene*, Arch. Museum, Inv. No. 229: see above, note 24; *Rhodes*, Arch. Museum. Inv. No. 1160: see above, note 26.

[80] *Athens* (Kerameikos): see above, note 15.

[81] *Knossos*, Herakleion, Arch. Museum, Inv. No. 46: see above, note 28; *Kisamos*, Arch. Museum, Inv. No. 694 and Λ 3210: see above, note 29.

[82] *Corinth*, Arch. Museum, Inv. No. S2385: see above, note 38; *Rhodes*, Arch. Museum, Inv. No. E517: Gualandi 1976: 223 no 238, fig. 295; Aristodemou 2012: 337, no 208, pl.33.3; *Athens*, National Museum, Inv. No. 1879 (relief): See above, note 66.

[83] Cf. *Messene*, Arch. Museum depot, Inv. No. 76 (seated Nymph, or Muse), see above note 21; *Gortyn*, Nymphaeum F25: Ghedini 1985: 155-164 no 30, fig.64; Aristodemou 2012: 367, no 341, pl.47.7.

[84] *Olympia*, Arch. Museum, Inv. No. Λ170 and Λ108: Bol 1984: 187-190 no 48, pl. 59-61 and 190-193 no 49, pl. 62-63, 68; Aristodemou 2012: 294, nos 41 and 42, pls.7.2-3.

[85] *Gortyn*, Nymphaeum F25, Herakleion Arch. Museum, Inv. No. 416 (Velletri type): Ghedini 1985: 190-192 no 43, fig.85; Aristodemou 2012: 368, no 349, pl. 48.5.

[86] Cf. *Gortyn, Nymphaeum F25*, Herakleion, Arch. Museum, Inv. No. 415 and Gortyn *Lapidarium*, Inv. No. 22: Ghedini 1985: 85-88 no 7, fig. 16 and 88-91 no 8, fig. 17; Aristodemou 2012: 366, nos 335-336, pls. 47.1-2.

[87] Cf. *Epidauros* (statue base): Kent Hill 1974: 108 note 8; *Mysaion* (Pausanias, 7, 27, 11): Kapossy 1969: 15; Aristodemou 2012: 288 no 16 and 293 no 39.

[88] *Herakleion, Nymphaeum at Grande Thermae*, Herakleion Arch. Museum, Inv. No. 351 (Zara type, Argos/Salamis variation): Romeo and Portale 1998: 135-138 no 31, pl. xɪɪc; Aristodemou 2012: 368 no 353.

Figure 14. Olympia, Archaeological Museum, Inv. No. Λ150. Marcus Aurelius cuirassed statue. Photo: G. Aristodemou

Figure 15. Berlin, Altes Museum, Inv. No. Ol.3 (Sk. 1399). Herodes Atticus' togate statue. Photo: G. Aristodemou

Figure 16. Olympia, Archaeological Museum, Inv. No. Λ156. Regilla's statue.
Photo: G. Aristodemou

the struggle with the Nemean Lion (Athens, Argos, Isthmia, and Corinth).[89] Statues of him are recorded by the ancient authors,[90] and he is also recognized in a relief, depicted as a reclining River God.[91] Other heroes, such as Theseus, are rarely found in nymphaea of the Greek provinces.[92] The Homeric episode of Odysseus and Diomedes stealing the armour of Philoctetes seems to be depicted in a relief plaque decorating a small fountain in Merenta, in ancient Brauron.[93] As ancient sources attest, the mythological pair of Pegasus and Bellerophon, as well as the mythological monster Scylla, both now lost, decorated the Peirene Fountain of Corinth.[94]

2. *Genre sculpture,* including child figures and everyday people (e.g. fishermen, shepherds, hunters, elderly people), as well as animal figures, are mainly found at the Asia Minor and Syrian provinces.[95] In the Greek provinces, we know only a few, among them being the child figure from Isthmia, the standing youth with the water vessel from Thessaloniki, the sleeping river god from Thasos and the sleeping hunter (?) from Patra (Figures 2, 3, 5).[96]
3. *Imperial and private portraits.* This category includes honorific statues of the Emperor and his family members, as well as portraits of Roman and local office-holders, priests, high ranked military officers and members of the local Aristocracy.[97] Imperial and private portraiture follow an almost standardized typology in order to reflect the Imperial power and the prestige of the local Elite.[98]

In the Greek provinces the cases of individually erected Imperial statues at nymphaea are not so many. Regarding the pre-hadrianic period, we only know of the existence of a bronze sculptural group depicting Augustus and his family erected at the Arsinöe fountain in Messene, however without any further information regarding the number or the statue types of the depicted figures.[99] Regarding the Hadrianic period, we know of three cases of monumental nymphaea that housed Imperial statuary groups (Nikopolis, Athens, Olympia): nothing is known -or found- regarding the twin Nikopolis nymphaea scholarly known as Πα and Πβ (apart from an Augustus portrait head lacking an exact provenance), or the SE Nymphaeum in Athens (apart from a colossal statue base, assigned either to Hadrian or Antoninus Pius).[100] The only -almost entirely- preserved group comes from the Olympia nymphaeum that housed the families of Marcus Aurelius and Herodes Atticus.[101] The Nymphaeum F25 in Gortyn also housed a group of portrait statues, male and female, probably both imperial and private, however not easily identified, due to the lack of attributes and their fragmentary preservation state.[102]

Male statues are portrayed in divine or heroic nudity (Figure 11), wearing cuirass (Figure 14), himation, toga (Figure 15) or chlamys.[103] Imperial male portraits are diffused throughout the nymphaea of the Roman East; however in the Greek provinces they are limited in the province of Achaia (Argos and Olympia).[104] Apart from the lost bronze statuary group of Augustus and his family at the Arsinöe fountain in Messene, no other imperial statue of the Julian/Claudian Dynasty is safely attributed to a nymphaeum. Hadrian is the more frequently presented Emperor (Athens, Argos, Olympia), due to the numerous public works he accomplished for his Greek subjects.[105] So far, the Antonines seem to be limited at the Herodes Atticus

[89] See above, note 44.
[90] Cf. Pausanias (2,3,2) for *Corinth.* Also, Aristodemou 2012: 291 no 26 (statue base).
[91] *Thebes,* Louvre, MA 36 (MNC649): Gais 1978: 368 note 47, fig.20; Aristodemou 2012: 289 no 19, pl. 4.1.
[92] *Athens.* See above note 14.
[93] Kakavogianni and Argyropoulos 2005a: 93-94; Kakavogianni and Argyropoulos 2005b:177-188; Kakavogianni and Argyropoulos 2006: 171-180.
[94] See above, note 48.
[95] Kapossy 1969: 41-45, 48-53; Aristodemou 2012: 159-164, 168; recently on genre sculpture, Kunze 1999 and 2002.
[96] *Isthmia,* Arch. Museum, Inv. No. IS 136: see above, note 17. *Thessaloniki,* Arch. Museum, Inv. No. 895: see above, note 11. *Thasos,* Arch. Museum, Inv. No. Λ100: see above note 12. *Patra,* Arch. Museum, no Inv. No.: see above note 20.
[97] Kapossy 1969: 46-47; Aristodemou 2012: 169-197; on Roman portraits in context, Smith 1998; Eule 2001; Fejfer 2008.
[98] As demonstrated at Aristodemou (2012: 275) at least 111 figures are recognized as portraits (imperial and private), which rises almost at the ¼ percentage of the total amount of the 400 sculpture studied (from nymphaea of the Eastern provinces). According to the statistics private portraits of the eastern nymphaea outnumber the imperial ones.
[99] Aristodemou 2012: 196. Also M. Trabucco della Toretta's paper in present volume, and Aristodemou 2018 (forthcoming).
[100] Walker 1979: 152, 173; Leigh 1997: 279-290, fig. 12; Longfellow 2011: 124-126; Aristodemou 2012: 285, no 2; Aristodemou 2018 (forthcoming).
[101] Bol 1984; Aristodemou 2012: 294-299, nos 40-65, pls. 7.1-11.3 with previous literature.
[102] Ghedini 1985; Aristodemou 2012: 366-368 nos 335-351.
[103] Aristodemou 2012: 169-183 (on male -imperial and private- portraits and their statue types deriving from nymphaea).
[104] Aristodemou 2012: 182.
[105] *Argos* (Larissa), Argos Arch. Museum, Inv. No. 242: Maderna 1988: 220 (UD 8), pl. 25.1; *Argos* (Agora), Argos Arch. Museum Depots, Inv. No. R71+2012+1265: Marcadé 1980: 179-181, no 255+55a, 256, 257, figs. 62-64. *Athens* (hypothesis): Walker 1979: 152; *Olympia*, Arch. Museum. Inv. No.

Nymphaeum in Olympia: Antoninus Pius[106] and Lucius Verus[107] are depicted once, Marcus Aurelius, the Philhellene Emperor, is depicted twice (Figure 14),[108] while his son, T. Aelius Antoninus, is sharing the same base with his sister.[109]

Private male portraits from nymphaea of the Greek provinces are limited in the province of Achaia (Corinth, Olympia) and Crete (Gortyn) (Figure 15).[110] The depicted individuals were usually financially involved as sponsors of the construction or the embellishment of the monument.[111]

The vast majority of the female statues wear chiton and himation, following, more or less closely, figure schemes originating from late Classical and Hellenistic statue types (e.g. Large and Small Herculaneum Women, *Pudicitia, Orans, in formam deorum* etc.).[112] The figure types were chosen according to the different poses and actions these women embodied (Figure 16).[113] Female members of the Imperial house are identified at only one major Nymphaeum, that of Herodes Atticus in Olympia.[114] Female private portraits are also limited at the provinces of Achaia (Corinth, Olympia)[115] and Crete (Gortyn)[116].Their presence at the nymphaeum is always associated with them belonging to the family of the sponsor (or, with being sponsors themselves).

Interpretation

The decorative program of a Roman nymphaeum can be understood as a constant and interactive relation between sculpture, water and the specific environment to which the monument belongs. Gods like Poseidon and Aphrodite, mythological figures and water deities like river gods and nymphs, or even every-day people like fishermen, are directly connected to water through their watery habitat.[117] A great number of these figures (Silens, Satyrs, Nymphs, Pan) function as water spouts.

Apart from the above, the majority of figures placed at nymphaea are associated with water indirectly, either through their relation to nature, or through their divine qualities and deeds.[118] Such figures are Dionysus and Apollo and their *Thiasoi*, Erotes, Olympian Gods (cf. Zeus), healing Gods (Asclepius, Hygeia), other semi-gods and heroes (e.g. Heracles, Theseus).

There are also figures whose presence at the monument is interpreted through local myths and cults.[119] These consist of major or minor Olympian and local gods, heroes and mythical city-founders (e.g.

Λ148: Bol 1984: 151-153 no 28, pl. 15-17. On the above, cf. Aristodemou 2012: 181, 285 no 2(?) (*Athens*), 287 no 9 and 10 (*Argos*) pls.2.3, 2.5, 294 no 43, pl. 7.4 (*Olympia*).

[106] *Olympia*, Arch. Museum, Inv. No. Λ165+2248(a-g): Bol 1984: 153-156, no 29, pl. 18-19. On the above, cf. Aristodemou 2012: 181, 295, no 44, pl. 7.5 (Olympia), and perhaps 285 no 2 (Athens).

[107] *Olympia*, Arch. Museum, Inv. No. A166: Bol 1984: 159, no 31, pl. 22-23; Aristodemou 2012: 295 no 46, pl. 8.2.

[108] *Olympia*, Arch. Museum, Inv. No. Λ150 and Λ149: Bol 1984: 157-159, no 30, pl. 20-21 and 194-195, no 51, pl.66-67; Aristodemou 2012: 181; 295 no 45, pl. 8.1 and 298 no 65, pl. 11.3.

[109] *Olympia*, Arch. Museum, Inv. No. 160: Bol 1984: 117-118, no 7-8, pl. 6: Aristodemou 2012: 296, no 51, pl. 9.1.

[110] *Olympia*, members of the Herodes Atticus' family: Bol 1984: 165 no 33, pl. 26 (Herodes Atticus), 169-171 no 35, pl. 30-31 (T. Claudius Atticus Herodes), 164 no 32, pl. 24-25 (Appius Annius Gallus), 165-169 no 34, pl. 27-29 (M. Appius Bradua), 186 no 47, pl. 56 (Atticus Bradua), 185 no 46, pl. 57-58 (Regillus), 193 no 50, pl. 64-65 (Herodes Atticus). *Gortyn:* Ghedini 1985: 85-88 no 7, fig. 16 and 105 no 13, fig. 27; On the above, cf. Aristodemou 2012: 183, 296-298 nos 53, 55-58, 63-64 (Olympia) and 366 no 335, pl. 47.2, 366 no 337, pl. 47.3 (Gortyn).

[111] Aristodemou 2012: 196.

[112] Aristodemou 2012: 169. In general, Aristodemou 2012: 184-197 (on the female imperial and private portraits and their statue types).

[113] Stewart 2003: 79-175; Aristodemou 2012: 250.

[114] Berlin, Altes Museum, Inv. No. Ol.2 (Sabina); Olympia, Arch. Museum, Inv. No. 155 (Faustina Major); Inv. No. 158 (Faustina Minor); Inv. No. Λ 160 (Lucilla, or Annia Galeria Aurelia Faustina); Inv. No. 161a+169 (Domitia Faustina, or Lucilla): Bol 1984, 175 no 38, pl. 38; 173-175 no 37, pl. 35-37; 179 sq., no 42, pl.46-47; 182 no 44, pl.51-53; 184 no 45, pl. 54-5. On the above, cf. Aristodemou 2012: 196, 295 no 47, pl. 8.3, 295 no 48, pl. 8.4; 296 no 49, 296 no 50, 296, no 52.

[115] *Peirene Fountain*, Corinth, Arch. Museum, Inv. No. 55+1180 (S-55; torso of a Small Herculaneum Woman statue): Robinson 2011: 215 fig.114, note 44. *Peirene Fountain*, Corinth Arch. Museum, Inv. No. 63 (female torso wearing chiton and himation): Johnson 1931: 65, no 112. *Peirene* Fountain, Corinth Arch. Museum, Inv. No. 62 (I-62, Regilla statue base): Robinson 2011: 78-79, fig.51; On the above, cf. Aristodemou 2012: 196, 290, no 20, pl. 4.2; 290, no 23, pl. 5.1; 291, no 25, pl. 5.3. *Olympia*, Arch. Museum, Inv. No. Λ156+163α (Regilla), Inv. No. Λ157 (Vivoulia Alcia Agrippina), Λ 162 (Atilia Caukidia Tertulla), Λ 165? (Elpiniki), Λ 159 (Athenais): Bol 1984: 171-173 no 36, pl 32-34; 124 sq, no 12 and no 39, pl. 39-41; 177 no. 40, pl. 42-43; 177-179 no 41, pl. 44-45; 180-182 no 43, pl.48-50. On the above, cf. Aristodemou 2012: 297 no 54, pl. 9.4; 297 no 59, pl. 10.3; 298 no 60, pl. 10.4; 298 no 61, pl. 10.5; no 62, pl. 10.6.

[116] *Gortyn*, Nymphaeum F25 (lapidarium, without Inv. No.): Ghedini 1985: 179-180 no 38 fig.79; 180-183 no 39, fig. 80. On the above, cf. Aristodemou 2012: 196; 368 no 346, pl. 48.3; 368 no 347, pl. 48.4. On a general discussion, Taliaferro-Boatwright 1991.

[117] Kapossy 1969: 70; Aristodemou 2012: 200-212.

[118] Kapossy 1969: 71; Aristodemou 2012: 213-237.

[119] Aristodemou 2012: 237-242. See also Robinson (2017: 181-192), about the connection between fountains, myths and the formation of cultural

Heracles, Theseus), whose connection to the city as patron gods is equally important to their connection with water and they are usually recorded at the monument's building or dedicatory inscription.[120] Their presence at the sculptural display program of a nymphaeum expresses the city's gratitude for the divine protection and underlines the belief that local gods support and sustain not only the Imperial family but the citizens too, providing them with all the necessary goods for a prosper life. Additionally, as is the case of the so called Poseidon Fountain in Corinth,[121] statues of local patron gods at nymphaea financed by local benefactors underline the city's confidence and power.[122] Imperial and private portraiture underlines both the Imperial power and the prestige of local Elite.[123]

Display programs

Only a few sculptural assemblages are known from fountain structures from cities and/or sanctuaries of the mainland and island Greece (Argos, Athens, Nikopolis, Olympia, Corinth, Gortyn).

Sculptures adorning the monumental facades of Roman nymphaea and deployed according to pre-scheduled iconographical programs, differ among regions and among monuments. Each nymphaeum is unique because its construction and embellishment are a combination of various important parameters: the financial status and the intentions of the sponsor, his educational and cultural background, the technological expertise of the architect, along with the local historical, cultural and geomorphologic particularities of the given region.

Different directions in the sculptural display programs reflect different purposes: either glorifying the Emperor and his House,[124] or honoring the local gods,[125] or, presenting the dynamic local Elite, which often enough contributed to the construction of the monument.[126]

Monumental nymphaea were financed either by Emperors or wealthy patrons.[127] Financing the construction and/or the decoration of a nymphaeum in order to project specific messages consists a politically motivated action and the monument's display program is used as a medium to promote the donor and his family.[128]

The nymphaea of Argos (Kriterion Hill, Larissa),[129] Athens (Lycabettus Nymphaeum[130] and SE Nymphaeum at the Athenian Agora[131]), Messene (Arsinöe Fountain)[132] and the complex of the twin nymphaea in Nikopolis,[133] are characterized by a specific focus on the Emperor, expressed through the central placement of his figure on their facades. The Emperor's statue is placed at the most prominent location

identity at Roman Corinth.

120 The Nymphaeum of Herodes Atticus in Olympia is dedicated to Zeus, Bol 1984: 109 no 1, pl.2.3. The same principle applies to the Asia Minor Nymphaea, e.g. the Trajanic Nymphaeum in Ephesus is dedicated to the Ephesian Artemis (just as the majority of the Ephesian nymphaea: IvE II, nr.424; also, IvE II, nr.424A, nr.413, nr.414). The South Nymphaeum [F2] in Perge is dedicated to Artemis Pergaia: IK Perge I, nr.195-196.

121 Scranton 1951: 60-61, fig. 41; Williams 1989; Williams and Zervos 1990: 354, fig. 5; Agusta-Bularot 2001: 179 no 10; Robinson 2001: 244-263; Mylonopoulos 2003: 153 no 6.5, fig. 28; Aristodemou 2018 (forthcoming).

122 The same applies for the majority of public edifices financed by local benefactors. For the civic pride and patriotism, see Macro 1980: 682-684. On the politics of munificence in Asia Minor, Zuiderhoek 2009:78-46.

123 Aristodemou 2012: 242-252.

124 Longfellow 2005: 88-140; Longfellow 2011, for an overall discussion upon imperial and private propaganda as applied in Roman nymphaea; Aristodemou 2012: 252-257.

125 Aristodemou 2012: 257-271.

126 Aristodemou 2012: 271-273, 276; Aristodemou (forthcoming).

127 On patronage and benefactions, cf. Gauthier 1985(4th-1st century BC); Winter 1996; Ando 2000; Cramme 2001; Halfmann 2001: 1-10; Eilers 2002; Longfellow 2011; Thomas 2007: 71-90. Aristodemou 2012: 65-82 (regarding nymphaea donors).

128 Dorl-Klingenschmidt 2006.

129 Vollgraff 1944-45: 397-400, no 7; Vollgraff 1958: 516-539 (Kriterion), 539-555 (nymphaeum), 555-570 (aqueduct); Walker 1987: 60-71; Longfellow 2011: 113-120; Aristodemou 2012: 70, 252-253, 287, fig. 2.1; Aristodemou (forthcoming). It was directly connected to the Northern Aqueduct, bringing water from the Kephalobryso springs, and functioned as its terminal point and principal consumer, Lolos 1997: 306.

130 Leigh 1997: 279 note 3, 290, fig. 6, 12; Longfellow 2011: 122-123, fig. 41; Richard 2012: 261.

131 Thompson 1955: 57-59; Walker 1979: 60-71; Glaser 1983: 109 nr 74; Leigh 1998; Leigh 2000: 122-130; Martin 2002; Longfellow 2011: 128-130; Aristodemou 2012: 54, 76, 255, 285; Aristodemou 2018 (forthcoming). More on S. Leigh's paper in present volume.

132 On the Arsinöe Fountain: Themelis 1990; 1995; 1999; 2013; Felten and Reinholdt 2001: 321 with previous literature; Rheinholdt 2009; Trabucco della Torretta 2006: 199; Aristodemou 2018 (forthcoming). Most recently, on this Nymphaeum, see M. Trabucco della Torretta's paper in present volume.

133 Zachos and Georgiou 1997: mainly 598. Just recently, regarding the West Gate and the Nymphaea, see Zachos 2015: 107-110 with figures, and in the present volume, Zachos and Leontaris with the most recent research results.

at the monument and his presence acquires special meaning in those cases where the imperial statue is directly connected with the water flow.[134] At least two of these monuments (the Argos and the Lycabettus Nymphaeum) were sponsored by the Emperor himself (Hadrian).

Other nymphaea, such as the Herodes Atticus Nymphaeum within the sanctuary of Olympia,[135] the Poseidon's Fountain in the Agora of Corinth,[136] and the nymphaeum F25 into the Praetorium of Gortyn,[137] are financed by private sponsors, all prominent members of their local societies. Their chosen display programs present a specific focus towards their wealthy benefactors whose statues occupy an important position at the monument's façade. The self-representation of the private donor and his family at prominent locations at the monument's façade reveals the confidence of the upper class which was motivated by their financial and political power.[138]

The sculptural display programs of the Greek nymphaea evolve around three axes: the Emperor, the local myths and cults and the local Aristocracy. As an imperial benefaction, Greek nymphaea proclaim the imperial power, virtue and culture. They focus mainly on the Emperor, and the main message sent is that the Emperor has the power to control the life of his subjects. His central position within the sculptural display program and his visual and, sometimes, organic connection with the water flow reveals the effort of the local population to attract his interest and his favor, and their respectfulness towards him.

As objects of private patronage they signaled the donor's wealth, fame, educational and cultural background, their high rank position in the Roman state and their connection to the Imperial regime.[139]

The presence of the local deities expresses the city's gratitude for the divine protection and underlines the belief that local gods support and sustain not only the Imperial family but the citizens too, providing them with all the necessary goods for a prosper life.

Obviously, fountain figures do not cover practical needs. A nymphaeum's water supply is ensured by the city's distribution network.[140] However, the selection of the iconographical subjects, the involvement of figures with the water flow and their placement at the monuments façade, underline the organic relationship between the building, the sculpture and the water source. The artistic composure that derives from this interaction exceeds the monument's architectural context and connects it with the natural environment or the cityscape where the monument belongs.[141] Monumental nymphaea and their decorative programs are a reflection of a city's image and transmit specific messages throughout the Roman world.[142] From the 1st century AD onwards, 'civic leaders and benefactors began to think about their cities differently: water supply became a larger and larger part of the identity of the city and its citizens'.[143] Cities advertised their abundance of water initially by depicting their local river gods on their coinage and later on by erecting large scale and imposing fountain structures that drastically changed their urban landscape.[144] Theses monumental nymphaea advertised the city's prosperity whereas their decoration consisted the medium with which they promoted their essential role within the province.[145] At the same time, the discussion of

[134] Aristodemou 2012: 252; Aristodemou (forthcoming).

[135] The monument was published by R. Bol (1984). Longefellow 2005: 260; Herodes Atticus is also associated with financing two other semicircular Nymphaea, the Nymphaeum at Alexandria in Troas and the SE Nymphaeum in the Athenian Agora, Aristodemou 2012: 76.

[136] Scranton 1951: 60-61, fig. 41; Williams and Zervos 1990: 354, fig. 5; Robinson 2001: 244-263. Also, Agusta-Bularot 2001: 179 no 10; Mylonopoulos 2003: 153, no 6.5, fig. 28; Aristodemou 2012: 74, 101. The SE part of the fountain was ruined prior to the construction of the superposed temple, while the water tanks were destroyed sometime in the Hadrianic period. The fountain's remains were covered by the foundations of the temples J and H, Robinson 2001: 252-254, 261. Also, Aristodemou 2018 (forthcoming).

[137] Ortega 1986-87: 131-174, pl.10. On the 2nd and 6th centuries AD phases of the Nymphaeum Block, Lippolis *et al.* 2013:250-253. On the Nymphaeum sculptural assemblage: Ghedini 1985: 63-248; Longfellow 2011: 138-139 fig. 48, 206-207 fig. 70; Aristodemou 2012: 529, 366-368, nos 335-351.

[138] Graindor 1930: 83; Bol 1984: 22, 96-97, 124; Walker 1987: 61; Havé-Nicolaus 1998: 40; Longfellow 2005: 260; On the family of Herodes Atticus, Tobin 1997: especially 29-32. On Roman Elite self-representation and self-projection, also Hingley 2005: mainly 74-88 and Flower 2011: 271-285, with literature.

[139] Wealthy local families used monumental architecture to compete for social prestige and promote their political ambitions: Häusle 1980; Schorndorfer 1997: 8; Dorl-Klingenschmid 2001: 120-121; Dorl-Klingenschmid 2006; Thomas 2007: 76; Aristodemou 2012: 52-59.

[140] On the connection of nymphaea to the water network, Richard 2012.

[141] Segal 1997; Recently, Rogers 2015.

[142] Chi 2002; Dorl-Klingenschmid 2006: 381-386; Aristodemou 2018 (forthcoming).

[143] Koloski-Ostrow 2001: 1-17. Lendon 2015: 134.

[144] Aristodemou 2018 (forthcoming): 317-332.

[145] Drerup 1966: 181-196; Dorl-Klingenschmid 2001: 149-150; Thomas 2007: 158; Aristodemou 2012: 52-53, 64; Aristodemou 2014: 527.

a nymphaeum's decoration reveals a lot about the identity and the intentions of their sponsors and adds greatly to the reconstruction of the social, political as well as the cultural role that sculptural assemblages played in their given context, especially in the Greek provinces of the Empire.

Bibliography

Agusta-Boularot, S. 2001. Fontaines et fontaines monumentales en Grèce de la conquête romaine à l'époque flavienne: permanence ou renouveau architectural? In: J.-Y. Marc and J.-Ch. Moretti (eds) *Constructions publiques et programmes édilitaires en Grèce entre le IIe siècle av. J.-C. et le Ier siècle ap. J.-C. : Actes du Colloque organisé par l'Ecole française d'Athènes et le CNRS, Athènes, 14-17 mai 1995, Ecole française d'Athènes* (Bulletin de Correspondance Hellénique (=BCH) Suppl. 39): 167-236.

Amedick, R. 2007. Stein und Wasser. Römische Skulpturen von Meerwesen und Wassergottheiten. In: H. v. Steuben, G. Lahusen and H. Kotsidu (eds) ΜΟΥΣΕΙΟΝ. *Beiträge zur antiken Plastik. Festschrift zu Ehren von Peter C. Bol*: 423-433. Bibliopolis: Möhneese.

Ando, Cl. 2000. *Imperial Ideology and Provincial Loyalty in the Roman Empire*. Berkeley: University of California Press.

Andreae, B. 1999a. *Odysseus. Mythos und Erinnerung. Kurze Anleitung.* Rom, 22.Feb-2.Sept. Katalog Rom, München 99/00. Mainz: Philipp von Zabern.

Andreae, B. 1999b. Ist die Hypothese von Polyphem Giebel in Ephesos bereits falsifiziert?. In: H. Friesinger-F. Krinzinger (eds) *100 Jahre Österreichische Forschungen in Ephesos*. Akten des Symposions, Wien, 1995: 531-533. Wien: Bohlau.

Ajootian, A. 1993. Silenus at Sparta. In: O. Palagia and W. Coulson (eds) *Sculpture from Arcadia and Laconia. Proceedings of an international Conference held at the American School of Classical Studies at Athens*, April 10-14, 1992: 251-256. Oxford: Oxbow Books.

Archontidou, A. and Acheilara, L. 1999. *Αρχαιολογικό Μουσείο Μυτιλήνης*. Mytilene: Ministry of Culture. Archaeological Receipts Fund.

Aristodemou, G. 2011a. Sculptured Decoration of Monumental Nymphaea at the Eastern Provinces of the Roman Empire. In: Trinidad Nogales Basarrate and Isabel Rodà de Llanza (eds) *XI Colloquium on Roman Provincial Art 'Rome and the Provinces: Models and Diffusion'. Mérida, Spain 18-21 May 2009* (Hispania Antigua, Serie Arqueológica 3): 149-160. Roma: L'Erma di Bretschneider.

Aristodemou, G. 2011b. Theatre Facades and Facade Nymphaea. The Link between. *Bulletin de Correspondance Hellénique* (=BCH) 135.1: 163-197.

Aristodemou, G. 2012. *Ο Γλυπτός Διάκοσμος Νυμφαίων και Κρηνών στο Ανατολικό τμήμα της Ρωμαϊκής Αυτοκρατορίας*. Thessaloniki: Kornelia Sfakianakis Editions.

Aristodemou, G. 2014. Monumental Fountain Structures: The Role of Nymphaea within the Urban Context of the Cities of the Graeco-Roman East. In: I. K. Kalavrouziotis and A. N. Angelakis (eds) *Wastewater and Environment: Traditions and Culture*, 1st IWA Regional Symposium on Water 22-24 March 2014, Hellenic Open University (Patras, 2014): 523-535. (e-proceedings, wwetc2014.env.uwg.gr).

Aristodemou, G. 2018 (forthcoming). Fountain culture in the Greek provinces before Hadrian. Introducing the concept of luxury. In: G. Wiplinger and W. Letzner (eds) *Wasserwesen zur Zeit des Frontinus. Bauwerke - Technik - Kultur. 40 Jahre Frontinus Gesellschaft. Tagungsband des Internationalen Frontinus-Symposiums Trier, 25.-29. Mai 2016* (BABESCH Supplements 32): 317-331. Bonn: Frontinus-Gesellschaft e.V and Leuven: Peeters.

Aristodemou, G. (*forthcoming*). Fountain Sculpture and personal propaganda in Roman Greece. In: Th. Stefanidou-Tiveriou - P. Karanastasi - D. Damaskos (eds) *Έργα Πλαστικής στη Ρωμαϊκή Ελλάδα: Καλλιτεχνικά Προϊόντα, Κοινωνικές Προβολές, Διεθνές Συνέδριο, Ρέθυμνο, 26-28 Σεπτεμβρίου 2014* // Sculpture and Society in Roman Greece: Artistic Productions, Social Projections, Rethymno, 26-28 September 2014.

Auinger, J. and Rathmayr, E. 2007. Zur spätantiken Ausstattung der Thermen und Nymphäen in Ephesos. In: A. F. Bauer and Ch. Witschel (eds) *Statuen in der Spätantike*: 237-269. Wiesbaden: Reichert Verlag.

Aurenhammer, M. 1990. *Die Skulpturen von Ephesos, Idealplastik I.* Forschungen in Ephesos, X/1. Wien: Verlag der ÖAW.

Bakalakis, G. 1966. Satyros an einer Quelle gelagert. *Antike Kunst* 9: 21-28, pl.7.1-6.

Bauer, F. A. 1996. *Stadt, Platz und Denkmal in der Spätantike. Untersuchungen zur Ausstattung des öffentlichen Raums in den spätantiken Städten Rom, Konstantinopel und Ephesos*. Mainz: Philipp von Zabern.

Becatti, G. 1970/71. Ninfe e divinità marine. Richerche mitologiche iconografiche e stilistiche. *Studi Miscellanei* 17: 17-58.

Berg, D. 1994. *Fountains and Artistic Water Displays in Classical Antiquity. Origins and Development from 700 to 30 B.C.* Arch.dissertation, University of Texas at Austin.

Brinke, M. 1996. Die Aphrodite Louvre-Neapel. *Antike Plastik* 25: 7- 64.

Brinkerhoff, D. M. 1965. New Examples of the Hellenistic Statue Group 'The Invitation to the Dance', and their Significance. *American Journal of Archaeology* 69: 25-37, pls. 3- 6.

Burrell, B. 2006. False Front: Separating the Aediculated Fassade form the Imperial Cult in Roman Asia Minor. *American Journal of Archaeology* 110: 437-469.

Bol, R. 1984. *Das Statuenprogramm des Herodes-Atticus Nymphaeums.* Olympische Forschungen, Band XV. Berlin: Walter de Gruyter.

Can Özren, A. 1996. Die Skulpturenausstattung kaiserzeitlicher Theater in der Provinz Asia, am Beispiel der Theater in Aphrodisias, Ephesos und Hierapolis. *Thetis* 3: 99-128.

Chi, J. 2002. Studies in the Programmatic Statuary of Roman Asia Minor. Unpublished Ph.D. dissertation, NY University.

Cramme, St. 2001. *Die Bedeutung des Euergetismus für die Finanzierung städtischer Aufgaben in der Provinz Asia.* Ph.D dissertation, Köln University.

Chrysostomou, P. and Kefallonitou, F. 2001. *Νικόπολις*. Athens: Ministry of Culture. Archaeological Receipts Fund.

Curtius, E. 1876. *Die Plastik der Hellenen an Quellen und Brunnen.* Abhandlungen der Akademie der Wissenschaften zu Berlin. Berlin: Dümmler in Comm.

Despinis, G., Stefanidou-Tiveriou, Th. and Voutiras, E. (eds) 1997. *Catalogue of Sculpture in the Archaeological Museum of Thessaloniki* I. Thessaloniki: National Bank Cultural Foundation.

Dorl-Klingenschmidt, Cl. 2001. *Prunkbrunnen in kleinasiatischen Städten. Funktion im Kontext.* Bayerische Akademie der Wissenschaften. Studien zur antiken Stadt, Band 7. München: Dr. Friedrich Pfeil.

Dorl-Klingenschmidt, Cl. 2006. Brunnenbauten als Medium der interkommunalen Wettbewerbs. In: G. Wiplinger (ed.) *Cura Aquarum in Ephesos. Proceedings of the 12th International Congress on the History of Water Management and Hydraulic Engineering in the Mediterranean Region*, DWhG, 12 (Selçuk, 2- 10 October 2004) (BABesch, Suppl. 42,) 381-386. Leuven: Peeters.

Drerup, H. 1966. Architektur als Symbol. Zur zeitgenössischen Bewertung der römischen Architektur. *Gymnasium* 73: 181-196.

Eilers, Cl. 2002. *Roman patrons of Greek cities.* Oxford: Oxford University Press.

Eule, J. C. 2001. *Hellenistische Bürgerinnen aus Kleinasien: Weibliche Gewandstatuen in ihrem antiken Kontext.* Istanbul: Tarih, Arkeoloji, Sanat ve Kültür Mirasini Koruma Vakfi.

Fejfer, J. 2008. *Roman Portraits in Context.* Berlin/New York: Walter de Gruyter.

Felten, F. and Rheinholdt, C. 2001. Das Brunnenhaus derArsinoë in Messene. In: V. Mitsopoulos-Leon (ed.) *Forschungen in der Peloponnes. Akten des internationalen Symposions anlässlich der 100-Jahrfeier des Österreichischen Archäologischen Instituts.* Zweigstelle Athen vom 5.3.-7.3. 1998 in Athen (ÖAI Sonderschriften Band 38): 307-323, pl. 30. Athens.

Fittschen, K. 1999. *Prinzenbildnisse Antoninischer Zeit.* Beiträge zur Erschließung hellenistischer und kaiserzeitlicher Skulptur und Architektur 18. Mainz.

Fleisher, R. 1982. Zwei eklektische Statuen aus Ephesos. In: *Pro Arte Antiqua, Festschrift für Hedwig Kenner, I. Denkschriften.* (Österreichische Akademie der Wissenschaften. Philosophisch-Historische Klasse Band 18): 123-129. Wien: Verlag der ÖAW.

Flower, H. 2011. Elite Self-Representation in Rome. In: M. Peachin (ed.) *The Oxford Handbook of Social Relations in the Roman World.* Oxford: Oxford University Press Handbooks Online.

Fuchs, M. 1987. *Untersuchungen zur Ausstattung römischer Theater.* Mainz: Philipp von Zabern.

Gais, R. 1978. River-God Iconography. *American Journal of Archaeology* 82: 355-370.

Gauthier, P. 1985. *Les cités grecques et leurs bienfaiteurs (IVe-Ier s. av. J.-C.). Contribution à l'histoire des institutions.* Bulletin de Correspondance Hellénique, Suppl., 12. Paris/Athens: EFA.

Geominy, W. 1999. Zur Komposition der Gruppe 'Die Aufforderung zum Tanz'. In: P. C. Bol (ed.) *Hellenistische Gruppen. Gedenkschrift fur Andreas Linfert.* Schriften des Liebighauses-Museum alter Plastik: 141-155. Mainz: Philipp von Zabern.

Ginouvès, R. 1998. Architecture de l'eau. In: R. Ginouvès (ed.) *Dictionnaire Méthodique de l'Architecture Grecque et Romaine, Tome III : Espaces Architecturaux, Bâtiments et Ensembles:* 92-100. Rome: École française de Rome.

Ghedini, F. 1985. Sculture dal Ninfeo e dal Pretorio di Gortina. *Annuario della Scuola archeologica di Atene e delle Missioni italiane in Oriente* (=AsAtene) 45 n.s.: 63-248.

Glaser, F. 1983. *Antike Brunnenbauten (KPHNAI) in Griechenland.* Denkschriften, Österreichische Akademie der Wissenschaften. Philosophisch-Historische Klasse Band 161. Wien: Verlag der ÖAW.

Glaser, F. 1987. Brunnen und Nymphäen, In: Frontinus-Gesellschaft (ed.) *Die Wasserversorgung antiker Städte, Band 2. Pergamon. Recht/Verwaltung; Brunnen/Nymphäen; Bauelemente*: 103-131. Mainz: Philipp von Zabern.

Graindor, P. 1930. *Un milliardaire antique. Herode Atticus et sa famille.* Societe Anonyme Egyptienne. Caire.

Gros, P. 1996. *L'Architecture Romaine du début du IIIe siècle av.J.-C. à la fin du Haut-Empire. Tome 1, Les Monuments Publics*: 418-444. Paris: Picard.

Gualandi, G. 1976. Sculture di Rodi, *Annuario della Scuola archeologica di Atene e delle Missioni italiane in Oriente (=AsAtene)* 54: 7-259.

Havè-Nicolaus, F. 1998. *Untersuchungen zu den kaiserzeitlichen Togastatuen griechischer Provenienz.Kaiserliche und private Togati der Provinzen Achaia, Creta (et Cyrene).* Trierer Beiträge zur Altertumskunde, Bd.4. Mainz.

Halfmann, H. 2001. *Städtebau und Bauherren im römischen Kleinasien. Ein Vergleich zwischen Pergamon und Ephesos.* Istanbuler Mitteilungen, Beiheft, 43. Tübingen: Ernst Wasmuth Verlag.

Häusle, H. 1980. *Das Denkmal als Garant des Nachruhms: Beiträge zur Geschichte und Thematik eines Motivs in lateinischen Inschriften.* Zetemata, 75. München: Beck

Hingley, R. 2005. *Globalizing Roman Culture: Unity, Diversity and Empire.* London: Routledge.

Hülsen, J. (ed.) 1919. *Das Nymphaeum. Milet: Ergebnisse der Ausgrabungen und Untersuchungen seit dem Jahre 1899. Bd. I. Teil 5.* DAI. Berlin/Leipzig: Jacket.

Horn, H. 1972. *Samos XII: Hellenistische Bildwerke auf Samos.* DAI. Mainz: Philipp von Zabern.

Johnson, F. P. 1931. *Corinth IX: Sculpture 1896-1923.* Princeton: American School of Classical Studies at Athens.

Kakavogianni, O. and Argyropoulos, V. 2005a. Κρηναίο οικοδόμημα στη Μερέντα. In: Β. Βασιλοπούλου (επιμ.) Αττική 2004. *Ανασκαφές, Ευρήματα, Νέα μουσεία*: 93-94. Νομαρχιακή Αυτοδιοίκηση Αν. Αττικής & B ΕΠΚΑ. Athens.

Kakavogianni, O. and Argyropoulos, V. 2005b. Αρχαίος ναός και κρήνη στη Μερέντα. In: Πρακτικά Συνεδρίου '*Από τα Μεσόγεια στον Αργοσαρωνικό. Β΄Ε.Π.Κ.Α.: Το έργο μιας δεκαετίας 1994-2003*': 177-188. Athens.

Kakavogianni, O. and Argyropoulos, V. 2006. Κρηναίο οικοδόμημα πλησίον του αρχαίου ναού στη Μερέντα, In: *Πρακτικά της ΙΑ' Επιστημονικής Συνάντησης Νοτιοανατολικής Αττικής*: 171-180. Spata.

Kaltsas, N. 2001. *Εθνικό Αρχαιολογικό Μουσείο. Τα Γλυπτά.* Athens: Kapon Editions.

Kapossy, B. 1969. *Brunnenfiguren der hellenistischen und römischen Zeit.* Zürich: Juris-Verlag.

Karageorghis, V. 1964. *Sculptures from Salamis, I.* Nicosia, Cyprus: Department of Antiquities.

Kent Hill, D. 1974. Nymphs and Fountains. *Antike Kunst* 17: 107-108.

Kent Hill, D. 1981. Some *sculpture* from *Roman Domestic* Gardens. In: W. F. Jashemski and E. B. Macdougall (eds) *Ancient Roman Gardens. Dumbarton Oaks Colloquium on the History of Landscape Architecture VII* (1979): 81-94. Washington: Maguire.

Klementa, S. 1993. *Gelagerte Flussgötter* des *Späthellenismus und der römischen Kaiserzeit.* Köln/Weimar/Wien: Böhlau Verlag.

Kolonas, L. and Stavropoulou-Gatsi, M. 2017. *Αρχαιολογικό Μουσείο Πατρών*. Athens: Ministry of Culture. Archaeological Receipts Fund.

Koloski-Ostrow, A. O. 2001. Water as a Symbol of Wealth? An Overview of the Roman Evidence. In: A. O. Koloski-Ostrow (ed.) *Water Use and Hydraulics in the Roman City*: 1-17. Archaeological Institute of America.

Konstantinopoulos, Gr. 1997. Το Αρχαιολογικό Μουσείο της Ρόδου. Rhodes: Pergamos Editions.

Kunze, Ch. 1999. Verkannte Götterfreunde. Zu Deutung und Funktion hellenistischer Genreskulpturen. *Mitteilungen des Deutschen Archäologischen Instituts, Römische Abteilung (= RM)* 106: 43-82.

Kunze, Ch. 2002. *Zum Greifen Nah. Stilphänomene in der hellenistischen Skulptur und ihre inhaltliche Interpretation.* München: Biring & Brinkmann.

Lange, E. 1920. Die Entwicklung der antiken Brunnenplastik. Unpublished Ph.D. dissertation, University of Göttingen.

Lattimore, St. 1976. *The Marine Thiasos in Greek Sculpture* (Monumenta Archaeologica 9). Los Angeles: Institute of Archaeology, University of California, .

Laurenzi, L. 1956. Sculture inedited del Museo di Coo. *Annuario della Scuola archeologica di Atene e delle Missioni italiane in Oriente* (=*AsAtene)* 17/18 n.s: 59-156.

Leigh, S. 1997. The reservoir of Hadrian in Athens. *Journal of Roman Archaeology* 10: 279-290.

Leigh, S. 1998. The Aqueduct of Hadrian and the Water Supply of Roman Athens. Unpublished Ph.D. dissertation, University of Pennsylvania.

Leigh, S. 2000. Interdisciplinary research on the Aqueduct of Hadrian in the Athenian Agora. In: Jansen, G. C. M. (ed.) *Cura Aquarum in Sicilia. Proceedings of the Tenth International congress on the History of Water Management and Hydraulic Engineering in the Mediterranean Region (Syracuse, May 16-22, 1998)*: 117-124. (BaBesch Supplement 6). Leiden: Peeters.

Lendon, J. E. 2015. Rhetoric and Nymphaea in the Roman Empire. *Chiron* 45: 123-149.

Letzner, W. 1990. *Römische Brunnen und Nymphaea in der westlichen Reichshälfte.* München: Charybdis.

Lippolis, E., Livadiotti, M., Calio, l., Baldini, I., Giatti, Ch. 2013. Gortys of Crete. The Excavation in the Nymphaeum Block. In: M. Ανδριανάκης, Π. Βαρθαλίτου, Ί. Τζαχίλη (επιμ.) *Αρχαιολογικό Έργο Κρήτης 2 (Πρακτικά της 2ης Συνάντησης, Ρέθυμνο, 26-28 Νοεμβρίου 2010 (2013)*: 246-264.

Lolos, Y. A. 1997. The Hadrianic Aqueduct of Corinth, with an Appendix on the Roman Aqueducts in Greece. *Hesperia* 66: 271-314.

Longfellow, B. 2005. Imperial Patronage and Urban Display of Roman Monumental Fountains and Nymphaea. Ph.D. dissertation, Michigan University.

Longfellow, B. 2009. The Reuse of Statues in Ancient Roman Fountains. A Public Lecture given at Creighton University (27 March 2009).

Longfellow, B. 2011. *Roman Imperialism and Civic Patronage Form, Meaning and Ideology in Monumental Fountain Complexes.* Cambridge: Cambridge University Press.

Macro, A. D. 1980. *The Cities of Asia Minor under the Roman Imperium. ANRW II.7.2:* 658-697.

Maderna, C. 1988. *Juppiter, Diomedes und Merkur als Vorbilder für Römische Bildnisstatuen. Untersuchungen zum römischen statuarischen Idealporträt.* Heidelberg: Verlag Archäologie und Geschichte.

Manderscheid, H. 1981. *Die Skulpturenausstattung der kaiserzeitlichen Thermenanlagen* (Monumenta Artis Romanae XV). Berlin/Wiesbaden: Reichert Verlag.

Marcadé, J. 1980. Sculptures argiennes (III). In: *Etudes Argiennes. Bulletin de Correspondance Hellénique, Supplements, VI:* 133-184. Ecole française d'Athènes. Athènes/ Paris: Diffusion E. de Boccard.

Marcadé, J. 1993. Éros endormi à Delphes. In: Études de sculpture et d'iconographie antiques, *Scripta Varia, 1941-1991*: 475-484. Paris: Publications de la Sorbonne.

Marcadé, J. (ed.) 1996. *Sculptures deliennes* (Sites et Monuments XVII. Ecole française d'Athènes). Athènes/ Paris: Diffusion E. de Boccard.

Marcadé, J. and Raftopoulou, E. 1963. Sculptures argiennes (II). *Bulletin de Correspondance Hellénique* (=BCH) 87: 33-187.

Markoulaki, St. 2009. Αριστοκρατικές αστικές επαύλεις στην ελληνορωμαϊκή Κίσαμο. In: Χ. Λούκος, Ν. Ξιφαράς, Κ. Πατεράκης (επιμ.) *Ubi dubium ibi libertas. Τιμητικός τόμος για τον καθηγητή Νικόλα Φαράκλα.* ΦΛΣ Πανεπιστημίου Κρήτης: 337-380. Rethymno: University of Crete Editions.

Martin, S. 2002. The Romanisation of the Southeast Corner of the Athenian Agora. Unpublished MPhil dissertation, Cambridge.

Merker, G. S. 1973. *The Hellenistic Sculpture of Rhodes* (Studies in Mediterranean Archaeology 40). Goteborg: Paul Astroms Forlag.

Meunier, M. and Launay, P. J. 1955. *Dans les pas des heros et des dieux.* Paris: Librairie Hachette.

Muthmann, F. 1931. Mitteilungen aus dem Kerameikos V: Fragmente einer Wiederholung des Satyrs mit der Fussklapper. *Mitteilungen des Deutschen Archäologischen Instituts, Athenische Abteilung (=AM) 56:* 87-89.

Muthmann, F. 1951. *Statuenstutzen, und dekoratives Beiwerk an griechischen und romischen Bildwerke: ein Beitrag zur Geschichte der romischen Kopisten Tatigkeit.* Heidelberg: C. Winter Univeritat Verlag.

Mylonopoulos, I. 2003. *Πελοπόννησος οἰκητήριον Ποσειδῶνος. Heiligtumer und Kulte des Poseidon auf der Peloponnes* (Kernos Supplements 13). Liege: Centre International d'Etude de la Religion Grecque Antique

Neudecker, R. 1988. *Die Skulpturenausstattung römischer Villen in Italien.* Mainz: Philipp von Zabern.

Neuerburg, N. 1960. *The architecture of fountains and nymphaea in ancient Italy*. Ph.D. dissertation, New York University.

Neuerburg, N. 1965. *L'architettura delle fontane e dei ninfei nell'Italia Antica.* MemAccNapoli, 5. Napoli: Gaetano Macchiaroli.

Ortega, A. 1986-1987. Gortina. Il Ninfeo presso il Pretorio, *Annuario della Scuola archeologica di Atene e delle Missioni italiane in Oriente* (=AsAtene) 64-65: 131-174.

Padgett, J. M. (ed.) 2001. *Roman Sculpture in the Art Museum Princeton University. Princeton.*

Papapostolou, I. 1967. *Archaeologikon Deltion, 22, Χρονικά, B' 1:* 34-36

Parra, M. C. 1976. Per la definizione del rapporto fra teatri e ninfei. *Studi Classici e Orientali* (=*StClOr)* 25: 89-118.

Rheinholdt, C. 2009. *Das Brunnenhaus Arsinoe in Messene. Nutzarchitektur, Repräsentationsbaukunst und Hydrotechnologie im Rahmen hellenistisch-römischer Wasserversorgung.* Wien: Phoibos-Vlg.

Ridgway, B. S. 1971. The Setting of Greek Sculptures. *Hesperia* 40: 336-356.

Ridgway, B. S. 1981. Greek Antecedents of Garden Sculpture. In: W. F. Jashemski and E. B. Macdougall (eds) *Ancient Roman Gardens. Dumbarton Oaks Colloquium on the History of Landscape Architecture VII* (1979): 7-28, pl.I-XII. Washington: Maguire.

Richard, J. 2008. Roman Nymphaea and Monumental Fountains in the Levant: Water Supply vs. Urban Aesthetics? In: C. Ohlig (ed.) *Cura Aquarum in Jordanien. Proceedings of the 13th International Conference on the History of Water Management and Hydraulic Engineering in the Mediterranean Region* 13 (Petra / Amman, 31 March – 09 April 2007: 263-284. Siegburg: DWhG.

Richard, J. 2012. *Water for the City, Fountains for the People. Monumental Fountains in the Roman East: an Archaeological Study of Water Management. Studies in Eastern Mediterranean Archaeology* (SEMA 9). Turnhout: Brepols.

Riedl, N. 2003. *Gottheiten und Kulte in der Dekapolis. Deities and cults in the Decapolis*, Ph.D. dissertation, Freien Universität Berlin.

Robinson, B. A. 2001. *Fountains and the Culture of Water at Roman Corinth.* Unpublished PhD Diss. dissertation, University of Pennsylvania.

Robinson, B. A. 2011. *Histories of Peirene: A Corinthian Fountain in Three Millennia.* Princeton: American School of Classical Studies at Athens.

Robinson, B. A. 2017. Fountains as Reservoirs of Myth and Memory. In: Gr. Hawes (ed.) *Myths on the Map: The Storied Landscapes of Ancient Greece*: 178-203. Oxford: Oxford University Press.

Rogers, D. K. 2015. Water-Display and Meaning in the High Roman Empire. Unpublished Ph.D dissertation, University of Virginia.

Romeo, I. and Portale, E. C. 1998. *Gortina III: Le Sculture* (Monografie della Scuola Archeologica di Atene e delle Missioni Italiane in Oriente VIII). Padova: Bottega d´Erasmo.

Salowey, C. A. 1995. The Peloponnesian Herakles: Cult and Labors. Ph.D dissertation, Bryn Mawr College.

Scranton, R. L. 1951. *Corinth, I.3: Monuments in the lower Agora and north of the archaic Temple.* Princeton: American School of Classical Studies at Athens.

Schorndorfer, S. 1997. *Öffentliche Bauten hadrianischer Zeit in Kleinasien.* München: Lit Verlag.

Schwingestein, C. 1977. Die Figurenausstattung des griechischen Theatergebäudes (Münchener archäologische Studien Band 8). München: Fink.

Segal, A. 1997. *From Function to Monument. Urban Landscapes of Roman Palestine, Syria, and Provincia Arabia,* Oxbow Monographs, 66. Oxford: Oxbow.

Settis, S. 1973. 'Esedra' e 'ninfeo nella terminologia architettonica del mondo Romano. Dall'età repubblicana alla tarda antichità. *ANRW* 1.4: 662-740.

Smith, R. R. R. 1998. Cultural Choice and Political Identity in Honorific Portrait Statues in the Greek East in the Second Century A.D. *Journal of Roman Studies* 88: 56-93.

Söldner, M. 1986. *Untersuchungen zu liegenden Eroten in der hellenistischen und römischen Kunst.* Frankfurt am Main: P. Lang.

Soles, M. E.C. 1983. *Aphrodite at Corinth: A study of the Sculptural Types.* Ph.D dissertation, Yale University.

Stefanidou-Tiveriou, Th. 1995. Σιληνός καθεύδων. *Archaeologike Ephemeris* 134: 67-75.

Stewart, P. 2003. *Statues in Roman Society. Representation and Response.* Oxford: Oxford University Press.

Sturgeon, M. C. 1987. *Isthmia, IV: Sculpture I:1952-1967.* Princeton: American School of Classical Studies at Athens.

Taliaferro-Boatwright, M. 1991. Plancia Magna of Perge: women's roles and status in Roman Asia Minor. In: S. B. Pomeroy (ed.) *Women's history and ancient history*: 249-272. University of North Carolina Press.

Themelis, P. 1990. Ανασκαφή Μεσσήνης. *Praktika tis en Athinais Arxaiologikis Etaireias* 1990: 56-103, mainly 72-76.
Themelis, P. 1995. Ανασκαφή Μεσσήνης. *Praktika tis en Athinais Arxaiologikis Etaireias* 1995: 56-86.
Themelis, P. 1999. *Η Αρχαία Μεσσήνη*. Athens.
Themelis, P. 2013. The 'Doryphoros' of Messene. In: V. Franciosi, P. G. Thémelis (eds) *Pompei /Messene. Il 'Doryphoro' e il suo contesto*: 125-210. Università degli Studi Suor Orsola Benincasa.
Thomas, E. 2007. *Monumentality and the Roman Empire; Architecture in the Antonine Age*. Oxford: Oxford University Press.
Thompson, H. A. 1955. Activities in the Athenian Agora: 1954. *Hesperia* 24: 50-71.
Thouvenot, R. 1974/75. Divinité des eaux. *BAParis* X/XI: 93-97.
Tobin, J. 1997. *Herodes Attikos and the City of Athens: Patronage and Conflict under the Antonines* (Archaia Hellas 4). Amsterdam: Gieben.
Tod, M. N. and Wace, A. J. B. 1906. *Catalogue of the Sparta Museum*. Oxford: Clarendon Press.
Trabucco della Torretta, M. 2006. Un edificio scomparso nella fontana Arsinòe di Messene? *Annuario della Scuola archeologica di Atene e delle Missioni italiane in Oriente (=AsAtene)* LXXXIV serie III, 6.I: 161-190.
Uğurlu, Nur Banu. 2009. *The Roman Nymphaea* in the *Cities* of *Asia Minor*. Saarbrücken: VDM Verlag.
Vermeule III, C. C. 1977. *Greek Sculpture and Roman Taste. The Purpose and Setting of Graeco-Roman Art in Italy and the Greek Imperial East*. Michigan: Ann Arbor.
Vollgraff, W. 1944-45. Inscriptions d'Argos. *Bulletin de Correspondance Hellénique* (=BCH) 68-69: 391-402.
Vollgraff, W. 1958. Fouilles et sondages sur le flanc oriental de la Larissa à Argos, *Bulletin de Correspondance Hellénique* (=BCH) 82 : 516-570.
Waldhauer, O. (ed.) 1928. *Die antiken Skulpturen der Ermitage, I* (Archäologische Mitteilungen aus Russischen Sammlungen I). Berlin: W. de Gruyter.
Walker, S. 1979. The Architectural Development of Roman Nymphaea in Greece. Unpublished Ph.D dissertation, University of London.
Walker, S. 1987. Roman nymphaea in the Greek world. In: S. Macready and F. H.Thompson (eds) *Roman Architecture in the Greek World*. Eighth Research Seminar, London 22nd March 1985: 60-71.Occasional Papers, 10. Society of Antiquaries of London. London: Thames and Hundson.
Weber, Th. 2002. *Gadara-Umm Qēs I. Gadara Decapolitana. Untersuchungen zur Topographie, Geschichte, Architektur und Bildenden Kunst einer »Polis Hellenis« im Ostjordanland* (ADPV 30). Wiesbaden: Otto Harrassowitz Verlag.
Williams, II, C. K. 1989. A Re-evaluation of Temple E and the West End of the Forum of Corinth. In: S. Walker and A. Cameron (eds) *The Greek Renaissance in the Roman Empire*. Papers of the Xth British Museum Classical Colloquium (Bulletin supplement 55): 156-162. London.
Williams, II, C. K. and Zervos, O. H. 1990. Excavations at Corinth, 1989: The Temenos of Temple E. *Hesperia* 59: 325-369.
Winter, E. 1996. *Staatliche Baupolitik und Baufürsorge in den römischen Provinzen des kaiserzeitlichen Kleinasien*. Asia Minor Studien, 20. Bonn: R. Habelt GmbH.
Woodward, A. M. 1926-1927. Excavations at Sparta, 1927: 2. The Theatre. Appendix: Sculptures from the Theatre and the Nymphaeum. *The Annual of the British School at Athens (=BSA)* 28: 6-36.
Xagorari-Gleisner, M. 2002. Tyche, Muse oder Nymphe? Brunnenfiguren aus Messene. *Thetis* 9: 75-80.
Zachos, K. and Georgiou, A. 1997. Νομός Πρέβεζας. *Archaeologikon Deltion* 52, Χρονικά, Β' 2: 576-600
Zachos, K., Karabelas, N. and Stork, M. (eds) 2007. Νικόπολις Β'. *Πρακτικά του Δεύτερου Διεθνούς Συμποσίου για τη Νικόπολη*: 11-15 Σεπτεμβρίου 2002 // *Nikopolis B'. Proceedings of the Second International Nicopolis Symposium*: 11-15 September 2002. Preveza: Actia Nicopolis Foundation.
Zachos, K. 2015. *Αρχαιολογικός Οδηγός Νικόπολης. Περιδιάβαση στο ιστορικό, ιερό και αστικό τοπία* (Μνημεία της Νικόπολης 10). Athens: Ministry of Culture. Archaeological Receipts Fund.
Zuiderhoek, A. 2009. *The Politics of Munificence in the Roman Empire: Citizens, Elites and Benefactors in Asia Minor*. Cambridge: Cambridge University Press.

Dr Georgia Aristodemou
School of Humanities,
International Hellenic University,
Thessaloniki-Greece
g.aristodemou@ihu.edu.gr

The monumental fountain in the Athenian Agora: reconstruction and interpretation*

Shawna Leigh

Abstract

In the last ten years the architecture of and various issues regarding the monumental fountain in the Athenian agora, a building not preserved above foundation level, leaving all possible reconstructions largely hypothetical, have undergone renewed scrutiny. Brenda Longfellow has briefly reviewed its evidence in her book on Roman monumental fountains and suggests that, like the more recent reconstruction of the Olympia nymphaeum, the Agora fountain had two stories, based on the thickness of its back wall. Additionally, Julian Richard has considered the monument in terms of siting and meaning in his monograph on Roman monumental fountains in the eastern empire. These studies leave questions regarding the monument that require a fresh look at the building.

In this article I will reconsider the evidence and possibilities for the architecture and decorative program of the Athenian fountain, and the degree to which its reconstruction based on the Olympia structure is likely. Additionally, I will postulate the possible water technology utilized in the structure, a topic largely ignored in previous studies. I will also discuss the meanings behind the siting of the 'nymphaeum', its imperial connections, and how the monument and its supply aqueduct visually changed the southeastern Agora space and the important ceremonial approach to the Acropolis, the Panathenaic Way. This focused restudy will allow the building to be better understood within its context in Imperial Greece as well as within Hadrian's program of Athenian euergetism.

Keywords: Fountain, Athens, Agora, Hadrian, Antonine.

During the 1954 American School of Classical Studies' Agora excavation season the remains of a curvilinear Roman monumental fountain (Nymphaion) were investigated in the southeast corner of site. Very little above the foundations of the building remained *in situ*, and the 11th century AD Church of the Holy Apostles sits on the western portion of the monument. Homer Thompson associated a few decorative architectural elements found in various nearby locations and a statue with the fountain's superstructure but stated that no reconstruction was possible because of the scanty remains at the site. He compared aspects of the building to the Nymphaeum of Herodes Atticus at Olympia and the monumental fountain at Tipasa in Algeria, and suggested that the Agora Nymphaion was covered by a semi-dome as was postulated for those buildings. As a part of her 1979 dissertation, Susan Walker went so far as to suggest that the architect for both the Olympia and Athenian structures was the same, presenting a reconstruction that resembled the contemporary reconstruction of the Olympia Nymphaeum (Figure 1);[1] at that time Walker supported the theory that both buildings had one story and were hypaethral. Three additional statues were associated with the Athenian structure.[2]

Recently the architecture of and various other issues regarding the monumental fountain in the Athenian agora have undergone renewed scrutiny. At the outset it must be stated that since the building is not preserved above foundation level all possible reconstructions are largely hypothetical. Brenda Longfellow briefly reviewed the evidence in her book on Roman monumental fountains and speculated that, like Bol's 1984 reconstruction of the Olympia Nymphaeum, the Agora fountain had two stories based on the fact that the thickness of the foundation trench for the back wall would allow the structure to support an extra level. Additionally, Julian Richard has considered the monument in terms of siting and meaning

*I would like to thank John M. Camp for suggesting that I work on this topic, the staff of the Agora Excavations of the American School of Classical Studies for their aid on the project, most particularly Sylvie Dumont, Craig Mauzy and Richard Anderson. I would also like to thank the chief editor of this volume, Georgia Aristodemou, for the opportunity to publish this paper and Susan Walker, without whose work this paper would not have been possible. Finally, I would like to thank William Wallace Roberts II for his support and comments on drafts of this project. All remaining errors are my own.

[1] For the one story reconstruction of the Olympia Nymphaeum, see Yalouris 1982: 107, ill. 43.

[2] Thompson 1955; Walker 1979.

Figure 1. Restored perspective drawing of the 'nymphaion' looking southeast by Susan Walker.
(Courtesy American School of Classical Studies at Athens: Agora Excavations)

in his monograph on Roman monumental fountains in the eastern Roman Empire. These studies leave questions regarding the monument that require a fresh look at the building.

In this paper I will reconsider the evidence and some possibilities for the architecture and decorative program of the Athenian fountain, and the degree to which its reconstruction based on the Olympia structure is likely. Additionally, I will postulate the possible water technology utilized in the structure as indicated by its supply aqueduct, a topic largely ignored in previous studies. I will briefly discuss the discernible meanings behind the siting of the Nymphaion, its imperial connections, and how the monument and its supply aqueduct visually changed the southeastern Agora space, functions and the important ceremonial approach to the Acropolis. This focused restudy will enable scholars to better understand the building within the context of the major 2nd century AD water embellishments of Imperial Greece both symbolically and technologically.

The remains

In situ remains

As mentioned above, the fountain sits in the southeast corner of the Agora, which, given its high elevation compared to the rest of the Agora (66.88m above sea level on the foundation for the floor), and the location of water sources brought in from the east of the city is topographically logical (Figure 2 a). It is

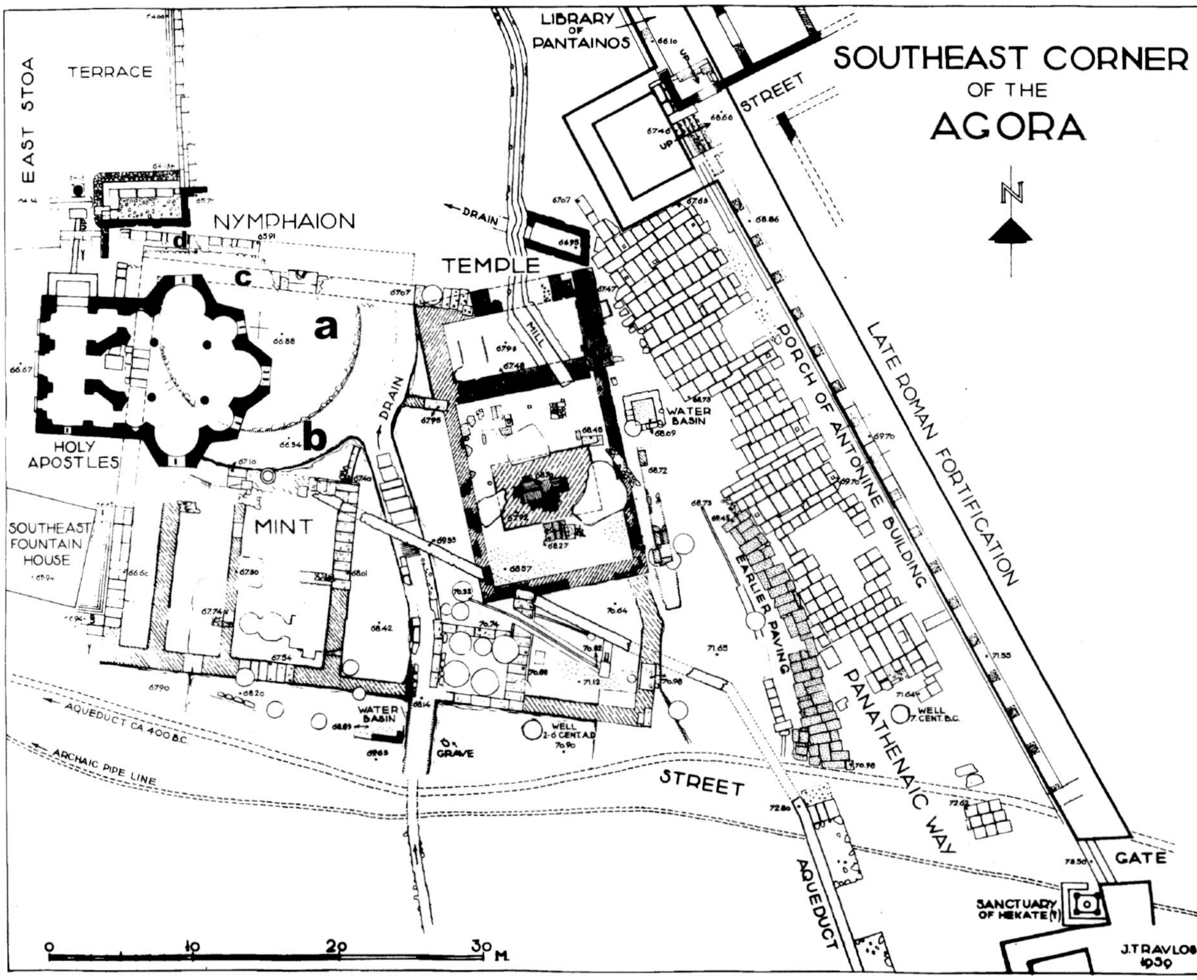

Figure 2. Actual State plan of the Southeast Corner of the Agora by John Travlos with added lettering. (Courtesy American School of Classical Studies at Athens: Agora Excavations).

the last of a series of hydraulic installations to occupy the area. For example, as can be seen in the center left side of the state plan, the building effectively blocks off access to the Southeast Fountain House, which dates to the 6th century BC (Figure 2). It is notable this Archaic structure was still functioning at the time Nymphaion was built, as was the 4th century BC fountain house in the southwest corner of the Agora.[3] Therefore the Roman building was not an attempt to renew ruined resources. Thompson's original excavation report described the foundations for the floor of the extant Roman fountain, which would have served as the bottom of the interior of the water basin (Figures 2 a; 3; 4) as somewhat larger than a semi-circle with a radius of 7.10m; Walker gives this measurement as 7.25m (Figure 5). The distance from the central axis of the rear of the flooring to the front of the basin is 8.30m. The concrete surface of the basin floor show grooves that record the locations of rectangular paving slabs. None of the slabs remain. A cutting for the foundations of the surrounding wall, 2.60m wide, bounds the outside of the floor (Figures 2b; 4). The extreme width of this trench is a point to which we will return below.

A conglomerate stretcher block topped by two orthostates remains *in situ* at the north part of the east side of the outer wall. Note that one of the blocks was cut back to allow the contemporary eastern extension of the Great Drain to pass (Figure 2, labeled 'drain' to the east of the Nymphaion). The drain did not communicate with the fountain at this point. There are also two blocks *in situ* at the western face of the outer wall near its north-east end, below the basin's floor level. The northernmost block is made of

[3] Leigh 1998: 15-17.

Figure 3. East end of the 'nymphaion' looking south.
(Courtesy American School of Classical Studies at Athens: Agora Excavations).

Figure 4. East half of the 'nymphaion' looking north.
(Courtesy American School of Classical Studies at Athens: Agora Excavations)

Architectural section	Measurement in meters	Measurement in Roman feet (.296m)
Concrete foundation for floor showing negatives of marble paving slabs.	7.10 m radius (Thompson); 7.25 m (Walker)	23.99 (Thompson) 24.49 (Walker)
Central axis depth of basin.	8.30m	28.04
Full building front length.	19.50m	65.88
Basin front length.	14.50m	48.99
Back wall trench width.	2.60 m	8.78
Frontal platform width.	3.60 m (Thompson)	12.16
Monument base.	1.65m x 1.70m	5.57x 5.74
Reconstructed monument base	2.20^2	7.43^2

Figure 5. Table showing the measurements of the 'nymphaion'. All dimensions from Walker 1979, except where noted.

poros and the southern one is of conglomerate. There is also a partial base slab made of blue Hymettian marble, found *in situ* on the bedding for the parapet that closed the basin, indicating the material that made up that element. A rectangular foundation for an approach to the building 3.60m wide (Figures 1; 2c; 3, right) abuts the straight, northern edge of the parapet bedding, and given the drop off of the area in front of this closing rectangular terrace, Thompson calculated that three steps led up to this level. Walker reported that part of the lowest step is preserved on the western side. (Figure 2d) At the rear of the rectangular foundation, on the central axis of the flooring, is the foundation for a base measuring 1.65m front to back and 1.70 m wide. Walker suggests that the width of the base might be restored to $2.20m^2$ without giving her reasoning. Both scholars restored a monumental imperial statue on this base. No remains of such statue have been found in the excavations.[4]

Remains associated with the superstructure

Other than the parts of the building discussed above, nothing remains *in situ*. A number of architectural elements found during the excavations, in the vicinity of the fountain and built into the Church of the Holy Apostles, have been identified as decorative pieces of the building (Figure 6). Some pieces have been associated based on their vertical curvature. Others have been joined in groups because of common dimensions and molding sequences.

Walker restored the four wall crowning moldings of the same dimensions as topping a plinth located around the inside of the basin on which an engaged Corinthian order sat. This location is implied by the water deposit on block A2552b. She considered one smaller scale crowning molding possible evidence for decorative framing niches on the interior of the back wall. Utilizing the pilaster capitals, Walker reconstructed a series of engaged Corinthian pilasters with a suggested height of 5.60m. As in the case of the restored dimensions of the monument base on the terrace, no explanation is given for this measurement. To explain the association of a similar full capital, Walker reconstructed free-standing columns fronting the ends of the wall enclosing the basin. Notably, Thompson also associated a section of curved architrave decorated with a guilloche pattern on its soffit with the building. He stated that the soffit of the piece showed that it sat on full columns and suggested that a free-standing Corinthian order fronted the fountain's enclosing wall.[5] However, Thompson had not associated the pilaster capitals with the building, thus this architrave could indicate that there were some free-standing columns inside the building. Walker used the partially preserved raking geison to postulate that there were also pediments on the wall ends. Finally, Walker assigned the entry point of the water to the center of the fountain's back wall, but this is not shown in her reconstruction (Figure 1). She stated that these elements of the reconstruction could be given 'with certainty'.[6] Note that Walker also mentioned that some marble facing plaques were found in the excavations, so there is evidence for marble veneer of the building.[7] The fact

[4] Measurements from Thompson 1955: 57-59 and Walker 1979: 151-156.
[5] Thompson 1954: 58.
[6] Walker 1979: 160.
[7] Walker 1979: 157.

Architectural Element	Dimensions	Notes
Wall crowning molding A2489	0.32m high	Found above the fountain levels immediately to the west of the building.
Wall crowning molding A2499	0.32m high	
Wall crowning molding A2552a	0.32m high	
Wall crowning molding A2552b	0.32m high	Upper surface has traces of water deposit.
Wall crowning molding A2737	0.32m high	
Wall crowning molding A2553	'Smaller scale'	
Water spout A2444		White Pentelic marble with a wreath. Curved channel cut in top with a heavy water deposit.
Water spout A2472	Maximum width of face 0.13m	White Pentelic marble lion's head. Found in the area of the fountain.
Lower piece of a plinth or pilaster carved with the upper torus of an Ionic base.	0.21m thick at base 0.41m wide at base 0.455m high Bearing surface: 0.58m by 0.31m	Found In the foundation rear wall trench of the fountain.
Corinthian pilaster capital A4073.	0.21m thick at base 0.41m wide at base 0.455m high Bearing surface: 0.58m by 0.31m	Reused in the late-Roman basilica built behind the library of Pantainos.
Corinthian pilaster capital A4474.	0.21m thick at base 0.41m wide at base 0.455m high Bearing surface: 0.58m by 0.31m at base	Reused in the late-Roman basilica built behind the library of Pantainos.
Corinthian pilaster capital uncatalogued.	0.21m thick at base 0.41m wide at base 0.455m high Bearing surface: 0.58m by 0.31m at base	Reused in the late-Roman basilica built behind the library of Pantainos.
Corinthian pilaster capital.	0.21m thick at base 0.41m wide at base 0.455m high Bearing surface: 0.58m by 0.31m	Stored in the southwest marble pile in the Library of Hadrian.
Corinthian pilaster capital.	0.21m thick at base 0.41m wide at base 0.455m high Bearing surface: 0.58m by 0.31m	Stored next to the Tower of the Winds.
Full Lotus and Acanthus capital.	Similar height to pilaster capitals	Reused in the Church of the Holy Apostles.
Fragmentary Dentil A2558.	Height restored to 0.16m	Cyma reversa and rebate on top. Found near the Church of the Holy Apostles.
Corner of a raking geison A2425.	Maximum height 0.16m Width 0.19m Angle of the cornice 14°	Recovered from the marble piles near the fountain.
Two uncatalogued undecorated bases.	0.915m on sides 0.54m high	One has a vertical cutting for a water pipe. Frantz 1971: 8 and plate 5,e.
Fragments of marble veneer.		Found in the excavations.
Curved piece of architrave with guilloche decoration on its soffit	Measurements unavailable	Associated with the building by Thompson 1954: 58. He states that the state of the soffit shows that this rested on full capitals.

Figure 6. Table with information on the pieces of superstructure associated with the 'nymphaion'. Information taken from Walker 1979, except where noted.

that some of the pilaster capitals share the same curvature as the enclosing wall of the Nymphaion does strongly suggest that they belong on the fountain, even though they were found elsewhere. But, if it does belong to the structure, the location of the corner of the raking geison could equally have been somewhere on the building's interior, perhaps crowning a niche cut into the surface of the enclosing wall.

Architectural Element	Moldings and decoration
Wall Crowns	From bottom to top: cavetto, bead and reel, egg and dart, bead nd reel, rebate. Some vertical faces are curved that would fit on the curve of the fountain's back wall.
Pilaster Capitals	On the frontal side are two palmettes with three leaves, flanking the central stem of a fleuron. The fleuron continues down the bell. Calyces sit diagonally. Helices flank the palmettes. Below the palmettes: ovolo, tongue, ovolo. Acanthus leaves appear on the bell's four corners. The sides carry diagonal palmettes.
Full Capital	Lotus and acanthus; the acanthus is the same as those on the pilaster capitals.
Fragmentary dentil	Topped by a cyma reversa and rebate.
Corner of raking geison	On top: ovolo and rebate. Below: hawksbeak
Curved piece of architrave with guilloche decoration on its soffit	Associated with the building by Thompson 1954: 58. He states that the state of the soffit shows that this rested on full capitals.
Bases	No decoration. Finished on all sides. Because of their measurements Waker does not include them in her reconstruction.

Figure 7. Table showing the decorative elements appearing on the architectural members associated with the fountain. All material from Walker 1979 except where noted.

Walker continued to suggest additional elements of superstructure based on comparison with the 2nd century AD monumental fountains at Olympia and Nikopolis. She reconstructed a total of ten Corinthian pilasters lining the enclosing wall of the water basin, presumably based on a comparison of the measurements of the internal perimeter of the curved wall and the largest width of the pilaster capitals (0.58m, see Figure 6). The width of the niches was calculated to be ca. 1m, and these were assumed to have been surrounded by architectural elements that are not preserved. She suggested that statues stood in these niches. The 2.50m width of the short ends of the surrounding wall would not give room for both free-standing columns and niches, so she stated that there were no niches at the ends of the walls, although these are shown in the reconstruction (Figure 1).Therefore, all sculpture must have been contained within the interior of the basin wall. Walker added a technical point, stating that the thickness of the back wall of the basin would allow a 'circulation channel' within its fabric, and she postulated that the water would then be divided inside the wall and exit through 'branch pipes' located 'below the level of the niches.[8] This contradicts her previous statement that the water entered the basin at the center of the back wall. This is a point to which we will return.

There is no direct evidence for the roofing of the building. While Thompson and Wycherley have suggested that the building was covered with a semi-dome,[9] Walker argues that, while such a roofing system would visually relate the Nymphaion to earlier Greek fountains located in actual grottos with religious functions, this was likely not the case. She based this conclusion on comparison with other large Roman semi-circular buildings that avoid having an apsidal form extending from their rear wall, as such a curve would normally be contained within outer walls or cut into a hill. Walker suggested two possible solutions for roofing the Athenian building. As roof tiles were found in the excavations there may have been a curved 'stoa-type' roof at the top, as that reconstructed by P. Aupert on the nymphaeum at Tipasa.[10] Alternatively, a plain attic may have topped the building. While Walker suggested that either roofing system is plausible, one must note that, unless extremely narrow, a 'stoa-type' roof would structurally need to bridge a space, presumably between the enclosing wall of the fountain and a free-standing colonnade, which does not match the engaged Corinthian order likely in the Athenian example.

From the time that the Nymphaion was excavated, Thompson associated a previously discovered 2/3 life-sized statue of the 'Venus Genetrix' type carrying a water pitcher with the decoration of the building.[11]

[8] Walker 1979: 161.
[9] Thompson and Wycherley 1972: 203.
[10] Aupert 1974.
[11] Thompson 1953: 53-54 and plate 19 a and b.

Additionally, he theorized that fragments of a similarly sized Amazon should belong to the Nymphaion and he hypothesized that the fountain displayed reproductions of 'adaptations' of famous earlier sculptures.[12] Because of the small size of the Venus statue, Walker questioned its association, as it would be difficult to place within the building. Nevertheless, given the decorative elements of other Roman monumental fountains, particularly the one at Olympia, Walker believed that an elaborate sculptural program must have adorned the building. She suggested that a statue of Hadrian, the original donor of the water supply system, would have stood on the monumental base on the Nymphaion terrace. Hypothesizing that an Imperial dedication would have featured a series of Imperial commemorative statues, Walker associated two additional sculpted heads found in the area with the building. One is a Pentelic portrait of Faustina the Younger. A larger head, possibly of Aelius Verus, was also associated with the decorative assemblage. However, Longfellow pointed out the tenuous nature of assigning these sculptures to the fountain. Her reasons are that over time, as has been demonstrated at other excavations, statues can move long distances from their original positions and that there is evidence in this area of the Agora for both marble workshops and other monuments on which such decorative elements could have been used. Longfellow also noted that the different sizes of the imperial heads imply that they did not come from the same location. She did suggest that the water pitcher that the Venus statue carries makes it the most likely of the group to have been associated with the building. She also pointed out that the under life-sized statue is too small to have occupied a niche in the same single story as Walker's reconstructed full pilaster Corinthian order, and suggests that there might have been two levels of sculptural decoration, with the upper more diminutive than that on the first level, similar to the arrangement at the Hydrekdocheion of Trajan in Ephesos.[13] This brings us to the next point of discussion – how many stories occupied the monumental fountain in the Athenian Agora?

One or two stories?

Roman construction processes

In a 1984 publication Renate Bol reconstructed a second story on the Olympia fountain based on the number of statues found in the excavations, which would fit in two rows of superimposed niches, along with the existence of supporting buttresses located at the rear of the upper basin.[14] Walker did consider the possibility of a second story for the Athenian building, although she was dubious that the resulting height would fit well within the surrounding physical context of the Athenian fountain.[15] Longfellow subsequently suggested that the Athens monumental fountain should be reconstructed with two stories based on the thickness of the back wall and similarities with the Olympia nymphaeum.[16] Indeed, Walker found the Olympia and Athenian monumental fountains to be so much alike that she suggested the earlier Athenian building influenced the design of the Olympia example, and that they might have been designed by the same architect.[17]

Before comparing the actual similarities of the two monumental fountains, some preliminary points must be addressed regarding the current understanding of the process of planning Roman structures and the functioning of Roman architectural workshops. The studies conducted on such organization have largely concentrated on major construction projects in Rome, but it is imperative this evidence be considered in relation to what may have been the situation in the province of Achaia, and in Athens and Olympia in particular. In her consideration of the Baths of Caracalla, Janet DeLaine discussed evidence for the building industry in Rome at the time of the Severans. Based on a meticulous study of the fabric of the Baths she estimates the number of both skilled and unskilled workers necessary for the various portions of the project, and concluded that there must have been a pre-existing force of workers in the city in order to make the construction feasible from its inception. To elaborate the point, DeLaine reviewed the

[12] Thompson 1954: 58.
[13] Longfellow 2011: 127.
[14] Bol 1984.
[15] Walker 1979: 162.
[16] Longfellow 2011: 126.
[17] Walker 1979: 163.

Imperial structures constructed and repaired during the 20 years before the building of Caracalla's Baths, strongly indicating that a large prepared workforce of both skilled and unskilled workers was available in the city.[18] Intriguingly for our study of waterworks outside of Rome, she also notes that a permanent group of approximately 700 slaves were responsible for maintaining the aqueducts.[19] This makes one wonder how such maintenance was carried out in the Provinces.

As a specific example of one group of workers, DeLaine considers the evidence for the population of 'firms' of wood-workers in Rome based on members of that *collegium*. Her results indicate a number of *c.* 10,400-13,000 men available for such work. There would have been of course other groups, including marble workers, iron workers, makers of pavement, etc.[20] In addition to the availability of workers, DeLaine pointed out that there was a tradition of organization for such large-scale projects in Rome that would enable them to be undertaken efficiently. Previous literature argued that there must have been use of the contract system to supplement standing forces of laborers,[21] which could help explain the organization of some of the workers.

In a particularly telling point for this study, DeLaine compares the construction apparatus in the city of Rome with the issues encountered completing large-scale building projects in the Provinces, notably the problems encountered by Nicomedia and Nicaea constructing buildings such as an aqueduct and a theater during the reign of Trajan.[22] DeLaine stated that the supply of materials for such large building projects would have required a permanent workforce, much like the situation of the *curatores aquarum.*[23] Presumably the Provincial cities lacked the resources available in Rome, such as access to the best materials and financial support. This evidence leads scholars to question what the arrangement was for both Imperial and local commissions in the Provinces.

Obviously, the mechanism of construction and the organization of workers in Athens would have differed from that of Rome in the 2nd century AD, as would that in Olympia. Athens would have been in a singular situation in the Provinces at the time when the Hadrianic/Antonine water supply and display system was being built. Camp has suggested that Hadrian might have instigated the aqueduct project in response to a drought in Greece for which he provided several pieces of evidence.[24] However, one must also consider that the Emperor inaugurated the Panhellenion in AD 131/2 with its capital in the city. This institution gathered together cities with Hellenic lineage into a cohesive and formal organization. Spawforth and Walker have discussed various architectural projects implemented in Athens as embellishing the apparatus of the Panhellenion, which they term both a 'political and cultural' body.[25] While a complete consideration of the construction processes in Athens during the initial years of the Panhellenion, of which the Agora monumental fountain was a part, is outside the scope of this paper, some hypotheses may be drawn from the scope of the building program compared with earlier Roman building in the city.

Several Roman projects had been implemented in Athens in the years between the fall of the city to Sulla in 86 BC and the reign of Hadrian from AD 117-138. These included the Odeion of Agrippa in the center of the Agora space, the Market of Caesar and Augustus to the east of the ancient Agora, and a stoa-lined street that joined the Agora and Market with an archway leading into the Agora. In addition, the Library of Pantainos was built with colonnaded stoas facing both the new street and the Agora proper. It is notable that these structures date to either the period of Augustus, or the early 2nd century AD.[26] The Library was financed by a local Athenian, while two of the earlier buildings had Roman backing. In light of the previously mentioned problems with non-imperially funded monumental building in the Provinces the divide between Roman and locally sponsored projects must have had some effect on the

[18] DeLaine 1997: 197-198.
[19] Frontinus, *De Aquaeductu Urbis Romae,* 116; DeLaine 1997: 199.
[20] DeLaine 1997: 199-200.
[21] Brunt 1980: 84-88; Pearse 1975: 36-56.
[22] Pliny the Younger, *Epistulae*, 10, 37, 39; Malacrino 2010: 118-119.
[23] DeLaine 1997: 203.
[24] Camp 1991: 108-109.
[25] Spawforth and Walker 1985: 92-98.
[26] Shear 1981: 358-372.

organization of both their planning and execution. One can only postulate how much labor, if any, in terms of designers and planners may have come to the city from Rome in these instances. The case of the Hadrianic and subsequent Antonine building in Athens was a different situation. Such a large-scale building program, which included such an intrinsically 'Roman' structure as a long-distance aqueduct, would require specialized planning not readily available in the Provinces, especially those where Roman legions, with their trained personnel, were not stationed. As there is no evidence that a Roman legion was ever posted in the Province of Achaia,[27] it is possible that a group of Roman architects, engineers and even some workmen may have been sent to work in Athens over a long period of time in order to plan and oversee the Hadrianic projects. As we have the dedicatory inscription of Hadrian's Athenian aqueduct, which states that the project was begun under Hadrian and completed under his successor, Antoninus Pius, we know that imperially funded work continued in Athens after Hadrian's death. The bulk of the work on the Agora Nymphaion and other outlets for the aqueduct's water would have taken place in the Antonine period.[28]

While the exact process of the administration of planning and building in the Roman provinces is unknown, as I have discussed elsewhere,[29] one can compare with Athens the situation of Nonius Datus in designing the construction of an aqueduct for the city of Saldae in Algeria. An inscription records that the legionary engineer stationed in another province was sent to Saldae to design the system.[30] After he had left it was requested that he return to the city twice when the actual construction process of the water supply system went awry. The aqueduct was not completed until around fifteen years after it was begun because of problems building the system's tunnel. This shows us that in the case of some projects one or more experts, probably associated with the army, would be sent to a Province to plan and oversee such an undertaking, at least until it was determined that the construction was safely underway. In Achaia, and particularly in Athens, this would lead to a two-tiered system of workers, those trained in a 'Roman' system outside of the province and those locally trained. In fact, Walker suggested that differences in the construction of the Athenian aqueduct and the Nymphaion might show a division between Imperial construction (the aqueduct) and local manufacture (the Nymphaion).[31] However, this equally might be because of differences in the date of construction, the fountain necessarily being built after the completion of the aqueduct.

Of course there was a long tradition of monumental building in Athens. We can only conjecture as to the nature of a local building apparatus during the Roman period. Given the widespread trade of building materials, especially marbles and colored stones, across the Empire, some observations regarding local practice as evidenced by the Agora monumental fountain are pertinent. Walker, in considering stone usage in Achaia, noted that in Athens there were difficulties in obtaining local marble pieces large enough for entablatures in Pentelic marble as it tends to have fissures that control the size of the blocks that can be quarried. She noted that in Athens such blocks were made from Hymettian marble or re-used materials.[32] As has been specified above, the extant stones used in the Agora monumental fountain included locally available materials: Pentelic and Hymettian marble for decorative features, and poros and conglomerate for foundations. Of course, most of the extant remains of the Nymphaion consist of concrete, the production of which would have to be introduced into the area. The methods necessary to make the quantity of concrete needed for the projects included in Hadrian's Athenian building program would have had to be taught to local workers; notably there is no evidence that the concrete as described by Walker included imported materials such as pozzolana.[33] While occasional pieces of tile have been found in the concrete mixture (as well as ground brick used to make the material waterproof), there is no

[27] Sherk 1957: 60-62; Gilliam 1965: 65-73; Leigh 1998: 79.

[28] *CIL* III, 549. The text, which appeared on the façade of the initial terminus of the aqueduct above Athens on the Lykabettos hill, gives the completion date of the system as AD 140.

[29] Leigh 1998: 79.

[30] *CIL* VIII, 2728 (=ILS 5795). See also, Cuomo 2011: 143-165.

[31] Walker 1979: 270.

[32] Walker 1979: 262; Dr Scott Pike 1993, personal communication.

[33] Walker 1979: 152-154.

evidence for brick in the superstructure.[34] The few *in situ* pieces of stone mentioned above, and sections of marble plaques found in excavation suggest a building in stone-faced concrete covered with veneer.

Comparison with the Olympia monumental fountain

A number of assumptions about the Agora structure have been made based on the better preserved one at Olympia. Without direct evidence for the reconstruction of the Athenian superstructure any consideration of this topic must be hypothetical. Walker left the question open, but, as mentioned above, she theorized the tall building that would result from a second storey would not fit the physical context of the site. As also mentioned above, Longfellow suggested that the Athenian building might have had a second story based on a comparison of the thickness of its back wall (2.60m or 8.78 Roman feet) and that of the monumental semi-circular upper apsidal basin at Olympia which varies from 1.33m-1.44m (4.49–4.86 Roman feet) thick. But we must note that this measurement excludes the seven buttresses that ring this part of the building. The buttresses extend ca. 1.15m (3.89 Roman feet) out from the face of this back wall.[35] If we accept that the buttresses plus the number of sculptures found that belong to the Olympia structure indicate a second storey, we must question whether the thickness of the wall of the Agora fountain had the same or a different purpose. Considering this issue, Walker felt that the excessive thickness of the back wall was meant to enclose a 'circulation channel' for the water to run from several points into the basin.[36] She did not elaborate as to how exactly how this would work. Since the water would have been running through closed pipes set within the concrete of the back wall it is doubtful that such a thickness would have been necessary to support the water pressure exerted on the pipes. There is no evidence for the material used to make the pipes that served the Agora fountain, but this would be true whether they were terracotta or lead.[37]

The second rational for both the initial reconstruction of the Athenian building as a single-level fountain and the subsequent suggestion for a two-story building has been a general comparison with the Nymphaeum of Herodes Atticus at Olympia, which was constructed around a decade after the completion of the Agora Nymphaion. In fact, Walker hypothesized that the Athenian building may have strongly influenced the structure at Olympia, so much so that they could have had the same architect.[38] But, as mentioned above, we must consider exactly how similar the two buildings were to determine if this is a valid strategy.

Other than the fact that they both have semi-circular basins, the extant architecture of the two structures is not all that similar. The total length of front of the Athenian fountain is 19.50m, versus the 31.20m of the Olympia building. The central axes of the two basins are similar: 8.30m for Athens, 8.45m for Olympia. But the shape of the Athens basin is greater than a semi-circle, whereas the Olympia basin is an actual semi-circle. The Olympia monument has two water basins; a long rectangular lower basin fronts the upper semi-circular basin, so that there was no frontal parapet as in the Athenian building. The upper basin at Olympia is set on a podium that elevates it above the frontal rectangular basin. Notably, while the evidence suggests that the Athenian fountain was built of stone-faced concrete, the Olympia building is made of brick-faced concrete. Some of the bricks were stamped with Herodes' Atticus' name, indicating that the material was locally made and came from one of the benefactor's properties.[39] Thus there are significant differences between the two fountains.

In her consideration of the Olympia Nymphaeum, Walker emphasized that the design of both the building and extant decorative elements was based on simple geometric principals based on multiples of

[34] Walker 1979: 152-153; Walker (1979: 268) describes brick construction in Greece as 'highly localized', but further on (1979: 171) she suggests that while the superstructure of the supply aqueduct was executed in poros, brick would have 'played a substantial supporting role'. This cannot be verified.

[35] Longfellow 2011: 126; For the Olympia Nymphaion measurements, Bol 1984: Beilage 1.

[36] Walker 1979: 161-163.

[37] Camp 1991: 109.

[38] Walker 1979: 162-163; Longfellow 2011: 125-126.

[39] For the Olympia Nymphaion measurements, Bol 1984: Beilage 1. On the architecture of the Olympia Nymphaion, Walker 1979: 185-189.

three Roman feet.[40] She points out that this would allow standard cut pieces to be ordered from marble suppliers. The building included Pentelic and Hymettian marbles that would have come from Athens, along with Cipollino.[41] The measurements of the Athenian fountain shown in Figure 5 prove that the Athenian monumental fountain was not planned in this manner as they are not exact multiples or divisors of the Roman foot. In a related type of argumentation, DeLaine considered the retention of construction plans over time in Rome. She discerned that the planning of the baths of Trajan, dating c. 100 years before the construction of the Baths of Caracalla, and that of the later building shared similar processes. Delaine also pointed out that evidence in Vitruvius and an inscription from Puteoli attest to the preservation of architectural designs over time.[42] Athens itself has an example of the retention of building plans across generations. The Arch of Hadrian in Athens was dedicated to the Emperor by 'the Athenians'. It has been dated to shortly before Hadrian's third visit to Athens in AD 131/2. This was the visit during which the ruler dedicated the completion of the Olympieion (which had been under construction for over 600 years) and inaugurated the Panhellenion.[43] The Panhellenes funded two replicas of this arch in the outer court of the Sanctuary of Demeter at Eleusis. The identity of the Emperor referred to in the dedicatory inscription is not certain. Antoninus Pius, Marcus Aurelius and Commodus have all been suggested. Spawforth and Walker accept a date in the AD 170s. Accepting this date means that the plans for these arches would have been retained for at least 30 years. This information strongly suggests that the plan of the Athenian monumental fountain would have still existed at the time of the construction of the Olympia Nymphaeum. Given the above discussion of what we can logically hypothesize about the local and Imperial construction processes in Athens and Achaia, the construction plans of the two buildings were clearly different. Thus there is no evidence that the Athenian and Olympian structures were planned by the same architect based on similarities in their design. While comparison with the many Roman monumental fountains that carried decorative architectural orders on their facades, along with the associated remains of superstructure found in the Athenian Agora and environs allows a very hypothetical reconstruction of the Nymphaion's elevation, that reconstruction should not hang on that of the Olympia building. Therefore the reconstruction of a second storey on the Athens structure is a possibility, but not a necessity. It remains to consider whether there is any other evidence that would influence the issue.

Technological evidence: the aqueduct supplying the Athenian Agora

The water supply to the Athenian monumental fountain might shed some light on the height of the building. Since its discovery, the Roman aqueduct feeding the southeast corner of the Agora has been associated with the Hadrianic/Antonine system, the main conduit of which ends in a dividing basin on the slope of the Lykabettos hill to the north and east of the city.[44] The agora branch of the aqueduct enters the area of the excavations from the east, runs under the Panathenaic Way and makes a right turn in a basin to run north along the street. At first the water channel ran on a continuous foundation wall, and then atop at least five piers, possibly connected by poros stone arches.[45] Based on some *in situ* concrete found behind the center point of the back wall of the Nymphaion Walker reconstructed an extension of five additional aqueduct piers ending at the fountain. This mass lies 1.50m south of the wall of the fountain, indicating that there was another element that connected the aqueduct channel to the building. Walker suggested that the water was supplied by an 'offset pipe' from this last pier.[46] She used the slope of the last part of the main channel Hadrian's aqueduct before it entered the Lykabettos hill basin to reconstruct the angle necessary for the water in the Agora channel to descend along the Panathenaic way before emptying into the back of the fountain, and hence to estimate the height of the

[40] Roman foot used = 0.296m (Wilson Jones 1989: 108. Note that other scholars use slightly different metric equivalents for the Roman foot, ex. 0.295m, Romano 2006: 65. The 0.296m equivalent has the advantage of being the same as the metric equivalent for the Attic foot, Senseney 2011: 107).

[41] Walker 1979: 185-186.

[42] *CIL* I, 577. Vitruvius, *De Aquaeductu Urbis Romae* 1,1,14. DeLaine 1997: 202-203.

[43] Adams 1989: 10-15.

[44] Thompson 1954: 59.

[45] Walker 1979: 163-171; Leigh 1998: 191-202.

[46] Walker 1979: 171-172.

last pier behind the building as 7.74m.[47] As I have noted elsewhere, the extant last portion of Hadrian's aqueduct before it enters the Lykabettos Hill building is a 19th century reconstruction, and therefore its slope is not applicable to the Agora aqueduct. Given the distance that the Agora aqueduct's above-ground continuous foundation and piers drop, Walker's suggested slope would result in a fall of only 0.034m across the 87m to the end of the fifth extant pier, which is probably too shallow to allow enough of a head of water to the fountain. Instead, using the steepest known slope on a short section of aqueduct, which is a ground-level loss of 28m per km in height, the final pier behind the fountain would have carried the bottom of the water channel at a height of 13.29m.[48] The full height of the water channel atop the pier would be somewhat higher. Of course, a less steep slope in the water line would result in a higher pier.[49] Logically, the level of the aqueduct channel at the point where the water would enter the Nymphaion should not be higher than the building itself, so this may give an idea of how tall the building was.

There is no evidence for the exact mechanism by which the water would pass from the Agora aqueduct's conduit into the monumental fountain. Extant archaeological evidence suggests that after the water in the main extramural tunnel of an aqueduct reached its first point of access in or on the edge of a city, known as a *castellum*, the water would continue to additional access points. The two possibilities for the Agora fountain are that either the water moved directly from the supply channel atop the aqueduct piers into the back wall of the building or that the water would be transferred from that channel into a pipe that led into the structure. These two situations would result in different hydraulic dynamics; the first is an 'open' conduit which would not run full and the water would move based on gravity. The second is 'closed', and therefore runs full and under pressure.[50] Examples of the first solution appear on the Larissa hill at Argos[51] and in the North Nymphaeum at Perge.[52] Similarly in the case of the supply to the Hydrekdocheion of C. Laecanius Bassus at Ephesos, terracotta pipes ended in a 'service room' behind the façade that would serve as a transition point for the water.[53]

Alternatively, the water could be divided in what A. T. Hodge called a 'secondary *castellum*'. This would be a basin from which pipes would carry portions of the water in different directions. Intriguingly, these basins could sit atop a pier. A number have been found at Pompeii and although there are few extant examples in Rome, Frontinus lists 247 in the city.[54] It is tempting to reconstruct such a secondary *castellum* atop the pier behind the Athenian Nymphaion. This solution would allow the water to connect directly with a vertical pipe in the back wall of the fountain, and then possibly to a horizontal pipe or channel that could have fed one set of marble water spouts, examples of which were found in the Agora excavations (Figure 6). Such a dividing basin could also accommodate the possibility that an additional supply line ran to the west behind the building, perhaps to feed the Southwest Fountain House and the industrial area beyond. Hydraulically, water running in a pressure pipe is a more satisfying solution than the earlier suggestion that a connection to the west ran between the Agora aqueduct and along the back wall of the ruined South Stoa II during the Antonine period.[55] In addition, Walker's reconstruction of the fountain shows no outlet for the water after it passed through the second set of spouts from the semi-circular basin to a low access channel in front of the parapet, although she does consider runoff in her text.[56] As there must have been a mechanism to gathering the runoff at the front of the monument, this could well have been channeled to feed structures to the north. Thus the Nymphaion itself, in addition to being a major access point in the water supply system of the city, would have itself fed other buildings. Channeling the water into a pressure pipe rather than an open drain would have allowed a better quality of water to reach higher elevations in the rest of the Agora.

[47] Walker 1979: 172 and 180, note 21.
[48] Leigh 1998: 212. Note that I use this slope to test what the shortest extant pier behind the fountain would be.
[49] Leigh 1998: 214-215.
[50] For discussions of the 'open' versus 'closed' pipe systems, see Hodge 2002: 216-245 and Richard 2012: 99-100.
[51] Walker 1979: 100.
[52] Longfellow 2011: 158.
[53] Richard 2012: 101.
[54] Frontinus, *De Aquaeductu Urbis Romae*, 2, 78-86; Hodge 2002: 291-294.
[55] Thompson 1959: 97 and plate 16. It is known that a water channel ran across the back wall of South Stoa II during the post-Herulian period, see the discussion in Leigh 1998: 215-216.
[56] Walker 1979: 172-173.

Thus, whether the Agora aqueduct's channel fed directly into the back wall of the fountain, or a secondary *castellum* communicated with the structure, the reconstructed height of the pier behind the central axis of the Athenian Nymphaion would indicate the approximate height of the back wall of the building. If the hypothetical figure of 13.29m for the bottom of the water channel is close to accurate, it indicates that the monument's total elevation was more than twice as tall as the 5.60 internal Corinthian order shown in the current reconstruction (Figure 1). Theoretically, some sort of sharp drop in water elevation between the conduit and the fountain could have existed in the Athenian example, such as that on the Larissa hill at Argos.[57] Whatever the exact system, a one-storey fountain would leave the aqueduct channel's masonry visible above the Agora building when seen from a distance. While this argument does not definitively prove that there were two stories on the structure, it adds weight to that solution. To fully appreciate the ramifications of reconstructing a second story, the visual and symbolic impact of the building must be addressed.

Roman Water display in a Greek context

While its location was controlled by access to water, few Roman buildings in Greece can claim a site as symbolically loaded as that of the Agora monumental fountain (Figure 8: 60). Previous publications have offered readings of the imagery presented by this area of the Agora. It was surrounded by colonnaded buildings, which has been interpreted as giving the appearance of the colonnaded streets known in the eastern Roman Empire or the type of forecourt that appeared in Imperial fora.[58] Notably, the fountain's orientation is not aligned with the nearby buildings or the Panathenaic way (Figures 2 and 8: 60), but rather toward the southwest corner of the Stoa of Attalos, thus giving a clear view of the building across the trapezoidal square (*plateia*) that fronts the area (Figure 8, between points 48, 52, 60 and 66). The neighboring Southeast Temple shares this orientation. The view of a fountain building that was so different from the old historic buildings in the surrounding Agora, with its curvilinear aedicular façade and probable Imperial statuary, must have drawn the eye. There were two static views of the Nymphaion when entering the space: one through the arch from the marble street between the Stoa of Attalos and the Library of Paintainos (Figure 8: 47) and one very acute view from the street between the Library of Pantainos and the Southeast Stoa (Figure 8, between points 48 and 51). However, the Roman architectural spectacle would be especially potent when one was proceeding up the Panathenaic Way and starting the climb up the path toward the Acropolis that had been followed in the Panathenaic festival for ca. 600 years.[59] Additionally the Ionic Antonine Southeast stoa ran south of the Nymphaion along the east side of the Panathenaic Way and continued the colonnaded frontage of the ceremonial street.

Richard attributed a 'high' rate of flow to the Agora monumental fountain as it received water from a 'large aqueduct' and the reconstructed method of water supply outlined above supports this conclusion.[60] As discussed above, Walker initially suggested an entry point for the water into the basin at the building's central internal axis, but in her reconstruction (Figure 1) the water is shown entering the basin through spouts in the bases of statuary standing in niches between the Corinthian pilasters. While the fact that two different types of marble water spout were found within the building makes this possible, the high rate of water flow coming out of the large aqueduct would be better emphasized by entering centrally through a cascade, or possibly with a steep drop, such as that in the Larissa fountain at Argos.[61] Such a rushing stream would add a louder and more obvious auditory experience for passersby. Of course both systems could have existed simultaneously.

Although it is an important part of this transitional space from the 'Greek' section of the Agora to the Roman-style trapezoidal platea and beyond, the plan of the Athenian Nymphaion is somewhat truncated because of its tight fit between the pre-existing Southeast Temple (Figure 8: 52), located next to the

[57] Walker 1979: 100.
[58] Walker 1979: 149-151, 278; Martin 2002; Longfellow 2011: 126-130.
[59] Leigh 1998: 219.
[60] Richard 2012: 163.
[61] Volgraff 1958: 551-554.

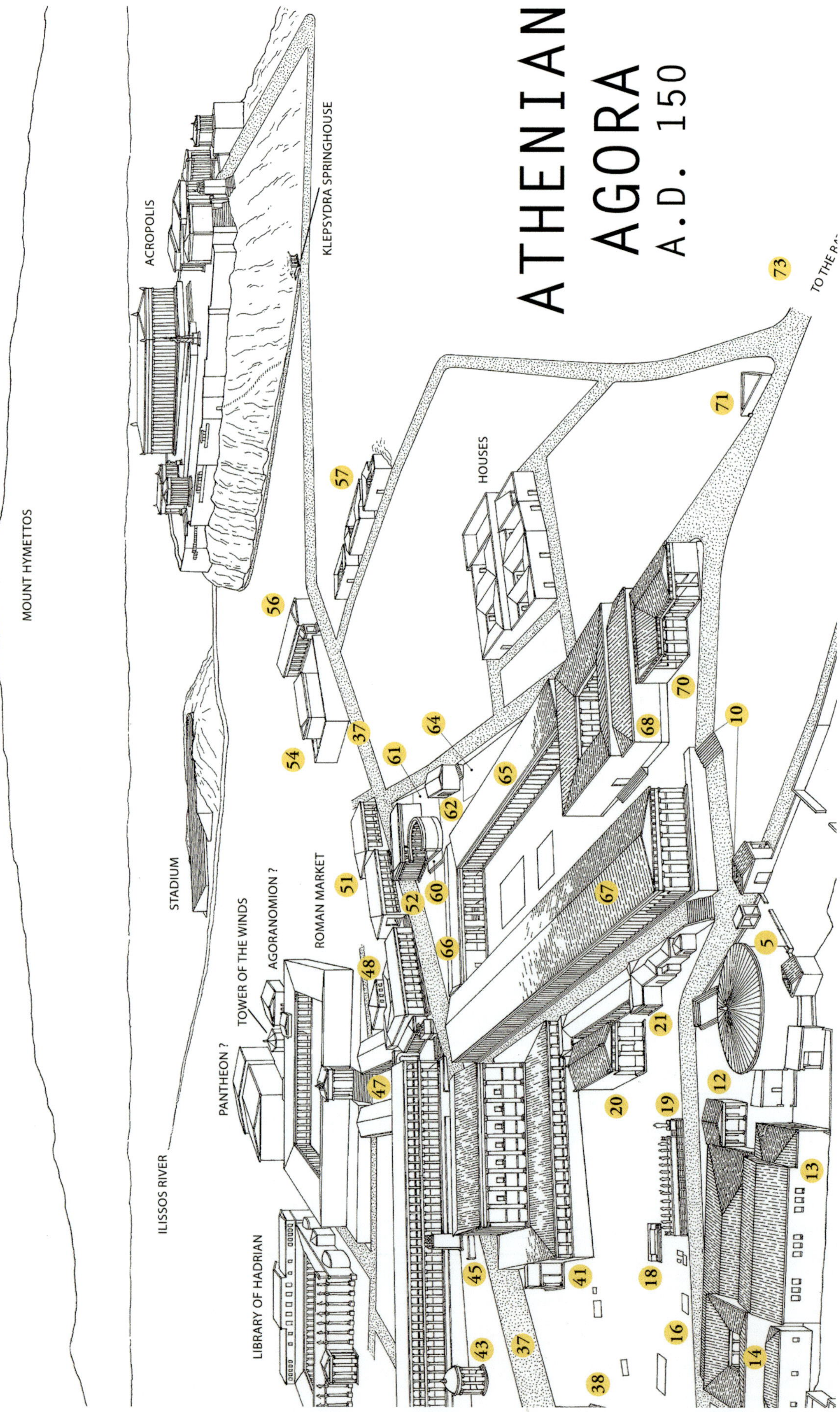

Figure 8. Partial view of the perspective drawing of the Athenian Agora c. 150 AD by W.B. Dinsmoor. Courtesy American School of Classical Studies at Athens: Agora Excavations.

Panathenaic Way and tentatively dated to the 1st century AD,[62] and the East Building (Figure 8: 66), originally constructed during the 2nd century BC and rebuilt in the mid-2nd century AD.[63] Thus the fountain completed this architectural circuit. Walker's one-storey reconstruction of the Nymphaion brings the top of the building roughly in line with the Doric frieze of the Southeast Temple (Figure 1), and while the Roman building notably differs in style, its visual impact would be somewhat lessened by the crowded area. Nonetheless, this juxtaposition of architecturally Roman and Greek styles would have been striking to the viewer, especially since the Doric columns of the prostyle Southeast Temple were actually re-used Greek-period building materials.[64] Still, should the monumental fountain have had two stories the resulting height of the display would have dwarfed everything around it. Fronted by a statue of the Emperor, and possibly holding other imperial statues the structure would emphasize the Roman gift of water to the city in a way that would be impossible to miss, underscoring the official Roman largess. Given the necessary height of its supply aqueduct and its thick back wall, it is very tempting to reconstruct a second story on the building as a truly monumental example of Hadrian and Antoninus Pius' Euergetism in Athens and an unmistakable visual commentary on the Roman control of this very potent Greek space.

Bibliography

Abbreviations used

CIL *Corpus Inscriptionum Latinarum* (Berolini 1863-).
ILS *Inscriptiones Latinae selectae*, ed. H. Dessau (Berlin 1892-1916)

Modern Sources

Adam, J. P. 1994. *Roman Building: Materials and Techniques*. London: Batsford.

Adams, A. 1989. The Arch of Hadrian at Athens. In: A. Cameron and S. Walker (eds) *The Greek Renaissance in the Roman Empire*: 10-15. London: Institute of Classical Studies.

Arafat, K.W. 1996. *Pausanias' Greece: Ancient Artists and Roman Rulers*. Cambridge: Cambridge University Press.

Aupert, P. 1974. *Le Nymphée de Tipasa et les Nymphées 'Septizonia' Nord-Africans*. Rome: Ecole Française de Rome.

Bol, R. 1984. *Das Statuenprogramm des Herodes-Atticus-Nymphäums*. Berlin: Verlag de Walter de Gruyer und Co.

Brunt, P. A. 1980. Free labor and public works at Rome. *Journal of Roman Studies* 70: 81-100.

Camp, J. M., II. 1991. Water Supply and its Historical Context. In: A. T. Hodge (ed.) *Future Currents in Aqueduct Studies*: 105-110. Leeds: Francis Cairns.

Camp, J. M., II. 1992. *The Athenian Agora: Excavations in the Heart of Classical Athens*. London: Thames and Hudson Ltd.

Clinton, K. 1989. Hadrian's contribution to the renaissance of Eleusis. In: A. Cameron and S. Walker (eds) *The Greek Renaissance in the Roman Empire*: 56-68. London: Institute of Classical Studies.

Cuomo, S. 2011. A Roman Engineer's Tales. *Journal of Roman Studies* 101: 143-165.

DeLaine, J. 1997. *The Baths of Caracalla: A Study of the Design, Construction and Economics of Large-Scale Building Projects in Imperial Rome* (Journal of Roman Archaeology Supplementary Series 25). Portsmouth.

Frantz, A. 1971. *The Church of the Holy Apostles*. Princeton: American School of Classical Studies at Athens.

Fullerton, M. 1985. Review of R. Bol, Das Statuenprogramm des Herodes-Atticus- Nymphäums. *American Journal of Archaeology* 89: 369-370.

Gilliam, J. F. 1965. Romanisation of the Greek East: The role of the army. *Bulletin of the American Society of Papyrologists* 2: 65-73.

Hodge, A. T. 2002. *Roman Aqueducts and Water Supply* (2nd ed). London: Duckworth.

[62] Thompson 1960: 343.
[63] Thompson 1972: 66, 71; Camp 1986: 177-179.
[64] Camp 1986: 186.

Holloway, R. R. 1966. Exploration of the Southeast Stoa in the Athenian Agora. *Hesperia* 35: 79-85.

Lancaster, L. 2000. Building Trajan's Markets 2: The Construction Process. *American Journal of Archaeology* 104: 755-785.

Leigh, S. 1998. The Aqueduct of Hadrian and the Water Supply of Roman Athens. Unpublished Ph.D. dissertation, University of Pennsylvania.

Leigh, S. 2000. Interdisciplinary research on the Aqueduct of Hadrian in the Athenian Agora. In: G. C. M. Jansen (ed.) *Cura Aquarum in Sicilia. Proceedings of the Tenth International congress on the History of Water Management and Hydraulic Engineering in the Mediterranean Region (Syracuse, May 16-22, 1998)*: 117-124. (BaBesch Supplement 6) Leiden: Peeters.

Longfellow, B. 2011. *Roman Imperialism and Civic Patronage: Form, Meaning and Ideology in Monumental Fountain Complexes.* New York: Cambridge University Press.

Malacrino, C. G. 2010, *Constructing the Ancient World: Architectural Techniques of the Greeks and Romans.* Los Angeles: J. Paul Getty Museum Publications.

Martin, S. 2002. The Romanisation of the Southeast Corner of the Athenian Agora. Unpublished M. Phil. dissertation, Cambridge University.

Mylonas, G. 1961. *Eleusis and the Eleusinian Mysteries.* Princeton: Princeton University Press.

Pearse, J. L. D. 1975. *The organisation of Roman building during the Late Republic and Early Empire.* Unpublished Ph.D. dissertation, Cambridge University.

Richard, J. 2012. *Water for the City, Fountains for the People: Monumental Fountains in the Roman East.* Belgium: Brepols Publishers.

Romano, D. G. 2006. Roman Surveyors in Corinth. *Proceedings of the American Philosophical Society* 150: 62-85.

Schleif, H. and Weber, H. 1944. *Das Nymphaeum des Herodes Attikos* (Olympische Forschungen I): 53-86.

Senseney, J. R. 2011. *The Art of Building in the Classical World: Vision, Craftsmanship and Linear Perspective in Greek and Roman Architecture.* New York: Cambridge University Press.

Shear, T. L., Jr. 1981. Athens: From City-State to Provincial Town. *Hesperia* 50: 356–377.

Sherk, R. K. 1957. Roman imperial troops in Macedonia and Achaea. *American Journal of Philology* 78: 52-62.

Spawforth, A. J. and Walker, S. 1985. The World of the Panhellenion. I. Athens and Eleusis. *The Journal of Roman Studies* 75: 70–104.

Thompson, H. A. 1953. Activities in the Athenian Agora: 1952. *Hesperia* 22: 25-56.

Thompson, H. A. 1955. Activities in the Athenian Agora: 1954. *Hesperia* 24: 50–71.

Thompson, H. A. 1959. Activities in the Athenian Agora 1958. *Hesperia* 28: 91-108.

Thompson, H. A. 1960. Activities in the Athenian Agora: 1959. *Hesperia* 29: 327-368.

Thompson, H. A. and Wycherley, R. E. 1972. *The Agora of Athens: The History, Shape and Uses of an Ancient City Center.* Princeton: American School of Classical Studies at Athens.

Volgraff, W. 1958. Le Flanc Oriental de la Larissa. *Bulletin Correspondance Hellenique* 82: 516-570.

Walker, S. *The Architectural Development of Roman Nymphaea in Greece.* Unpublished Ph.D. dissertation, University of London.

Wilson Jones, M. 1989. Principals of Design in Roman Architecture: The Setting Out of Centralised Buildings. *Papers of the British School at Rome* 57: 106-151.

Yalouris, J. 1982. *Οι ολυμπιακοί αγώνες στην αρχαία Ελλάδα: Αρχαία Ολυμπία και ολυμπιακοί αγώνες.* Athens: Ekdotiki Athinon.

Dr Shawna Leigh
Adjunct Assistant Professor
Hunter College, Department of Classical and Oriental Studies
New York, USA
sle0100@hunter.cuny.edu

New water from old spouts: the case of the Arsinoe fountain of Messene

Mario Trabucco della Torretta

Abstract

In the proposed contribution, a new analysis will be proposed of the two Roman phases of the Arsinoe fountain in Messene, an urban monumental water facility originally built at the end of the 4th century BC and transformed many times later in its life, using the same pieces several times and adding new ones.

The first Roman transformation of the building during the second half of the 2nd century BC will be examined in its architectural features and set in the broader context of the so called 'Romanisation' of the province of Achaia, a process accomplished throughout the Roman world also through the acquisition and new elaboration or Roman architectural models in the newly acquired provinces. The evidence from the fountain Arsinoe will be connected with other similar cases in Greece and in the other provinces (Asia Minor, Sicily), looking for a model of architectural Romanisation of the provincial landscape.

Lastly the second general refurbishment of the building, occurred in the second half of the 1st century BC, will be analysed underlining the place of Messene in the long story of refunctionalization of urban monumental fountains from mainly utilitarian buildings to places of display and propaganda, and its role in the transition from the fountain to the *nymphaeum*.

Keywords: monumental fountains, nymphaeum, Messene, Romanisation, civic architecture

Situated between the big square of the *agora* and the theatre of Messene, the Arsinoe monumental fountain-house is one of the longest-living buildings of the entire city. It was originally built with great probability at the end of the 4th century BC or at the beginning of the 3rd century BC in the shape of a 36m long Doric stoa, following the fashion of the period, and it was remodelled at least three more times during its long history, before ending up as a water mill in the Late-Antique phase of the city.[1]

The archaeological excavation of the monument began in 1988 under the direction of P. Themelis and the auspices of the Greek Archaeological Society; later, in 1990, F. Felten and C. Reinholdt of the University of Salzburg joined the project for the technical study of the building and the excavated finds. Once cleared from the earth that covered it for centuries, the building revealed itself unmistakeably as a large monumental fountain-house and was identified with the fountain dedicated by the Messenians to the mother of Asklepios, Arsinoe, and mentioned by Pausanias (3, 31, 6) in the record of his visit of the city between AD 155 and 160.[2]

Given the chronological scope of this volume, I will discuss here the third phase of the life of the Arsinoe fountain, that is the refurbishment of the building during the age of emperor Nero, putting this intervention in the broader context of the acts of civic patronage that along with the direct intervention of the emperors permitted many monuments of the past throughout the Greek world to survive the injuries of time and retain their functions until the final collapse of the ancient cities.

[1] The archaeological remains are presented in Felten-Reinholdt 2001 and thoroughly published in Reinholdt 2009. Given the fragmentary state of the evidence the interpretation of the remains is debated. The Austrian excavators identify only three phases of the life of the building (1. late Hellenistic; 2. early-Imperial 3. age of Diocletian), to which must be added a fourth late-Antique phase as a watermill during the 6th century AD. They assume that in the first phase the building was created reusing the Doric elements from another unlocated and unspecified building (Felten-Reinholdt 2001; Reinholdt 2009); I have challenged this interpretation, sustaining that the unspecified building is actually a very early phase of the fountain-house itself, in its late-Classical original configuration as a two-nave Doric stoa (Trabucco 2007). In this text I will follow this reconstruction and the phases will be numbered accordingly.

[2] On the identification of the complex with the Arsinoe Fountain, Muth 2007: 52-53, esp. notes 355 and 356.

Figure 1. Messene, Arsinoe fountain-house, present state (after Trabucco 2007: fig.1).

The architectural remains

The Arsinoe fountain-house is a wide rectangular building, approximately 36 x 14m, divided longitudinally in two main areas: on the rear (North) portion of the building there is a wide narrow basin covered with a thick layer of reddish hydraulic plaster, with a semicircular exedra at the centre; the front (South) section is divided into a square paved area at the centre corresponding to the exedra, and two subsidiary basins to the right and left of it (Figure 1). At the two extremities of the front side there are two little square projecting *podia.* The whole building is poorly preserved; apart from the back terrace wall in isodomic masonry, only a couple of courses of the other walls survive in their original position, and a minimal amount of the other architectural members has been recovered from the dismantling of later structures.

All the decorative architectonic members found have been divided in three groups. The first is a group of column shafts and capitals in the Doric order, along with part of a Doric entablature and a series of stylobate blocks; these pieces, relying on their stylistic and technical features, can be assigned to the last years of the 4th century BC or the beginning of the 3rd. Along with the Doric members there is a second consistent group constituted by columns, pillars with half-columns and entablature in the ionic order; parallels for the form of the Ionic basis and capital (so called 'ionischen Festlandkapitelle') suggest for this group a chronology in the first half of the 2nd century BC. The third major group of architectonic members is constituted by the pieces of a wall in isodomic masonry pierced by three openings crowned by an equal number of round arches, each decorated by three progressively projecting bands surmounted by a moulded cyma. This last group is dated by the excavators to the 1st century AD. This brief survey of the decorative elements of the fountain can be completed with the addition of a fourth group, made of the parts of the exedrae and the anathemata that were dedicated in the area in front of the fountain façade.[3]

After a careful analysis of the blocks and consideration of their position, it became clear that the fountain had several construction phases, at least five from the late 4th century BC to the 6th century AD; during these phases the same architectonic members were re-employed and re-arranged, in order to renew the layout of the monument with the least of the expense, only adding some new pieces to the ones already in place in order to give the building a more fashionable look.

In this study we will examine more closely the central phase of the life of the building, that is the partial refurbishment of the late-Hellenistic fountain-house at the beginning of the Roman Imperial period.

Since the Roman period interventions were altering the previous late-Hellenistic configuration, some words must be spent on how the fountain actually looked like before the restorations. At the beginning of our era, the fountain-house was a fairly traditional Ionic stoa divided longitudinally in two naves by a shallow parapet surmounted by a row of piers with ionic half-columns on the front side.[4] These inner piers supported in turn a simple entablature with double-fascia epistyle, plain frieze with dentilated cornice and geison, all in one piece. The presence of mortises in the flanks of the piers leads to imagine that the space between them was filled by some sort of screens, probably wooden, placed in order to protect the water basin beyond them.[5] This big basin (*lacus*) occupied the whole rear nave of the building and was fed by a channel coming out from the rear terracing wall and covered with a thick layer of reddish hydraulic plaster.

This kind of layout was fairly common for the fountain-houses of late-Hellenistic period and we can find parallels for the whole conception -or for some of its features- in Greece (Athens, Corinth and Delos) and in Asia Minor (Pergamon, Ephesus). The characteristics of the Ionic order, with the capitals belonging to the group of the so-called 'ionischen Festlandkapitelle', are also similar to those of the Palaestra of Olympia and point out to a chronology between the end of the 3rd century and the beginning of the 2nd century BC. This was a particularly good period for the city of Messene, seeing intense building activity and the creations of the famous sculptor Damophon.

The reddish hydraulic plaster of the main basin is not the only layer of plastering in it. During the 1st century AD a semi-circular exedra (4.28m wide) was placed in the centre section of the basin and its poros foundations pierced through the plaster, so that a new layer of *opus signinum* was laid out over the previous one.[6] (Figure 2) On these foundations only the first course of the limestone base of the exedra, decorated with a *cyma recta*, is preserved.

The positioning of an exedra with statues in the centre of the rear basin states the beginning of the alteration program and will constitute the main focus of the new form of the edifice. The Ionic façade was demolished, reducing the depth of the covered portion of the building by half, while the stylobate of the colonnade was moved to the south redefining the southern part of the building as an open strip *c.* 7.20m deep. The strip is divided in three sections: one central square area covered with slabs in front of the exedra and two more rectangular basins built of reused pieces from the numerous honorary monuments that once stood in front of the building.

What was once the dividing line between the front and the rear nave of the late-Hellenistic building is now the new façade of the fountain-house and the regular repetition of the traditional trilithic scheme of the colonnade is now interrupted by the insertion of a solid wall perched by three fornices, whose poros elements have been found reused in the later stages of the monument. The foundation for this new

[3] For a careful technical analysis of these founds, Reinholdt 2009.

[4] On the wide spatial and chronological diffusion of this type of building, Dorl-Klingenschnid 2001: 33-34; Longfellow 2011: 67 and 108-109 (in particular on the Arsinoe fountain-house).

[5] For a similar feature in the Greek Phase 7 (1st half of the 2nd century BC) of the Peirene fountain at Corinth, Robinson 2011: 148-150.

[6] Reinholdt 2009: 33-36 (exedra) and 39-40 (two layers of hydraulic plaster).

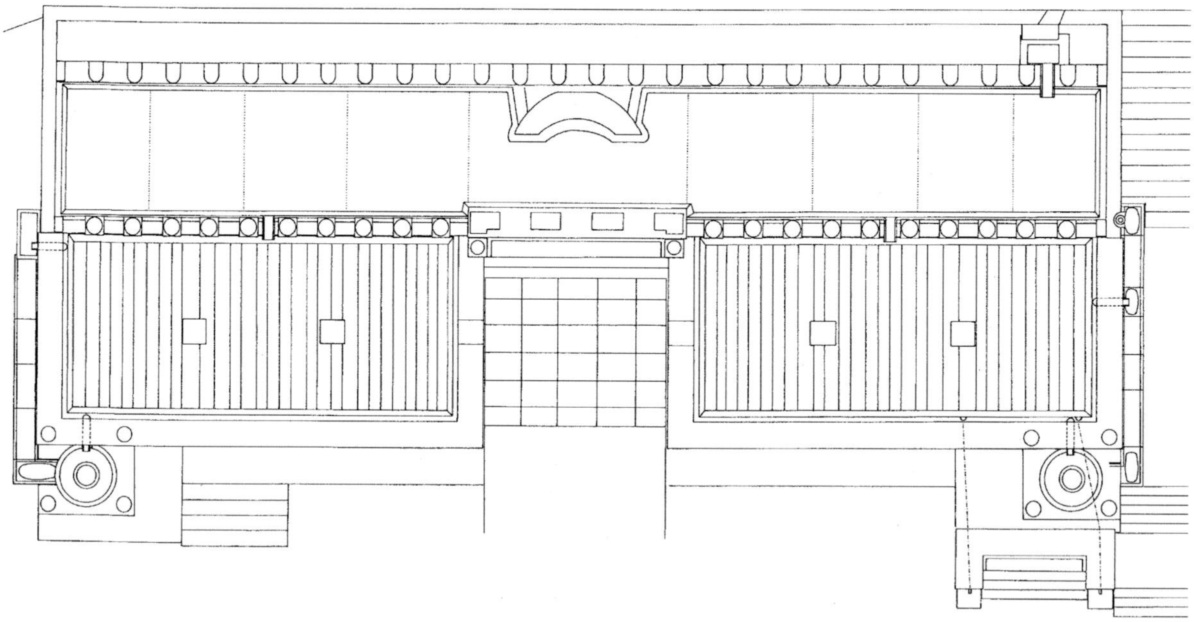

Figure 2. Messene, Arsinoe fountain-house, reconstructed plan of the Roman phase (after Reinholdt 2009: fig. 144)

installation is constituted by an enlargement of the central section of the median foundation wall crafted in *opus testaceum* in front of the exedra and for a length of 7.05m.[7] The use of the wall with arched openings in connection to a fountain-house is a direct reference to the Peirene Fountain of Corinth, where in the first decades of the Imperial period a screen wall was erected in front of the spring, decorated with arches and half-columns over two storeys.[8]

Even more close to the Corinthian example is the other major modification of the Messenian fountain, that of transferring the water outside of the fountain-house, *en plein* air, creating the two rectangular basins to the left and right of the square in front of the arched wall. At Corinth the creation of a *hypaithros krene* in the court in front of the spring is attested during the late Augustan period, 'no later than the reign of Tiberius'.[9] In our case the south boundary of the building was moved further south, in order to create a square area (7.2m x 7.2m) in front of the arched wall to emphasize the monumentality of the composition. The blocks of the stylobate of the demolished front colonnade were moved south and the parapets for the two basins were built upon them, reusing material from the Hellenistic honorary monument bases that were scattered all over the area in front of the building - probably also being in a state of great disrepair. The monumental effect was completed by adding to the front side of the two basins two platforms, each one sustaining a marble *labrum* (diam. 2.10m) that received the overflow from the basin behind (Figure 3).[10]

The insertion of the wall with the three arches and the addition of the two open-air basins with the platforms and the *labra*, apart from being a bold statement of Roman style, is a radical transformation of the initial conception of the building. If the late-Hellenistic fountain-house was just a functional building with no precise focal point, where all the parts of the façade were -almost democratically- the same, now the gaze of the passer-by is attracted to the centre of the building and forced to look through the arches to see the statues on the exedra. The purpose of the fountain-house is no more the mere supply of fresh water to the citizens; the building has now become also a display of specific propagandistic messages.

[7] Reinholdt 2009: 87-102.
[8] Robinson 2011: 181-182 and fig. 99 (Roman Phase 2, post. 30 BC).
[9] Robinson 2011: 188-189.
[10] On the use of *labra* in Roman baths and fountains, Ambrogi 2005 (the Arsinoe *labra* are no. 119 of her catalogue).

Figure 3. Messene, Arsinoe fountain-house, reconstructed elevation of the Roman phase (after Reinholdt 2009: plate 12)

Restorations of monuments and civic patronage at Messene

One of the most interesting inscriptions from Messene is a catalogue of donations made by private citizens for the restoration of civic buildings, found by A. Orlandos at the NE corner of the Agora with another fragment near the entrance of the Sebasteion.[11] The chronology of the inscription, at first generically fixed between the period 31 BC – AD 14 by L. Migeotte, has now been restricted to the years AD 11-14 by D. Baldassarra.[12] In the inscription a certain *Epinicos*, the secretary of the city council, records the names of twenty-nine individuals from twenty-two families that have provided different amounts of money for the restoration works. Part of the contributions are mentioned only with the name of the donor(s) and the amount, some other record also the particular building that should be restored with the money promised. From the preserved part of the stela a total amount of 4720 *denarii* is recorded but a grand total of 6000/6200 *denarii* has been estimated. These amounts of money were devoted to permanently resolve the problem of the state of disrepair of some public monuments on behalf of the city during a moment of weakness of the public finances. The reward for this act of generosity was that the benefactors gained both the gratitude of the city and a wide publicity by being recorded on official stelae, placed next to the Sebasteion and next to the restored monuments.

Though the Arsinoe fountain is not mentioned in this document, some observations on this text could provide us a more precise idea of the context of the intervention made on the fountain-house.

The biggest contribution specified in the text is the 1000 *denarii* promised or given by a *L. Bennius Glycon*, while the lowest one is the 50 *denarii* given by a *T. Ninnius Philippion*, an amount twenty times lower than the first one. If Migeotte is right in his interpretation that the appeal for contributions was made only to the wealthiest families of the city, we must assume that a gift of 50 *denarii* was a reasonable amount allowing the donor to stay within the budget without making a poor impression and, consequently, that a sum twenty-times greater was a very generous gift. Along with these specified amounts there were also five 'blank cheques', pledges to restore some specific monuments – or parts of them – whatever the cost of the repairs. One of these open promises could be related to the recent find of an inscribed epistyle from the Doric stoa in the Agora, recording works of an estimated cost of 1875 *denarii* covered by a private citizen.[13]

As already noted by L. Migeotte, great emphasis is given in the text regarding the link between this restoration program, the Roman Emperor and the Roman people; in two instances (ll. 3 and 36-37) the text makes reference to some obligations towards the Emperor and the people of Rome, and prescribes that the text itself should not be placed at the Agora or near the *bouleuterion*, as we could expect, but at the Sebasteion.[14] Going even further we could add that the appeal is made in a moment of 'weakness' of the city finances: the contributions recorded are not really voluntary ones – they were specially prompted in order to allow the city to fulfil a requirement that could not be avoided. We do not know the reason for such an urgent need, but whatever the reason was we should notice the absence from this list both of the fountain Arsine and of the money spent for its restoration.

According to P. Themelis the most probable cause for the damages to all those civic buildings was an earthquake that must have struck the Peloponnese during the last years of the reign of Augustus. As a proof of it Themelis recalls an inscription dated to the year AD 14 recording an embassy sent to the new Emperor Tiberius to ask for his 'mercy' following the *atopon* of Poseidon.[15] If this was indeed the case what should we think of the absence of the Arsinoe from the list? Was the fountain-house left untouched by the earthquake – or was it in such a state of disrepair that even the most generous of the benefactors decided to look elsewhere?

---

11 Orlandos 1959: 68 and 170-172; *SEG* 23, 205 and 207; *SEG* 35, 343; Migeotte 1985; Baldassarra 2007: 29-30.

12 Baldassarra 2007: 29, with discussion of the chronology at note 26.

13 Themelis 2012: 57, Inv. No. 17781. The author suggests that this unnamed citizen could be a certain *Ti. Claudius Mystes* (I) portrayed in a bronze statue whose base was found in the stoa (60).

14 Migeotte 1985: 604.

15 Themelis 2010b: 97.

It is also possible that the earthquake of AD 14 did not severely damage the city and its buildings, and that the repairs were only minor ones due only to the old age of the monuments and some possible lack of maintenance. This would explain the apparent contradiction of the Messenians first trying to solve the problem by themselves and then appealing to the mercy of the Emperor, and lead us to agree with C. Grandjean in her supposition that the object of the Imperial mercy was the restitution of the Dentheliatis region, that Augustus had taken away from the control of Messene as a revenge for their support to Antonius before Actium.[16]

Since during the reign of Nero the fountain-house was laying in ruins, we must look for another event as the reason of its state. A suitable candidate could be another earthquake, the same that damaged Sparta and forced its inhabitants to call for the help of Vespasian, as documented by other epigraphical evidence.[17] Given the notorious lack of sympathy of Nero for the Spartans, it is well possible that the earthquake had taken place some years before, but that the Spartans were able to secure the Imperial help only with Vespasian.

Whatever the reason that produced the destruction of the Arsinoe fountain-house, the monument was the perfect object for an act of *euergetism*. Its position at the NW corner of the Agora, on the path between the Agora and the theatre, its value as a utilitarian building fundamental for the supply of fresh water to the citizens, made it the perfect monument to be restored by a wealthy member of the elite in search for a chance to show his power, his connections and to gain the gratitude of the people. Through the expenditure of a relevant amount of money the benefactor demonstrates his patriotism, his care, his sense of duty for 'what is right for the *polis*' (as stated in the benefactors' inscription); he increases his status and visibility and proves to be worth of the honours he aspires to. And now that the *fons honorum* is Rome, it is absolutely important to demonstrate fidelity to Rome and to the Emperor.

Profiling the donor: the Ti. Claudii in Messene

During the 1995 excavation season and not far from the fountain, an inscription was found recording the dedication by a certain '*Tiberios...* ', of the statues of the Sebastoi and the restoration works on the fountain that was in ruins (*ten krenen kateageisan epeskeuasan*), paying for all this 3800 *denarii* of his own money.[18] P. Themelis, who promptly published the inscription, initially proposed to identify the dedicant as *Ti. Claudius Dionysiou Aristomenes*, already attested as the dedicant of another statue of the Emperor Nero dated probably in AD 55 (IG V 1, 1450). However, the poor state of preservation of the stone along with the great number of *Ti. Claudii* attested at Messene in the 1st century AD are sufficient grounds to be cautious and avoid accepting this identification straight away.[19]

The *praenomen* Tiberius allows us to assume that we should look for the Arsinoe fountain dedicant amongst the ranks of Greek citizens that received the privilege of the Roman citizenship during the Julio-Claudian period and in particular those who were in good terms with Claudius or Nero. Of these two emperors, Claudius was more favourable towards the Spartans, while Nero was particularly friendly with the Messenians. We will briefly consider here the two most suitable candidates for the role of the benefactor in relation to the fountain-house.

Ti. Claudius Aristomenes was a member of one of the most prominent families of Messene (if not the most prominent at all), whose members enjoyed the burial in the splendid mausoleum K3 in the civic Gymnasium. His father, *Dionysios*, was celebrated by the city as *heros*. The *genos* made consistent reference in the member's onomastics to Messenian heritage through the repetition of the name *Aristomenes*, the hero king of the Messenians, every other generation. The *Aristomenes* of the early Imperial period, the son of *Dionysios*, was awarded the Roman citizenship and appointed priest of the Imperial cult at least

[16] Grandjean 2003: 250.
[17] Themelis 2010b: 97; IG V 1, 691; *SEG* 11, 1950, 848.
[18] Themelis 1995: 56-57; Reinholdt 2009: 133-134.
[19] Rizakis *et al.* 2004: MES 336; Baldassarra 2007: 33. On the *Ti. Claudii*, Rizakis 2007.

from 55 AD, when a dedication of a bronze statue of Nero by him and his wife *Gemonia* is recorded on an inscription.[20] This official recognition of both local prestige and fidelity to Rome was the start of a new phase in the history of the family: his son, *Ti. Claudius Dionysius Crispianus*, though maintaining a strict link with his city and being called 'new Epaminondas', was appointed priest of the Imperial cult for life, *Helladarch* of the Achaian League, *prefectus* of the *Cohors prima Bosphoriniana* and *tribunus* of the *Legio XII Fulminata certa costans* and was welcomed in the equestrian order.[21] At the same time he maintained the link with his city holding high local offices, like the priesthood of the Mysteries of Andania.

The second most suitable candidate is *Ti. Claudios Saithidas*, appointed as High priest of Nero from AD 62 onwards, who should have received the Roman citizenship during the reign of Claudius. He is the author of the dedication of another bronze statue of Nero, upon the inscription of which the dedicant is defined as *philokaisar* – a clear declaration of fidelity towards the Emperor and the Roman state, and at the same time a claim purely made to attract the Imperial favour. Furthermore, the *Saithidai* had a very rich ancestry line that went back at least to the 1st century BC – and they were immensely wealthy as mentioned by Pausanias (4,32,2). The descendants of this *Ti. Claudius Saithidas* had a spectacular career, starting from the *Ti. Claudius Saithidas Caelianus,* who was Imperial priest for life and *Helladarch* of the Achaian League under Trajan, *to Ti. Claudius Frontinus,* who was *consul suffectus* during the reign of Antoninus Pius and the first Messenian to enter the Senate. Once reached the senatorial rank, the interest of the family gradually shifted to their new Italian base in the territory of Abellinum.[22]

The identification of the Arsinoe benefactor with *Aristomenes*, as suggested by P. Themelis, is indeed tempting. His family demonstrated throughout all its history a great attachment to the civic heritage of Messene, from the historical and ideological values preserved and transmitted in the very name of the various *Aristomenai* to the priesthood of the Mysteries of Andania. Whenever the city asked for help, as in the case of the decree proposed by *Epinicos*, they were there to do their part: the father of *Aristomenes*, *Dionysios*, provided along with his mother 500 *denarii* for the restoration of the temple of Demetra and of the so called 'stoa of Nikaios'.[23] This same Dionysios was so popular that the citizen of Messene after his death called him *heros* and dedicated a statue of him in the city Gymnasium, as an example for all. It is also probable that Pausanias in his journey enjoyed the friendship and hospitality of the *Aristomenai*, a family that constituted the perfect guide through the region with their local prestige, extensive knowledge of the local traditions and their evident connections with Rome.[24] Nonetheless, there are some unclarified issues regarding the possibility of *Aristomenes* financing such an expensive restoration project that is characterized by extensive Roman features and values.

At first, the amount of money spent; how should one evaluate the expenditure of 3800 *denarii* for the bronze statues and the repairs of the fountain-house? Considering the data discussed above it seems to be an extremely generous donation. With such an offer the donor goes beyond the simple statement that he was caring for the good of the city, as all the contributors of the Augustan inscription did. The investment of this huge amount of money had a dual purpose; on one hand to please his fellow citizens by restoring for them the fountain-house. On the other hand to demonstrate to the Romans how faithful a subject he was by installing the statues of the *Sebastoi* and remodelling the architectonic form of the whole fountain as to be more 'Roman'. In this broader context, the amount of 3800 *denarii* is not merely a gift; it is an investment in prestige and a powerful *captatio benevolentiae*. The amount of money invested by *Dionysios Aristomenous* one generation earlier was indeed worthy of admiration, but it was neither the open amount set by some of the donors, nor the maximum among the fixed amounts promised.

Secondly, we should also consider the effect of such an effort on the perspective of the family sustaining it. It is legitimate to presume that a wealthy individual from a prominent family making a lavish gift

[20] IG V 1, 1450. Rizakis *et al.* 2004: MES 131, 205; Baldassarra 2007: 32; Luraghi 2008: 320, n. 104. On the imperial cult at Messene, Deshours 2004.
[21] Baldassarra 2007: 34-35.
[22] Baldassarra 2007: 38, particularly n. 65.
[23] Migeotte 1985, ll: 28-29.
[24] Habicht 1985: 115-135; Baldassarra 2007: 37.

to his city and enhancing the Roman presence in the topographical appearance of the city would not go unnoticed – especially when this person is the High Priest of the Imperial cult. While the son of Aristomenes, though holding high offices in the Roman administration, did not go further than obtaining the admission to the equestrian order, the nephew of Saithidas was able to climb up the ladder to the very top, the Senate. Did the large investment of Romanising the Arsinoe fountain contribute to this success? One cannot say for sure, but it is my opinion that Saithidas fits better the profile of the *euergetes* that restored the fountain-house in comparison with their local contenders, the Aristomenai family.[25]

Conclusion

The *polis* of Messene had contacts with Rome at least since the year 210 BC, as attested by Livy (27, 30,13). After the destruction of Corinth, the city was proclaimed *libera et immunis* as a reward for not taking part in the war moved by the Achaian League against Rome; this allowed the city to enjoy a greater degree of economic freedom without the imposition of a tribute, and to keep the right to mint its own coin. Nonetheless the circulation of Roman *denarii* in the region gradually increases and the Roman coin is used more frequently in the everyday activities, as an accounting currency and for savings hoarding.[26] This numismatic evidence, along with the early presence of Romans in the city in such numbers to justify the creation of a tribe of foreigners, leads us to imagine a city that is very open to the Roman influence. During the Augustan period the very elite of the city was of mixed origin, as attested by the appeal of *Epinicos* for financial help, addressed to both 'the Greek and Roman residents'.

The Greek elite of the city, like in all cities, was divided. On one side there were those who accepted the Roman presence wholeheartedly, welcoming the change to a new way of life; on the other side there were those who resisted this change, remaining bound to the local traditions and values though cooperating with the Roman administration. Unsurprisingly, the first group was the most successful one – as the case of Messene may confirm once again in the event of an epigraphical confirmation of the proposed identification of the benefactor who repaired the Arsinoe fountain-house with *Ti. Claudius Saithidas*. Integrating Roman elements in the traditionally Greek language of the *euergetism*, both in the material details and in the concept underlying the whole realisation, was the key to pass from the level of the local elite to that of the provincial and, evidently, the imperial one.[27]

Under this respect also, the Roman domination added a totally new level to the ever-lasting strive for honours and prestige that characterizes the internal dynamics of all elites; being the best was no more enough for the members of these restricted local groups. In the new perspective of an *orbis Romanus* they wanted to excel, and in order to achieve this goal exceptional efforts were needed. One aspect of these efforts was being engaged in the celebration of the newly introduced cult of the deified Emperors; another aspect was contributing through their family wealth to the creation of a new – Roman – world in the old Greek territories.[28] This transformation of the landscape is one aspect of the Romanisation process – the process of transforming a civic landmark that 'embodied the symbolic and social ideas of the community', as B. Longfellow stated it, from a functional building whose appearance was harmonized into the architectonic landscape surrounding it into a monumental declaration of principles, values and self-promotion.[29]

This process that we have observed, was applied to the Arsine fountain-house of Messene and constitutes one of the first stages of the long transition from the Hellenistic *krene* to the Roman *nymphaeum.*

[25] Same opinion on the identity of the donor is expressed in Felten-Reinholdt 2001: 319 and Luraghi 2008: 307, but not discussed.
[26] Sidiropoulos (2011: 1031) commented that 'It is perhaps superfluous to go on and argue in favour of the fact that such people were familiar with the denarius'.
[27] On this evolution on the scale and aim of the euergetic activities, Rizakis 2009. The value of high offices, priesthoods and public works for the building of a public figure is underlined also by Longfellow (2011: 61).
[28] See Camia 2008, for the role of the Imperial Priesthood as a key way to the highest positions in the Empire.
[29] Longfellow 2011: 1.

Bibliography

Abbreviations used

IG V,1	*Inscriptiones Graecae, V,1. Inscriptiones Laconiae et Messeniae*, ed. Walter Kolbe. Berlin 1913.
SEG	*Supplementum Epigraphicum Graecum.* Vols. 1-11, ed. Jacob E. Hondius, Leiden 1923-1954. Vols. 12-25, ed. Arthur G. Woodhead. Leiden 1955-1971. Vols. 26-41, eds. Henry W. Pleket and Ronald S. Stroud. Amsterdam 1979-1994. Vols. 42-44, eds. Henry W. Pleket, Ronald S. Stroud and Johan H.M. Strubbe. Amsterdam 1995-1997. Vols. 45-49, eds. Henry W. Pleket, Ronald S. Stroud, Angelos Chaniotis and Johan H.M. Strubbe. Amsterdam 1998-2002. Vols. 50- , eds. Angelos Chaniotis, Ronald S. Stroud and Johan H.M. Strubbe. Amsterdam 2003-.

Agusta-Boularot, S. 2001. Fontaines et fontaines monumentales en Grèce de la conquête romaine à l'époque flavienne: permanence ou renouveau architectural ? In: J.-Ch. Moretti – J.-Y. Marc (eds) *Constructions publiques et programmes édilitaires en Grèce entre le IIe s. av. J.-C. et le Ier s. ap. J.-C. Actes du colloque organisé par l'École Française d'Athènes et le CNRS, Athènes, 14-17 mai 1995* (Bulletin de correspondance hellénique, Supplément 39): 167-236. Paris- Athènes: EFA.

Ambrogi, A. 2005. *Labra di età Romana in marmi bianchi e colorati.* Rome: L'Erma di Bretschneider.

Baldassarra, D. 2007. Famiglie aristocratiche a Messene nella prima età imperiale. Il contributo dell'epigrafia. In: G. Cresci Marrone and A. Pistellato (eds) *Studi in ricordo di Fulviomario Broilo. Atti del Convegno (Venezia, 14-15 ottobre 2005)*: 25-62. Padova: S.A.R.G.O.N. Editrice e Libreria.

Camia, F. 2008, Imperial priests in second century Greece: a socio-political analysis. In: A.D. Rizakis and F. Camia (eds) *Pathways to Power: Civic Elites in the Eastern Part of the Roman Empire. Proceedings of the International Workshop held at Athens, 19 December 2005*: 23-41. Athens: Scuola archeologica italiana di Atene.

Dorl-Klingenschmid, C. 2001. *Prunkbrunnen in kleinasiatischen Städten: Funktion im Kontext.* München: Dr Friedrich Pfeil.

Felten, F. and Reinholdt, C. 2001. Das Brunnenhaus der Arsinoë in Messene. In: V. Mitsopoulos-Leon (ed.) *Forschungen in der Peloponnes. Akten des internationalen Symposions anlässlich der 100-Jahrfeier des Österreichischen Archäologischen Instituts.* Zweigstelle Athen vom 5.3.-7.3. 1998 (ÖAI Sonderschriften 38): 307-323. Athens.

Grandjean, C. 2003. *Les Messéniens: de 370/369 au Ier siècle de notre ère. Monnayages et histoire* (Bulletin de correspondance hellénique Supplément 44). Paris/Athènes: EFA.

Habicht, C. 1985. *Pausanias und seine Beschreibung Griechenlands.* München: C. H. Beck Verlag.

Longfellow, B. 2011. *Roman Imperialism and Civic Patronage: Form, Meaning and Ideology in Monumental Fountain Complexes.* New York: Cambridge University Press.

Luraghi, N. 2008. *The Ancient Messenians. Constructions of Ethnicity and Memory.* Cambridge: Cambridge University Press.

Migeotte, L. 1985. Réparation de monuments publics à Messène au temps d'Auguste. *Bulletin de correspondance hellénique* 109: 597–607.

Muth, S. 2007. *Eigene Wege. Topographie und Stadtplan von Messene in spätklassisch-hellenistischer Zeit* (Internationale Archäologie 99). Rahden/Westf: Verlag Marie Leidorf.

Orlandos, A. 1959. Ανασκαφή Μεσσήνης. *Praktika tēs en Athēnais Archaiologikēs Hetaireias* 1959 (1965): 162-173.

Reinholdt, C. 2009. *Das Brunnenhaus der Arsinoë in Messene: Nutzarchitektur, Repräsentationsbaukunst und Hydrotechnologie im Rahmen hellenistisch-römischer Wasserversorgung.* Wien: Phoibos-Verlag.

Rizakis, A. D. 2007. Les Ti. Claudii et la promotion des élites péloponnésiennes. In: Y. Perrin (ed.) *Neronia VII - Rome, l'Italie et la Grèce: Hellénisme et philhellénisme au premier siècle ap. J.-C.*: 183-195. Bruxelles: Éditions Latomus.

Rizakis, A. D. 2009. Supra-civic landowning and supra-civic euergetic activities of urban elites in the Imperial Peloponnese. In: Chr. Gallou (ed.) *Being Peloponnesian. Cohesion and diversity through time*

(International conference, University of Nottingham, 31 march-1 april 2007): chap. 6: 1-15. Univesity of Nottingham: accessed from http://www.nottingham.ac.uk/csps/ events/pelo09_abstracts.php.

Rizakis, A. D., Zoumbaki, S., Lepenioti, C. (eds) 2004. *Roman Peloponnese II, Roman Personal Names in their Social Context (Laconia and Messenia).* The National Hellenic Research Foundation/ Research Centre for Greek and Roman Antiquity (Ethnikon Hidryma Ereunōn/Kentron Hellēnikēs kai Rōmaïkēs Archaiotētos [EIE/KEPA]). Athens/Paris: Diffusion E. de Boccard.

Robinson, B. A. 2011. *Histories of Peirene: a Corinthian Fountain in Three Millennia.* Princeton: American School of Classical Studies at Athens.

Sidiropoulos, K. 2011. A hoard of denarii and early Roman Messene. In: N. Holmes (ed.) *Proceedings of the XIV International Numismatic Congress,* Glasgow 2009. International Numismatic Council: 1025-1036. Glasgow: SPINK.

Themelis, P. 1995. Ανασκαφή Μεσσήνης, *Praktika tēs en Athēnais Archaiologikēs Hetaireias* 150: 56-86.

Themelis, P. 2010a. Die Agora von Messene. In: H. Frielingshaus and J. Stroszeck (eds) *Neue Forschungen zu griechischen Städten und Heiligtümern: Festschrift für Burkhardt Wesenberg zum 65. Geburtstag*: 102-125. Möhnesee: Bibliopolis.

Themelis, P. 2010b. The economy and society of Messenia under Roman rule. In: A. D. Rizakis and C. E. Lepenioti (eds) *Roman Peloponnese III. Society, Economy and Culture in the Imperial Roman Order: Continuity and Innovation, Innovation.* Meletemata 63: 89-110. The National Hellenic Research Foundation/ Research Centre for Greek and Roman Antiquity (Ethnikon Hidryma Ereunōn/Kentron Hellēnikēs kai Rōmaïkēs Archaiotētos [EIE/KEPA]). Athens/Paris: Diffusion E. de Boccard.

Themelis, P. 2012. Ανασκαφή Μεσσήνης, *Praktika tēs en Athēnais Archaiologikēs Hetaireias* 2012 (2015): 53-62.

Trabucco, M. 2007. Un edificio scomparso nella fontana Arsinòe di Messene? *L' Annuario della Scuola Archeologica di Atene e delle Missioni in Oriente* 83 serie III.6: 197-214.

Dr Mario Trabucco della Torretta
Independent Researcher
London
dr.m.trabucco@gmail.com

Reflecting the past: the nymphaeum near the so-called Praetorium at Gortyn

Brenda Longfellow

Abstract

The single-story Π-shaped nymphaeum near the so-called Praetorium in Gortyn was built by an unknown patron in the second half of the 2nd century AD. At this time, it was adorned with an extensive sculptural group that included seated muses, fountain sculptures of gods, and portrait statues that presumably portray the patron of the monument and members of his family. Standing in the civic center and facing the so-called Praetorium across a large street that traverses the city, the fountain was highly visible in the cityscape, where it provided a convenient source of potable water as well as an eye-catching artistic water display. Sometime between the 5th and 8th century, the open-air settling basin was enclosed and the edifice transformed into a barrel-vaulted reservoir. Even though the barrel vault of the cistern effectively blocked from view the lower part of the statues from the view of passers-by, the 2nd-century sculptural group was re-erected along the colonnaded façade wall of the transformed monument.

This paper examines the ancient and late-antique phases of the monument in to order to answer the question of why a late-antique patron would re-erect the sculptural group at all. Economics is a tenacious explanation for the use of *spolia,* or repurposed materials, in late Antiquity yet seems to play no role here. More likely, the sculptural group's resurrection can be attributed to the ties of the late-antique patron to the community as well as late-antique imperial legislation that repeatedly restricted the removal of statues and other architectural elements from public buildings and urban centers across the empire.

Keywords: Gortyn, sculpture, patronage, *spolia*, Theodosian Code

Monumental civic fountains were built by cities, emperors and elite patrons throughout the ancient Roman Empire. Typically fed by aqueducts and positioned in heavily trafficked areas of cities, these imposing edifices stood one to three stories high. They showcased a variety of sculptures, from donor portraits to divinities to local heroes, all arranged to highlight the prestige of the community and benefactor. Such sculptural groups often interacted with flowing water to create innovative kinetic displays. In most cases, the sculptural display was created for the fountain and reflects deliberate choices on the part of the patron. Exceptions to this general rule of a sculptural group created for the monument, then, are particularly notable – the architecture of the fountains and the public amenity they provided of flowing water in urban centers were no less remarkable or costly as acts of public munificence, but the re-use of sculptural elements suggests the interchangeability of sculptural types for certain patrons. This essay considers the sculpture that has been associated with the nymphaeum near the so-called Praetorium in Gortyn in its antique and late-antique phases, both of which seem to incorporate sculptures originally intended for other contexts.

The nymphaeum near the so-called Praetorium in Gortyn (F 25) was built by an unknown patron in the second half of the second century AD, probably during the Antonine period (Figure 1).[1] Standing in the civic center and facing the so-called Praetorium across a large street that traverses the city, the fountain was highly visible in the cityscape and provided a convenient source of potable water.[2] Fed by a massive aqueduct that carried water 35km from the springs at the foot of Mount Ida, this fountain is just one in a series of civic fountains known in Gortyn.[3] Its topographical placement, size, architecture, and decoration are fairly standard for monumental civic fountains of the period, including other fountains

[1] The dating of the monument is based on the style of the capitals and sculpture that is presumed to belong to the original phase of the nymphaeum. For the excavation history and reconstructions of this monument see Bendinelli 1914; Ghedini 1985; Maiuri 1914; Ortega 1986-87.
[2] The so-called Praetorium was first built as a gymnasium in the early 1st century AD. By the Trajanic period, it had been transformed into a monumental bath complex, Di Vita 2010.
[3] See Taramelli (1902: 118-140) and Giorgi (2016) for a discussion of the aqueduct, which was actually one of two that fed the imperial city. See Perali (1914); Ortega (1986-87: 164-174); and Giorgi (2008) for discussions of other civic fountains in Gortyn.

Figure 1. Nymphaeum near the Praetorium, Gortyn. Photo taken with permission of the Ephorate of Prehistoric and Classical Antiquities of Herakleion.

in Gortyn, particularly the nymphaeum near the Large Bath. What is remarkable about this particular edifice is the preservation of the 2nd-century sculptural program despite a late-antique renovation that transformed the monument's appearance and function.

Approximately 16m long and 10m wide, the 2nd-century fountain's predominant architectural feature is a hypaethral rectangular settling basin framed at the back and lateral sides by a single-story columnar façade (Figure 2). The central axis of each façade wall is emphasized by an apsidal sculptural niche, the base of which features four steps for water to cascade down before tumbling into the settling basin below.[4] The rest of the façade walls are enlivened with engaged brick pilasters aligned behind freestanding marble Ionic columns. The columns carry a marble architrave, and the façade walls and engaged pilasters are reveted in marbles imported from Greece, Asia Minor, and Africa.[5] Somewhat surprisingly, similar marble revetment did not cover the interior walls of the settling basin, which instead had exposed *opus signinum* below the water line and fresco above. The few surviving fragments of fresco consist of wavy blue and red lines, probably in imitation of marble veining, and hence provide evidence that some cost cutting measures were taken in the finishing touches of the monument.[6]

A total of fourteen statues in fragmentary condition and exhibiting a wide range of marbles, styles, and quality have been associated with the nymphaeum, both during the original excavation of the edifice (1911-1913) and subsequent investigations (Figure 3). Three of the statues recovered in the edifice are colossal in scale and thus universally discounted from display in the fountain due to their incongruent size with the

[4] The main conduit entered the fountain at the northwest corner and was channeled to the central apsidal niche on each façade wall. Pipes passed the water through the front wall of the settling basin and into the long, narrow draw basin at the front of the monument. Secondary channels lead from the back of the lateral niches to the front of the monument to fill the small basins at the corners.

[5] The marble for the revetment, columns, and architrave comes from the local quarry, Caristo (Eubaea), and Teos and Phrygia (Asia Minor), Ortega 1986-87: 158.

[6] Ortega (1986-87: 157-158) suggests that the fresco might have been a marine scene rather than marble veining.

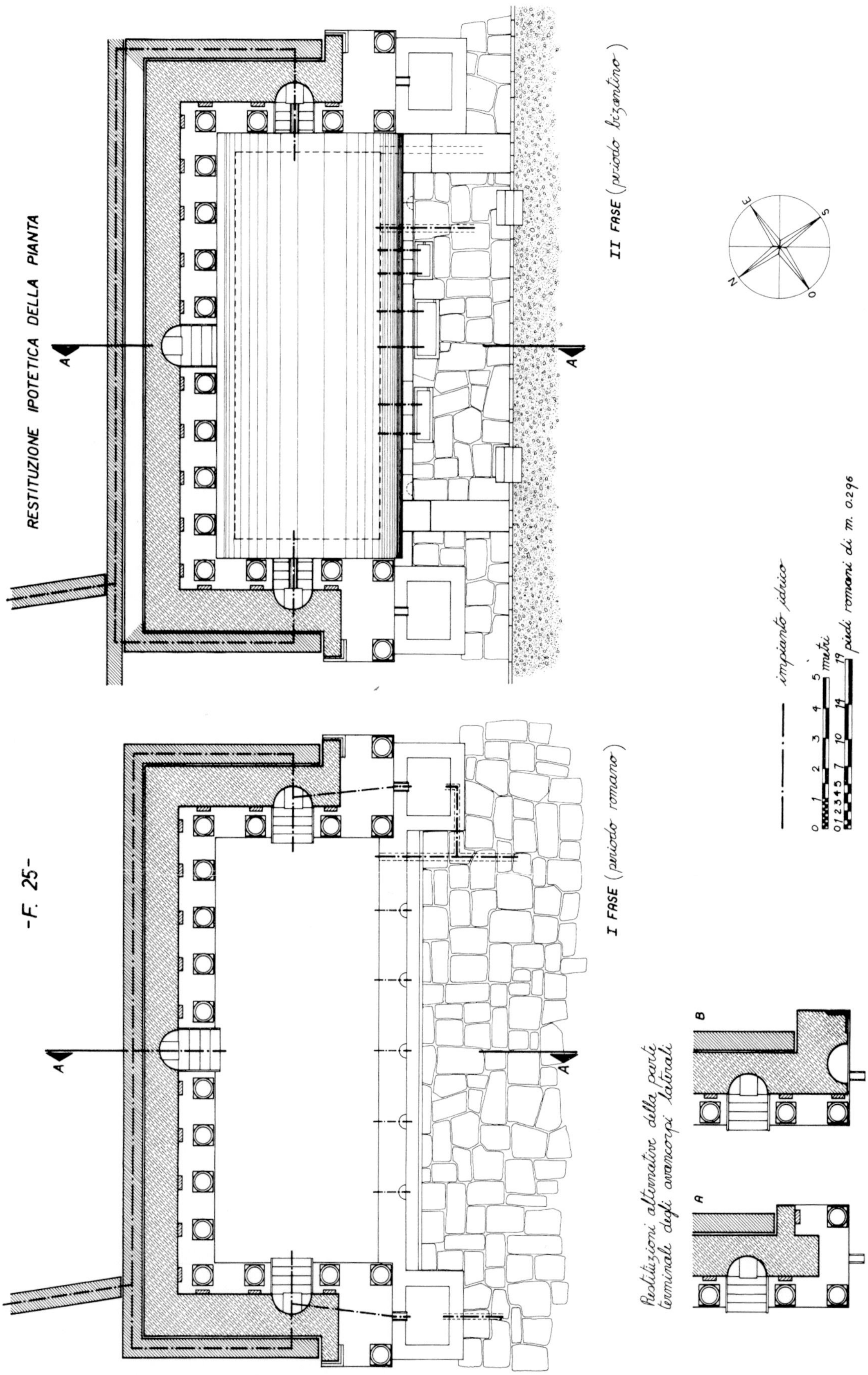

Figure 2. Plan of the ancient and late antique phases of the nymphaeum near the Praetorium, Gortyn (after Ortega 1986-1987: fig. 14/ reproduced with permission of the Scuola Archaeologica Italiana di Atene).

Figure 3. Sculptural and architectural elements found during the 1911 excavation. SAIA, neg. 145. (reproduced with permission of the Scuola Archaeologica Italiana di Atene).

façade wall.[7] Of the remaining statues, the following are generally dated by stylistic analysis to the second half of the 2nd century, and, according to Ortega, securely assigned to the nymphaeum: an Aphrodite of the Louvre-Naples type, a male torso with chlamys, a now-lost nymph holding a shell (Figure 4), the muse Thalia (Figure 5), a Small Herculaneum Woman, a Large Herculaneum Woman, and a male torso variously identified as Hermes or Heracles.[8] The two *togati* found in the edifice but rejected by Ortega because of their typology should be added to the nymphaeum group.[9] These statues probably stood in the axial niches and intercolumniations of the façade walls, where Ortega suggests they may have been joined by other muses, nymphs, or sileni. Due to the lack of solid evidence for dating either the architecture or sculpture

[7] Fragments of 3 colossal statues were recovered from the nymphaeum: a female statue possibly dating to the 4th century BC (Ghedini 1985: no. 32), a 1st century BC or AD female bust (Ghedini 1985: no. 33), and a mid-2nd century AD head of Athena Velletri (Herakleion Museum, Inv. No. 416; Ghedini 1985: no. 43).

[8] For reasons that aren't quite clear, Di Vita (2010: 227) only associates three of these statues with the nymphaeum: the Aphrodite of the Louvre-Naples type, the nymph, and Thalia.

[9] Ortega (1986-87: 155) suggests that the togate figures associated with the edifice do not belong to this monument due to their typology. However, portrait statues of donors and their family members were typically mixed with divinities and their entourages for such displays. In addition, the Large and Small Herculaneum women were two of the most common statue types for female portraits in the 2nd century.

Figure 4. Nymph associated with the nymphaeum near the Praetorium, Gortyn (now lost). Scuola Archeologica Italiana di Atene, neg. 141 (reproduced with permission of the Scuola Archaeologica Italiana di Atene).

Figure 5. Thalia associated with the nymphaeum near the Praetorium, Gortyn. Herakleion SAIA, neg. 18.949 (reproduced with permission of the Scuola Archaeologica Italiana di Atene).

associated with the fountain, it is difficult to say which statues originally belonged to the monument, which were slightly later additions, or which were placed there only in late antiquity. Indeed, the range in quality and styles indicates that different workshops produced the sculpture. Because this essay emphasizes the late-antique phase of the sculptural display, the 2nd-century statues of a size appropriate for display in the niches of the monument will be discussed as a single group.[10]

Based on their smaller, rather uniform size, three of these statues can be identified as occupants of the axial niches: an Aphrodite of the Louvre-Naples type that presumably held a water jug rather than the traditional apple, a male torso with chlamys that has been identified as Hermes but may be better identified as a youthful Dionysus due to the grape clusters on the support, and a nymph holding a shell at her waist, which has been pierced through for a water pipe.[11] Both the Aphrodite and the nymph with shell were relatively standard fountain figures in the 2nd century, although the pipe hole in the nymph's shell adds a rather unique dimension to the display.[12] Perched on a pedestal above the water staircase and holding the spewing shell at the level of her genitalia, the nymph suddenly became an active generator of water, reprising the traditional role of nymphs in Greek religion as protectors of freshwater springs.

[10] Ghedini (1985) suggests that a nymph of the Nemesis type that was found in the triple apsidal hall of the so-called Praetorium belonged to the nymphaeum, but discounts a number of statues found in the nymphaeum itself, including the togate statues and the Large and Small Herculaneum Women, suggesting that they were not appropriate for the display.

[11] Aphrodite: Herakleion Museum, Inv. No. 325; Ghedini 1985: no. 23; male torso with chlamys: Herakleion Museum, Inv. No. 415; Ghedini 1985: no. 7; nymph with shell (now lost): Ghedini 1985: no. 28. These three statues have a reconstructed average height of 1.20m, compared with 1.80m for the other statues (Ortega 1986-87: 154)

[12] Another nymph holding a shell fed by water pipe stood in the House of the Ephebe, in Pompeii (Naples Archaeological Museum, Inv. No. 144276).

The six larger sculptures that Ortega associates with the edifice and stood along the façade walls were a mix of divinities and portraits. The divinities include Thalia, the muse of Comedy, and the nude torso of a youth that lacks any identifying characteristics but Ghedini tentatively identifies as Hermes and Ortega identifies as Hercules.[13] Ortega has convincingly argued that Thalia faced the street from the loggia occupying the terminal end of the eastern lateral wall and probably formed part of a pendant group.[14] The muse sits on a rocky outcrop, from which water may have spouted into the small square basin below, a suggestive setting that, like the nymph with the shell, evoked the mountain spring origins of the water feeding the fountain.

As is to be expected of such a civic display in the 2nd century, the assemblage is rounded out by male and female portraits, presumably of the elite family that donated the edifice. The fragmentary togate statues found in the nymphaeum are balanced by a Large Herculaneum Woman and a Small Herculaneum Woman.[15] In the 2nd century, both the Large and the Small Herculaneum Women were the most common statue types used for civic portraits of local and regional elite women, with the popularity of the types peaking during the Antonine period – which is the probable date of these two examples. Trimble has convincingly demonstrated that these richly draped statues of elite women with restrained gestures showcased the importance, wealth, and high social status of an elite family and, because this type was used by elite families across the Empire in the 2nd century, emphasized its ability to engage in an imperial-wide dialogue with peers of the same social status.[16] Placed in monuments that usually provided a practical benefit – such as potable water – and that typically were located in highly trafficked areas of a city – such as this fountain –, these statues also emphasized the symbiotic relationship between elite families and cities. For, local elites who funded public projects like the nymphaeum near the so-called Praetorium not only improved the general quality of civic life but also elevated their own social status in the community.

Unlike the homogenous sculptural groups produced by single workshops for contemporary nymphaea like the nymphaeum of Regilla and Herodes Atticus at Olympia, the sculptural group associated with the nymphaeum near the so-called Praetorium in Gortyn appears to be an amalgamation of imported and local work, all of varying quality. Among the more elegantly crafted pieces is the figure of Thalia, and among the works with simplified elements, perhaps due to inferior workmanship, is the youthful Dionysus figure. With the probable exception of the Thalia, which is made of Italian Luna marble, and the possible exception of the Large Herculaneum Woman, which is made of Attic Hymettian marble, Ghedini has suggested that the sculptures were produced by more than one local sculptural workshop, which either crafted them in their entirety or refined them from roughed out imported pieces of marble.[17] All statues, regardless of quality, were crafted from imported stones, a choice necessitated by the unsuitability of the local gray stone for sculpture.

To summarize: it is interesting to note that the 2nd-century sculptures associated with the fountain did not form a stylistically cohesive group, even in antiquity. Given the appropriate size of the Aphrodite, the nymph with a shell, and Dionysus for the axial niches with their water cascades, these statues probably were created specifically for the fountain. Moreover, the incorporation of a water pipe indicates that the nymph with a shell was originally carved to decorate a fountain. As Maiuri has noted, the varying proportions of the other statues suggest that some may not have been intended for the fountain but rather gathered together from across Gortyn. Regardless of their origins, by the time the fountain was transformed in late Antiquity, these sculptures seem to have become associated with it.

Gortyn served as the religious metropolis of Crete from the 4th century to the end of the 7th century, and, sometime during this period, the settling basin of the fountain was converted into a reservoir (Figure 6).[18]

---

13 Torso of a youth: Haghioi Deka, Phylakion Inv. No. 22; Ghedini 1985: no. 8; Thalia: Ghedini 1985: no. 30.
14 Ortega 1986-87:155.
15 Togate statue: Ghedini 1985: no. 13 (late 2nd-early 3rd century); Large Herculaneum Woman (mid-2nd century): Ghedini 1985: no. 38; Small Herculaneum Woman (mid-2nd century): Ghedini 1985: no. 39.
16 Trimble 2011.
17 Ghedini 1985: 243.
18 Ortega (1986-87: 160) suggests the devastating earthquake of July 21, AD 365 provides a *terminus post quem* for the renovation.

Figure 6. Reconstructed elevation of the late-Antique phase of the nymphaeum near the Praetorium (after Ortega 1986-1987: tav. 12 /reproduced with permission of the Scuola Archaeologica Italiana di Atene).

This was done by building a cistern within the settling basin and covering it with a barrel vault.[19] The cistern was fronted by a taller façade, with ten large reused limestone blocks and sixteen cornice fragments placed on top of the original parapet wall of the settling basin. Circular holes for water pipes were drilled near the top of the limestone blocks, often cutting through the existing inscription, and three Roman sarcophagi were set in front of the monument to serve as retrieval basins.[20]

The blocks were reused from an honorific statue base dating to the 1st century BC and incorporated into the façade without regard for the existing inscription.[21] In the 6th century, another inscription was added to the cornice blocks. Traversing the length of the façade wall, this inscription refers to a renovation of a watery place by a certain Georgios, perhaps the 6th-century bishop of Gortyn.[22] Sometime between AD 618 and AD 638, the upper parts of four column shafts were carved with cyclical acclamatory inscriptions that honored the Emperor Heraclius I and his family and referenced an Imperial intervention following an earthquake of AD 618.[23] These were placed high on the columns, in order to be seen above the barrel vault; therefore they provide a *terminus ante quem* for the final phase of the monument. Although these five inscriptions are incorporated into the new monument and the acclamatory inscriptions provide a *terminus ante quem* for the monument, the actual transformation of the monument cannot be securely attributed to any of them, as it is very possible that they were added to the monument long after its transformation into a reservoir.[24]

This transformation of the fountain into a reservoir and the upward migration of the pipes feeding the retrieval basins suggest that the city had acquired a need for greater amounts of stored water. Even though the aqueduct was still supplying the fountain, water no longer entered the niche in the back facade and the side basins were no longer in use, suggesting that less water was flowing to the monument. The reduction in water flow is indicative of either a decrease in the amount of water available at the aqueduct's source, an inability or lack of desire to maintain the aqueduct at the same level as it had been maintained during the Imperial period.

Yet the 2nd-century sculptural group was erected along the colonnaded façade wall of the transformed monument, even though the taller parapet wall and even taller barrel vault of the cistern effectively blocked from view the lower part of the colonnade and statues.[25] This raises the question of why the late-antique patron would bother with re-erecting the sculptural group at all, especially given the placement of the water pipe for the nymph with a shell, a sexual image that seems to be at odds with the public appearance of a Christian community. Ortega suggests that the decorative scheme remained in place due to a lack of specific reasons for moving it.[26] But the complete architectural transformation of the edifice from a hypaethral fountain to a covered cistern makes it surprising that the ancient sculptural group was retained. Economics is a tenacious explanation for the use of *spolia* in late Antiquity, and the reduction in water flowing to the fountain makes it a possible explanation in this case.[27] Yet, there is a much better explanation, and that is that the sculptural group's resurrection can be attributed to late-antique Imperial legislation that repeatedly restricted the removal of statues and other architectural elements from public buildings and urban centers across the Empire.

---

[19] Ortega 1986-87:164. Prior to these renovations, the pipe feeding the central niche in the back wall had been cut, but water continued to flow to the two niches at the center of each lateral façade wall.

[20] The sarcophagi include an Attic sarcophagus of Pentelic marble (Ghisellini 1989: no. 4) and two sarcophagi of local production (Ghisellini 1989: nos. 25 and 28).

[21] These blocks came from the honorific base of the quaestor Larcius Lepidus Sulpicianus (AD 67/8).

[22] IC IV, 461 and 465. Maiuri (1914: 132) reconstructs it as follows: ... ὁ πρὸς εὐσέβειαν ἐκτρέχων Γεώργιος *(6-7)*: καὶ προσεπιβάλλων τῇ *πόλει* τὴν *ο (3-5)*... ἐξανιστᾷ τὸν τόπον ⁝ δι *(1-2) [— — πρ]*οσερριμμένον *(15-16)*.
Maiuri thinks the inscription dates to the same time period as the other ones in the monument.

[23] IC IV, 512 (A-D).

[24] According to Ortega (1986-87), a sufficient part of the architectural material of the 2nd-century nymphaeum, including its columns and capitals, were used in the construction of the 5th-century basilica in Mavropapa, suggesting that, by this time, the nymphaeum had been transformed into a cistern.

[25] This addition to the parapet wall raised the façade by 2.45 m, with the crest of the barrel vault visible behind.

[26] Ortega 1986-87: 159.

[27] Other equally persistent explanations for late antique *spolia* would be the shortage of skilled workmen and the lack of time to carve a new sculptural group.

On March 26, AD 429 the Emperor Theodosius II announced to the Senate of Constantinople that a compendium of Roman Imperial law would be gathered in sixteen books and named after him.[28] Completed in AD 437 and written in Latin, the Theodosian Code preserves more than 2,500 Imperial decrees dating from AD 320 – 420. Addressed to officials stationed in both the East and West and legislating everything from the grain supply of Constantinople, the use of torture during trials, or even the return of an actor to the stage after taking the last Christian sacraments, these laws provide unique insight into the central administration of the Empire.

Fifty-three decrees in the Theodosian Code legislate civic building projects, both new constructions and older edifices. Most of these encourage the maintenance and restoration of older buildings, and many restrict magistrates from launching new projects until dilapidated civic structures are repaired.[29] A few decrees protect civic buildings from private annexation,[30] and a handful regulates the reuse of building materials and decoration. The last group provides evidence that spoliation in Rome and the provinces could be carried out legally only by Imperial permission and under strict surveillance.[31] Moreover, if a building was characterised as irreparable and destined for demolition, the construction and decorative material had to be reemployed on public buildings, usually in the same city. The Emperor Constantius emphasized this stance in a decree addressed to the proconsul of Africa in AD 357: 'No man shall suppose that municipalities may be deprived of their own ornaments [i.e. public monument], since indeed it was not considered right by the ancients that a municipality should lose its embellishments, as though they should be transferred to the buildings of another city'.[32] Such directives emphasize the link between a city and its civic monuments, suggesting that, from the central administration's viewpoint, the buildings were intrinsic components of a city and thus had to be preserved.[33]

Such Imperial mandates controlling the reuse of public building materials continued throughout the following centuries. In AD 365, the Emperors Valentinian and Valens prohibited the moving of statues and other decorations from one town to another, a ban that was meant to stop the predatory practices of governors and other high-ranked administrative officials, who apparently had been stripping the more isolated cities of their civic ornaments.[34] In AD 376, the Emperors declared to the Senate in Rome that new buildings in the city could not use spoliated pieces: 'If any person should wish to undertake any new building in the City, he must complete it with his own money and labor, without bringing together old buildings, without digging up the foundations of noble buildings, without obtaining renovated stones from the public, without tearing away pieces of marble by the mutilation of despoiled buildings'.[35] Materials were not to be collected even from demolished public buildings, and this emphatic and absolute denial of spolia to magistrates suggests that enough office holders over the past two generations had modeled their building practices on those of Constantine to warrant the judicial chastisement of the entire Senate.

In AD 398, the Emperors Arcadius and Honorius forbade all new building projects and spoliation without their explicit approval: 'No judge shall burst forth into such rash lawlessness as to suppose that he should begin any new public work without consulting Our Piety, or that he should dare to tear from any structure any ornament of bronze or marble or any other material, which can be proved to have been in serviceable use or to constitute an ornamentation in any municipality, or that he should dare to transfer such material to some other place without the order of Your Sublimity...'[36] This decree places an emphasis on the necessity for permission to move materials around, and the usefulness of the construction. Bronze and marble ornaments are explicitly mentioned, indicating that the Emperors recognized that the more mobile parts of the building were in danger of being removed.

[28] The Theodosian Code was presented to the Senate in Rome on December 25, AD 438; the surviving minutes from this meeting include the original decree of Theodosius II. See Pharr (1952: 3-7) for an English translation.
[29] The repetition of this decree indicates that the Imperial concern regarding the plunder of civic edifices had little effect.
[30] *Cod. Theod.* 15.1.25.
[31] *Cod. Theod.* e.g. 15.1.37, 40, 43.
[32] *Cod. Theod.* 15.1.1. Translated in Pharr (1952: 423).
[33] Alchermes 1994: 169.
[34] *Cod. Theod.* 15.1.14
[35] *Cod. Theod.* 15.1.19. Translated in Pharr (1952: 425).
[36] *Cod. Theod.* 15.1.37. Translated in Pharr (1952: 427).

Such decrees continued to be issued long after the publication of the Theodosian Code. In AD 458 the Emperors Leo and Majorian legislated that no ancient building in Rome was to be destroyed or its decorations removed by anyone, either private citizen or public official. They then equivocated by stating that if a civic building was characterised irreparable by Imperial authorities, it could be demolished and *only* its decoration was transferred to another public edifice in the same city.[37] In issuing this decree, Leo and Majorian made clear that their goal was to preserve the civic landscape and thus protect the appearance and heritage of the city. Indeed, Alchermes has pointed out that the 4th- and 5th-century laws regulating reuse suggest an underlining Imperial policy of architectural conservation, the driving force for which appears to be the desire to visually and physically preserve the civic heritage of individual municipalities. The cumulative effect of these laws seeking the preservation of civic heritage was to preserve fragments of the community's physical monuments that were always being renewed and refreshed as they were incorporated into new settings over the centuries. Long after an original monument collapsed, fragments of it would remain on public display, transferred and incorporated into a new monument that, with this fragment, spoke to the longevity and continuity of the municipality.[38]

Scattered among these fragments of a city's urban landscape are its statues, which are implicitly protected in some decrees and explicitly in others. In Rome, these judicial pronouncements were supported by civic offices likes those of the urban prefect and *curator statuarum*, or curator of statues, the latter of which particularly demonstrates that civic statuary was considered an important component of Rome´s urban heritage. Constantine seems to have created the office of *curator statuarum* to inventory, protect, and maintain the bronze and marble sculptures erected in the civic spaces throughout Rome. This esteemed position may have been particularly attractive to former consuls, in large because the *curator statuarum* was able to augment his presence in the urban center by inscribing his name on statues erected or repaired during is tenure in office.[39]

The combination of Imperial decrees about reusing statues and the creation of the office of *curator statuarum* indicate that statues were to remain in the civic sphere, regardless of their possible pagan origins or associations.[40] With the exception of cult statues, this is true even when the statues were in pagan temples. In AD 381, Emperors Gratian, Valentinian, and Theodosius decreed that a temple and its statues inside were to remain accessible to the public, although for civic purposes rather than religious: '...We decree that the temple shall continually be open that was formerly dedicated to the assemblage of throngs of people and now also is for the common use of the people, and in which images are reported to have been placed which must be measured by the value of their art rather than by their divinity.' Thus, the temple and presumably the art, became disassociated from its original religious connotations. In AD 399, the separation of statues from their religious contexts is made explicit (16.10.15, 18). Other than cult statues and altars (16.10.19), no action was to be taken against the ornaments of public works.

This extraction of pagan statues from their original religious contexts continues into the 5th century. Even from the Christian standpoint, pagan statues were important because of their civic connotations, and thus all but cult statues could be saved. In a 5th-century poem addressed to Rome and its pagan leaders, the Christian poet Prudentius (AD 348-413) – who had earlier twice served as a governor of a province -, suggested that if the pagan statues could be stripped of their religious associations, their aesthetic qualities and civic associations would make them worthy of admiration and preservation.[41]

In the 4th and 5th centuries we have the beginnings of a legislated tradition of reusing statues in a civic landscape as a means of providing a visual connection to the classical past and civic heritage. And I think this is what we are seeing in the nymphaeum in Gortyn. The reuse of the 2nd-century pagan sculptural program – despite the fact that Gortyn is now a Christian city and despite the fact that the group is

[37] *NMaj* 4.1. Pharr 1952: 553-554.

[38] Alchermes 1994: 178.

[39] *CIL* VI, 31906. See Chastagnol (1960: 33-42) and Kalas (2015: 116) for discussion of some of the epigraphic testimony for this office and for the practice of reusing statues.

[40] E.g. *Cod. Theod.* 16.10.8.

[41] *Contra Symmachum*, 1.501-505. This poem was about Symmachus' attempt to return the Altar of Victory to the Senate House.

partially obscured by the new architectural format of the edifice -- highlights the legislated necessity of maintaining the city's classical past.[42] The unknown patron of the Gortyn nymphaeum transferred the sculpture to a new context where the group retained an echo of its original purpose of adorning the city, but in a way that was relevant to contemporary citizens, who could see it as a vestige of their own classical heritage. I am suggesting that the reuse of the sculptural group reflects both legal reality and a nostalgic desire to remember and parade the fact that Gortyn was a significant city in the Roman world, even at the same time as the city was doing relatively well, serving as the seat of a bishop on Crete and receiving the attention of Byzantine Emperors like Heraclius I.

Conclusion

The reuse of older sculpture and architecture elements in new contexts is labeled as *spolia* by art historians. The most frequent use of spolia, the re-used materials themselves, is as space filler in walls and other places where it will never be seen. The Porticus Octaviae in Rome, which was first built sometime after 27 BC and rebuilt around AD 200, provides one example of this common use of spolia (Figure 7). The southern entrance took the form of a double pronaos.[43] On the inner pediment, you can see blocks of stone and pieces of fluted columns, spoliated elements that in antiquity would have been covered by a revetment and thus never visible, the once diligently worked architectural details reduced to raw building material. Another use of spolia, and one that was more common in antiquity than has been previously recognized, is what has been called a programmatic use of spolia.[44] This is when sculpture or architectural elements are transferred to new contexts and, in these new contexts, might be seen and recognized as reused

Figure 7. Inner pediment of the southern entrance of the Porticus Octaviae, Rome.

[42] A similar desire to maintain ancient statues can be seen in 4th- and 5th-century inscriptions from Rome that seem to record the re-erection of pagan statues in the same place where they fell, either from neglect, vandalism, or the damages of war (Ward-Perkins 1984). Such a reinstatement of local statues that had been damaged or neglected suggests a renewed pride in tradition and a desire to associate the current era with the Roman past.

[43] The remains of the south end of the porticus now serve as the entrance to the church of S. Angelo in Pescheria.

[44] Cf. Hansen 2003.

pieces. A good example of this comes from the northern Acropolis wall in Athens, where column drums and other architectural elements are displayed as a visual reminder of the Persian invasion, as kind of a political statement that it will not happen again.[45]

In the Roman world, the reuse of statues and architectural monuments has been an area of particular interest for scholars of Late Antiquity and the Middle Ages, who often identify the reigns of Diocletian and Constantine as the moment when recycled materials lacking military associations began to be deliberately incorporated into imperially-sponsored monuments; the Arcus Novus of Diocletian and the Temple of Romulus are discussed as the first examples of this practice and the Arch of Constantine is often heralded as the monument inaugurating the widespread use of programmatically repurposed materials.[46] The second-century reliefs reused in the Arch of Constantine include scenes of hunting, sacrifice, imperial address, justice, and munificence, all actions associated with potent emperors. In the spoliated reliefs, many of the heads of Hadrian and Marcus Aurelius have been reworked to represent Constantine. The transfer of these reliefs to this arch, the substitution of Constantine's portrait for those of the second-century 'good' Emperors, and the juxtaposition of old reliefs with new ones where Constantine is acting as a strong and beneficent emperor, all these factors relentlessly establish Constantine as rightful heir to this imperial tradition and legacy. This is an important point for him to be making, as this arch commemorates his victory over his co-ruler Maxentius, who had been particularly popular in Rome. The choices made regarding which images and which emperors would be included on the arch was deliberate and meaningful for Constantine. However, the spoliated material probably was not meant to be seen and recognized by the average Roman as fragments from specific monuments or specific emperors – but rather as generic pieces from the imperial past that lend the weight and authority of history to Constantine's claims of bringing peace, prosperity and legitimate succession to the empire. As Kinney notes, the fragmentary nature of the spoliated pieces allow them to occupy a space outside of a specific historical moment, where they are able to trigger a variety of memories and associations that could never come about from newly carved reliefs standing alone.[47] This difference between patron's intent and the general reception is one that can help us make sense of the retention of the statues in the nymphaeum near the Praetorium in Gortyn.

As discussed above, the *act* of re-using the second-century sculptural group as a backdrop to the Cretan reservoir was at once practical and cost-saving but also had the potential to carry an ideological charge; from the modern perspective, this complex process makes most sense in later antiquity – a period in which even today some think that the artists responsible for imperial commissions were far less skilled and nuanced in their presentation. But in part this phenomenon has been seen as so prevalent in late antiquity because that is where we have thought to look for it. In point of fact, the practice of reuse is evident much earlier in antiquity, and we may be seeing this reuse in the second-century nymphaeum near the so-called Praetorium in Gortyn. Moreover, because we do not know how ancient patrons or audiences would themselves have described the practice of the re-use of earlier materials, we need to pay particularly attention to the vagaries of context that necessarily differ from case to case.

Until recently suggestions have been given for interpreting the reuse of sculpture in late Antiquity, outside the bounds of Christianity or economics.[48] One reason for this is the focus on late Antique examples of spolia and the resultant thinking of spolia as a late-Antique phenomenon. Another reason relates to our current consumer mindset, in which it is easy to dismiss reuse as something done for reasons of economy, a lack of skilled workmen, or time constraints. In the late-antique phase of the nymphaeum at Gortyn, however, civic pride, probably with legislative underpinnings, seems to be a strong factor in the reuse of the second-century statues.

[45] Wrede 1996.
[46] Cf. Deichmann 1975: 5-7; Elsner 2000: 153; Hansen 2003: 15-17, 41; Kinney 1995: 56.
[47] Kinney 1995.
[48] A recent exception is Kristensen and Stirling 2016.

Bibliography

Alchermes, J. 1994. *Spolia* in Roman Cities of the Late Empire: Legislative Rationales and Architectural Reuse. *Dumbarton Oaks Papers (=DOP)* 48: 167-178.
Bendinelli, G. 1914. Sculture rinvenute nel ninfeo presso il pretorio di Gortina (Creta). *L'Annuario della Scuola Archeologica di Atene e delle Missioni in Oriente* 1:137-148.
Chastagnol, A. 1960. *La préfecture urbaine à Rome sous le Bas-Empire*. Paris: Presses Universitaires de France.
Deichmann, F.W.1975. *Die Spolien in der spätantiken Architektur*. Munich: Beck.
Di Vita, A. 2010. *Gortina di Creta: quindici secoli di vita urbana*. Rome: 'L'Erma' di Bretschneider.
Elsner, J. 2000. From the Culture of Spolia to the Cult of Relics: The Arch of Constantine and the Genesis of Late Antique Forms. *Papers of the British School at Rome (= PBSR)* 68: 149-184.
Ghedini, F. 1985. Sculture dal ninfeo e dal pretorio di Gortina. *L'Annuario della Scuola Archeologica di Atene e delle Missioni in Oriente* 63: 63-248.
Ghisellini , E. 1989. *Sarcofagi Romani di Gortyna*. Rome: 'L'Erma' di Bretschneider.
Giorgi, E. 2008. Water Technology at Gortyn in the 4th-7th c. A.D.: Transport, Storage and Distribution. *Late Antique Archaeology* 4: 287-320.
Giorgi, E. 2016. *Archeologia dell'acqua a Gortina di Creta in età protobizantina*. Oxford: Archaeopress
Hansen, M.F. 2003. *The Eloquence of Appropriation: Prolegomena to an Understanding of* Spolia *in Early Christian Rome*. Rome: 'L'Erma' di Bretschneider.
Kalas, G. 2015. *The Restoration of the Roman Forum in Late Antiquity: Transforming Public Space*. Austin: University of Texas Press.
Kinney, D. 1995. Rape or Restitution of the Past? Interpreting *Spolia*. In: S. C. Scott (ed.) *The Art of Interpreting*: 52-67. University Park, Pennsylvania State University.
Kristensen, T. M. and Stirling, L. (eds). 2016. *The Afterlife of Greek and Roman Sculpture: Late Antique Responses and Practices*. Ann Arbor: University of Michigan Press.
La Torre, G. F. 1988-1989. Contributo preliminare alla conoscenza del territorio di Gortina. *L'Annuario della Scuola Archeologica di Atene e delle Missioni in Oriente* 66-67: 277-323.
Maiuri, A. 1914. Un ninfeo presso il pretorio di Gortina (Creta). *L'Annuario della Scuola Archeologica di Atene e delle Missioni in Oriente* 1: 119-136.
Ortega, A.A. 1986-87. Gortina. Il ninfeo presso il pretorio. *L'Annuario della Scuola Archeologica di Atene e delle Missioni in Oriente* 64-65: 131-174.
Perali, P. 1914. Un ninfeo presso le grandi termi e altri ruderi di fontane in Gortina (Creta). *L'Annuario della Scuola Archeologica di Atene e delle Missioni in Oriente* 1: 149-159.
Pharr, C. 1952. *The Theodosian Code and Novels and the Sirmondian Constitutions, a Translation with a Commentary, Glossary and Bibliography*. Princeton: Princeton University Press.
Romeo, I. and Portale, E. C. 1998. *Gortina* III*: le sculture*. Padova: Bottega d'Erasmo.
Taramelli, A. 1902. Cretan Expedition. Gortyna. *American Journal of Archaeology* 6: 101-165.
Trimble, J. 2011. *Women and Visual Replication in Roman Imperial Art and Culture*. Cambridge: Cambridge University Press.
Ward-Perkins, B. 1984. *From Classical Antiquity to the Middle Ages: Urban Public Building in Northern and Central Italy, AD 300-850*. Oxford/New York: Oxford University Press.
Wrede, H. 1996. Waffen gegen die Perser. *Archäologischen Anzeiger* 1: 37-41.
Zanini, E. 2006. Indagini archeologiche nell'area del 'quartiere bizantino' di Gortina: terza relazione preliminare (campagna 2004). *L'Annuario della Scuola Archeologica di Atene e delle Missioni in Oriente* 82: 751-68.

Dr Brenda Longfellow
Associate Professor
The University of Iowa, School of Art & Art History
Iowa, USA
brenda-longfellow@uiowa.edu